BSAVA
MANUAL OF REPTILES

General Editor

Peter H Beynon
BVSc MRCVS

Scientific Editors

Martin P C Lawton
BVetMed CertVOphthal FRCVS

and

John E Cooper
BVSc CertLAS DTVM FRCVS MRCPath FIBiol

Iowa State University Press / Ames

Published by the
British Small Animal
Veterinary Association,
Kingsley House, Church Lane,
Shurdington, Cheltenham,
Gloucestershire GL51 5TQ.

Published in the
United States by
Iowa State University Press,
Ames, Iowa 50014.

Printed by J. Looker Printers,
Poole, Dorset.

Copyright BSAVA 1992. All rights reserved.
No part of this publication may be
reproduced, stored in a retrieval
system or transmitted in any form
or by any means electronic, mechanical,
photocopying, recording or otherwise
without prior written permission of the
copyright owner.

The publishers and contributors cannot take any responsibility
for information provided on dosages and methods of
application of drugs mentioned in this publication.
Details of this kind must be verified by individual users in the
appropriate literature.

First Published 1992
Reprinted 1994

ISBN 0-8138-2296-3

CONTENTS

Contents .... 2 and 3

Foreword 4

Dedication 5

Chapter One
    Introduction 7
    John E Cooper and Martin P C Lawton

Chapter Two
    Management in Captivity 14
    Andrew A Cunningham and Claudia Gili

Chapter Three
    Examination and Diagnostic Techniques 32
    Oliphant F Jackson and Martin P C Lawton

Chapter Four
    Post-Mortem Examination 40
    John E Cooper

Chapter Five
    Laboratory Investigations 50
    Elliott R Jacobson

Chapter Six
    Radiological and Related Investigations 63
    Oliphant F Jackson and Anthony W Sainsbury

Chapter Seven
    Integument 73
    John E Cooper

Chapter Eight
    Cardiovascular System 80
    David L Williams

Chapter Nine
    Respiratory System 88
    Lynne C Stoakes

Chapter Ten
    Gastrointestinal System 101
    Robin D Bone

CONTENTS

Chapter Eleven
Urogenital System 117
Peernel Zwart

Chapter Twelve
Neurological Diseases 128
Martin P C Lawton

Chapter Thirteen
Nutritional Diseases 138
Peter W Scott

Chapter Fourteen
Miscellaneous 153
Martin P C Lawton

Chapter Fifteen
Ophthalmology 157
Martin P C Lawton

Chapter Sixteen
Anaesthesia 170
Martin P C Lawton

Chapter Seventeen
Surgery 184
Martin P C Lawton and Lynne C Stoakes

Chapter Eighteen
Therapeutics 194
Mark A Pokras, Charles J Sedgwick and Gretchen E Kaufman

Appendix One
Useful Addresses 214

Appendix Two
Clinical Examination and Post-Mortem Sheets 217

Appendix Three
Haematological and Biochemical Data 219

Appendix Four
Legal Aspects 221

Appendix Five
Conversion Tables 224

Index 225

FOREWORD

Reptiles have fascinated the human race for thousands of years. In some cultures they have been revered and protected, in others despised and destroyed. They have certainly never been ignored.

The importance of reptiles is widely recognised today. Many play a significant ecological role in the wild. Large numbers are kept in captivity – as pets, as exhibits, for food and for research – and there is increasing concern over their health, welfare and conservation. The veterinary profession has begun to take an interest and involve itself in this field, building largely on the foundation laid by a small number of people with broad interests and considerable vision. It is appropriate, therefore, that the BSAVA should respond to requests for more information by commissioning a *Manual of Reptiles*. The *Manual of Exotic Pets* does, of course, contain basic information on reptiles but there is a need for the more detailed advice which this manual has sought to provide. I am in no doubt that it will succeed and I congratulate all those involved.

The BSAVA is indebted to the editors, Peter Beynon, Martin Lawton and John Cooper for steering the manual to completion. Lynne Hill played a key role in formulating and guiding the project in the early stages. Clare Knowler skilfully prepared the drawings for the cover and for one of the chapters.

The majority of the authors submitted manuscripts by the required date and all responded positively and cheerfully to requests for changes or additions to their contributions. The authors are busy people, authorities in their own field, and we owe a debt of gratitude to them.

Thanks are also due to a number of colleagues who commented on chapters and provided advice. Sally Dowsett and Sandie McConnachie undertook much of the supporting secretarial work. The Wellcome Library of the RCVS provided copies of papers and located relevant publications.

The editors are grateful to Paddy Beynon, Lynne Stoakes and Margaret Cooper for their support and encouragement, and to J. Looker Printers and in particular Matthew Poulson, for their help and advice in the production of this manual. Thanks are also due to Simon Orr and Harvey Locke who, as successive Chairmen of BSAVA Publications Committee, nurtured and supported this project.

I am delighted to be able to launch this new addition to the BSAVA series of manuals. Its appearance is timely and I am confident that it will prove of enormous value to practitioners. The profession is increasingly associated in the public eye with the care of all animals. The *Manual of Reptiles* will reinforce this association.

Michael E Herrtage MA BVSc DVR DVD MRCVS
President, BSAVA

DEDICATION

This manual is dedicated
to the memory of
Oliphant Fairburn Jackson PhD MRCVS,
a friend and colleague
and pioneer in the field
of reptile medicine.

LIST OF B.S.A.V.A. PUBLICATIONS

THE JOURNAL OF SMALL ANIMAL PRACTICE
An International Journal Published Monthly   Editor W. D. Tavernor, BVSc, PhD, FRCVS
Fifteen Year Cumulative Index published 1976
Available by post from: B.S.A.V.A. Administration Office, Kingsley House, Church Lane, Shurdington, Cheltenham, Gloucestershire GL51 5TQ

Manual of Parrots, Budgerigars and other Psittacine Birds
Edited by C. J. Price, MA, VetMB, MRCVS
B.S.A.V.A. Publications Committee 1988

Manual of Laboratory Techniques
Third Edition
Edited by D. L. Doxey, BVM&S, PhD, MRCVS
and M. B. F. Nathan, MA, BVSc, MRCVS
B.S.A.V.A. Publications Committee 1989

Manual of Anaesthesia for Small Animal Practice
Third Revised Edition
Edited by A. D. R. Hilbery, BVetMB, MRCVS
B.S.A.V.A. Publications Committee 1992

Manual of Radiography and Radiology in Small Animal Practice
Edited by R. Lee, BVSc, DVR, PhD, MRCVS
B.S.A.V.A. Publications Committee 1989

Manual of Small Animal Neurology
Edited by S. J. Wheeler, BVSc, CertVR, PhD, MRCVS
B.S.A.V.A. Publications Committee 1989

Manual of Small Animal Dentistry
Edited by C. E. Harvey, BVSc, DACVS, DAVD, MRCVS
and H. S. Orr, BVSc, DVR, MRCVS
B.S.A.V.A. Publications Committee 1990

Manual of Small Animal Endocrinology
Edited by M. F. Hutchinson, BSc, BVMS, MRCVS
B.S.A.V.A. Publications Committee 1990

Manual of Exotic Pets
New Edition
Edited by P. H. Beynon, BVSc, MRCVS
and J. E. Cooper, BVSc, DTVM, CBiol, FIBiol, CertLAS, MRCPath, FRCVS
B.S.A.V.A. Publications Committee 1991

Manual of Small Animal Oncology
Edited by R. A. S. White, BVetMed, PhD, DVR, DACVS, FRCVS
B.S.A.V.A. Publications Committee 1991

Manual of Canine Behaviour
Second Edition
V. O'Farrell, PhD
B.S.A.V.A. Publications Committee 1992

Manual of Ornamental Fish
Edited by R. L. Butcher, MA, VetMB, MRCVS
B.S.A.V.A. Publications Committee 1992

Manual of Reptiles
Edited by P. H. Beynon, BVSc, MRCVS
J. E. Cooper, BVSc, DTVM, CBiol, FIBiol, CertLAS, MRCPath, FRCVS
and M. C. P. Lawton, BVetMed, CertVOphthal, CertLAS, CBiol, FIBiol, FRCVS
B.S.A.V.A.. Publications Committee 1992

Manual of Small Animal Dermatology
Edited by P. H. Locke, BVSc, MRCVS,
R. G. Harvey, BVSc, DVD, CBiol, MIBiol, MRCVS
and I. S. Mason, BVetMed, CertSAC, PhD, MRCVS
B.S.A.V.A. Publications Committee 1993

Manual of Small Animal Ophthalmology
Edited by S. M. Petersen-Jones, DVetMed, DVOphthal, MRCVS
and S. M. Crispin, MA, VetMB, BSc, PhD, DVA, DVOphthal, MRCVS
B.S.A.V.A. Publications Committee 1993

Manual of Small Animal Arthrology
Edited by R. W. Collinson, BVMS, CertSAO, MRCVS
J. E. F. Houlton, MA, VetMB, DVR, DSAO, MRCVS
B.S.A.V.A. Publications Committee 1994

Manual of Feline Behaviour
V. O'Farrell, PhD
P. Neville, BSc
Edited by C. St. C. Ross, BVM&S, MRCVS
B.S.A.V.A. Publications Committee 1994

An Introduction to Veterinary Anatomy and Physiology
By A. R. Mitchell, MA, VetMB, MRCVS
and P. E. Watkins, MA, VetMB, DVR, MRCVS
B.S.A.V.A. Publications Committee 1989

Proceedings of the B.S.A.V.A. Symposium "Improved Healthcare in Kennels and Catteries"
Edited by P. H. Beynon, BVSc, MRCVS
B.S.A.V.A. Publications Committee 1991

Practice Resource Manual
Edited by D. A. Thomas, BVetMed, MRCVS
B.S.A.V.A. Publications Committee 1991

Members Information Service
Edited by D. A. Thomas, BVetMed, MRCVS
B.S.A.V.A. Publications Committee 1993

Practical Veterinary Nursing
Third Edition
Edited by G. M. Simpson, BVM&S, MRCVS
B.S.A.V.A. Publications Committee 1994

Small Animal Formulary
B. J. Tennant, BVSc, PhD, CertVR, MRCVS
B.S.A.V.A. Publications Committee 1994

B.S.A.V.A. VIDEO 1 (VHS and BETA)
Radiography and Radiology of the Canine Chest
Presented by R. Lee, BVSc, DVR, PhD, MRCVS
Edited by M. McDonald, BVSc, MRCVS
B.S.A.V.A. Publications Committee 1983

AVAILABLE FROM BOOKSELLERS
Canine Medicine and Therapeutics
Third Edition
Edited by E. A. Chandler, BVetMed, FRCVS
D. J. Thompson, OBE, BA, MVB, MRCVS
J. B. Sutton, MRCVS
and C. J. Price, MA, VetMB, MRCVS
Blackwell Scientific Publications 1991

Feline Medicine and Therapeutics
Second Edition
Edited by E. A. Chandler, BVetMed, FRCVS
C. J. Gaskell, BVSc, PhD, DVR, MRCVS,
and R. M. Gaskell, BVSc, PhD, MRCVS
Blackwell Scientific Publications 1994

An Atlas of Canine Surgical Techniques
Edited by P. G. C. Bedford, PhD, BVetMed, FRCVS
Blackwell Scientific Publications 1984

Jones's Animal Nursing
Fifth Edition
Edited by D. R. Lane, BSc, FRCVS
Pergamon Press 1989

CHAPTER ONE

INTRODUCTION

John E Cooper BVSc CertLAS DTVM FRCVS MRCPath FIBiol
Martin P C Lawton BVetMed CertVOphthal FRCVS

The Class Reptilia consists of approximately 6,547 species (Hare and Woodward, 1989). Therefore, it greatly outnumbers the Class Mammalia. The Class is divided into:-

1. Order Squamata (6,280 species)

    Sub-orders:-
    a) Sauria (lizards) - 3,750 species
    b) Serpentes (snakes or ophidians) - 2,400 species
    c) Amphisbaenia (worm lizards) - 140 species
    d) Sphenodontia - 1 species, the Tuatara (*Sphenodon punctatus*)

2. Order Chelonia (244 species)

    Sub-orders:-
    a) Pleurodira
    b) Cryptodira

3. Order Crocodylia (22 species)

For further divisions reference should be made to Tables 1 – 6.

Reptiles are found in many different habitats, in areas ranging from both the wet and dry tropics to the Arctic Circle. They may be terrestrial, aquatic or semi-aquatic, as well as arboreal or burrowing.

Although many species of reptiles are still extant, the Class has declined in numbers in recent millennia. The so-called "Age of Reptiles" lasted for about 215 million years, during which period an array of different species, some very large, others with bizarre morphological features, appeared - to survive for a while - and then to become extinct.

Reptiles remain integral members of many ecosystems and at present, when "biodiversity" is very much the vogue, it is increasingly important that their role is not overlooked. Many species of reptiles are threatened in the wild, largely on account of the same factors that have been instrumental in the decline (and in some cases, the disappearance) of other taxa. These include alteration of habitat, pollution, introduced predators/competitors and intentional human persecution. Indeed, in the UK the smooth snake (*Coronella austriaca*) and the adder (*Vipera berus*) are both threatened due to the destruction of their habitat by house and road building.

The role of disease in the decline of free-living (wild) reptiles is not fully understood and much work is needed on this subject. Nevertheless, it seems likely that both infectious and non-infectious diseases may, under certain circumstances, contribute to the decline in populations or limit their ability to spread. The availability of food undoubtedly plays a large part in the survival and fecundity of the local wild population. For example, sea temperature and, therefore, the availability and type of algae have a marked effect on the population of marine iguanas (*Amblyrhynchus cristatus*) in the Galapagos. (Cooper and Laurie, 1987).

Pathogens, at any rate, may perhaps be considered under the same heading as introduced predators and competitors. The veterinary profession is increasingly likely to be approached for advice by biologists who are concerned with morbidity and mortality in free-living reptiles and who require guidance or assistance over such matters as clinical examination, chemical immobilisation or pathological investigation, including *post-mortem* examination.

However, at the present time the veterinary profession's main contribution to reptiles is the provision of advice to owners and attention to specimens that are kept in captivity. This is, of course, important from a welfare point of view but it may also contribute indirectly to conservation of vulnerable species, because improved survival and successful reproduction of captive reptiles help to limit the numbers that have to be imported from the wild.

It is difficult to obtain accurate figures on the numbers of reptiles that are kept in captivity in the UK. This is because, with the exception of certain species covered by the Dangerous Wild Animals Act (see "Appendix Four – Legal Aspects"), there is no requirement for these animals to be licensed or registered. Some data are available from HM Customs and Excise relating to importation, especially those species covered by the Convention on International Trade in Endangered Species of Wild Fauna and Flora (Washington Convention or CITES). In addition, the pet trade bodies can provide a certain amount of information on the numbers imported for sale, but are often reluctant to do so.

In order to deal adequately with reptiles the veterinary surgeon needs knowledge of their basic biology. Reptiles and other ectotherms are commonly called "cold blooded". This and the term poikilothermic suggest that the animal is not able to regulate its own body temperature and is entirely dependent on the external environmental temperature. This concept is erroneous. Reptiles are more correctly referred to as ectothermic because they are thermoregulators, ie. they regulate their body temperature by means of behavioural strategies rather than by internal means. Thus, lizards bask in the sun until they start to overheat, when, either by hiding under a rock or changing the angle of their body to the sun or entering water, they are able to lower their body temperature. When they need to become warmer they return to bask in the sun. Burrowing species usually behave differently. For example, sand boas (*Eryx* spp.), when their core body temperature starts to rise, burrow deeper into the sand and thus prevent overheating.

All reptiles have a **preferred body temperature** (PBT) which varies between species, fluctuates with seasons and may even be different at certain times of the day. The PBT is the optimum temperature range for that species. A reptile will try to maintain its body within its PBT range so that its metabolic rate is optimum, its digestive enzymes are functioning correctly and it is able to digest properly. Antibody production is also optimum at a reptile's PBT.

Bellairs (1970) pointed out that when a python was kept at 28°C and fed a rabbit, digestion took five days, while if the environmental temperature was 18°C then it took 15 days to digest the rabbit. One can conclude from this and other studies that a reptile that is not able to achieve its PBT is unlikely to utilise its food adequately even if fed correctly; therefore, it may suffer from malnutrition (see "Nutritional Diseases"). Other bodily functions, eg. growth, sloughing and reproduction, tend to be optimum within a reptile's PBT range and will be adversely affected if this is not provided.

A reptile with a core body temperature below its PBT will have reduced metabolism. Similarly, a reptile with a core body temperature above its PBT may have an increased metabolism or, at very high temperatures, be in a critical state. This is relevant to the reptile's ability to deal with drugs, including anaesthetics. If the metabolism is altered the pharmacokinetics of a drug will also be affected. Lawrence (1983) reported that a fall of 6°C will double the half-life of certain antibiotics. Therefore, in this manual drug dosages are usually followed by the temperature at which the reptile should be kept (if it is different from their PBT) and this regime should be rigorously followed.

Mammals with bacterial infections become pyrexic. This increases the ability of white blood cells and antibodies to attack and "stick" to antigens. Although reptiles are not able to develop pyrexia as such, a sick reptile will seek heat so that it can raise its core body temperature to the upper limit or slightly beyond its PBT. Therefore, the veterinary surgeon dealing with a sick reptile should consider keeping it in an environment so that at least its PBT is attained. Ideally, a heat range or temperature gradient should be provided which allows the reptile to increase (and select) its core temperature.

Reptiles, being ectothermic, can only maintain their body temperature within their PBT range by using external heat sources and behavioural means. The **preferred optimum temperature zone** (POTZ) is the temperature range within which a reptile is able to maintain its core body temperature within its PBT. In order to achieve this, however, the reptile must be allowed to carry out its normal behavioural activities and thus regulate its temperature. The POTZ of most reptiles is between 22°C and 27°C and, as a general rule, the herpetologist should ensure that the vivarium provides this. Vivaria that are kept at a constant temperature can prove stressful as the inmates may not be able to reach or maintain their PBT by behavioural means. This is one of the main causes of maladaptation syndrome (see "Miscellaneous").

When a reptile is exposed to a temperature below its POTZ and is unable to maintain its PBT, it will undergo a physiological change, known as **hibernation**, which is a cold-induced torpor (Gregory, 1982). The animal's metabolism slows to a level that is sufficient only to maintain life and at which energy expenditure is reduced to a minimum. Although all reptiles may, in theory, undergo periods of hibernation, only a minority use this strategy. These are temperate species. The survival of hibernating reptiles depends upon fat body reserves and other factors.

In some species hibernation appears to be a physiological trigger and may be required to stimulate breeding. In corn snakes (*Elaphe* spp.) a cold spell followed by a short hibernation encourages reproduction in captivity.

Aestivation is a physiological state of stupor. It occurs when a reptile is exposed to a temperature above its POTZ and the PBT is exceeded. If the reptile's core body temperature continues to rise, it may reach a critical point and result in the death of the reptile. Reptiles kept in captivity in a vivarium should, therefore, be given a temperature gradient. If the heating element is wired into a thermostat, overheating of the environment can be avoided.

The anatomy of reptiles shows considerable diversity. Reptiles are vertebrates and have at least the vestiges of pentadactyl limbs. The most striking differences from other animals are seen in the skin (see "Integument"), heart (see "Cardiovascular System") and eyes (see "Ophthalmology"). There are also important differences in the number and types of bones, especially in the skull: indeed, these characteristics are often used in classification.

From time to time taxonomists change the scientific names of reptiles, based on new information on classification. Throughout this manual the editors have endeavoured to use current nomenclature which may differ from that given elsewhere. Thus, in this manual the red-eared terrapin is referred to as *Trachemys scripta elegans*: its previous names were *Chrysemys scripta elegans* and *Pseudemys scripta elegans*. Similarly, the desert tortoise, *Xerobates agassizi*, was formerly known as *Gopherus agassizi*. Even though scientific names change, they remain a consistent and internationally recognised way of identifying species.

It is clear from the number of people who belong to herpetological societies, ie. societies concerned with reptiles and amphibians, and the numbers of reptiles that are presented to practising veterinary surgeons, that substantial numbers of reptiles are kept in the UK and that veterinary attention for these is regularly requested. There would appear also to be a need for potential or new owners of such animals to receive advice on their management and care: this is the role that some veterinary surgeons are already playing.

Although, as mentioned earlier, the keeping of most reptiles does not require a licence in the UK, this does not mean that there are no legal requirements relating to the care of these species. A number of pieces of UK legislation - and their equivalent in other European countries - restrict the importation of certain reptiles and impose upon the keeper duties in respect of the health and welfare of his/her charges. Further information on the legal aspects is available from standard texts (Cooper, 1987; Cooper, 1991) and in "Appendix Four – Legal Aspects".

It is only in recent years that reliable information on reptiles has become generally available to those who keep or study them. Many books have appeared, ranging from those directed at the hobbyist to more professional texts on anatomy and physiology, reproduction and breeding, and health and disease. Examples of these are given in the list of References and Further Reading. The advent of books on these subjects has made it easier for the veterinary surgeon to obtain information which he or she can use or impart to clients. But in addition, particularly over the past five years, there has been an unprecedented increase in the amount of available veterinary information. This has included tape-slide programmes, videos, information and care sheets (some produced by veterinary surgeons, others by herpetological societies) and scientific papers, many of them in refereed journals. The veterinary surgeon who requires information on reptiles has no excuse for pleading total ignorance. A glance at the index of the *Veterinary Record* or the *Journal of Small Animal Practice*, or their equivalent in Europe or in North America, will usually yield some initial references. Preferably, the enquiring veterinary surgeon should consult the Royal College of Veterinary Surgeons (RCVS) Wellcome Library who will be able, for a modest fee, to perform searches and produce a printout of recent publications or, if necessary, to obtain copies.

What then is the aim of this manual? Does it offer anything that is not already catered for as described above? The manual is intended to provide, in an easily handled and accessible form, information on aspects of reptiles that are most likely to be of concern and relevance to the practising veterinary surgeon. The manual does not seek to compete with authoritative texts on reptile diseases or surgery. Indeed, references are listed at the end of each chapter as often it is necessary to refer to these for more detailed information. Nor is it intended to compete with herpetological books, some of which provide expert advice on housing, management and breeding. The editors of this manual have attempted to bring together readily available information from veterinary and herpetological quarters in one volume, and thus avoid the need for the busy practitioner to consult two or more books in order to cope with a particular case or problem.

However, the veterinary surgeon using this manual must do so wisely. First, it is best to consult the Index, remembering that a subject may be dealt with under a number of different headings. Secondly, even if a particular author does not go into the detail that the reader requires, it may be that a reference is given which will provide this. The reader should be prepared, if necessary, to consult that reference, eg. through the RCVS Wellcome Library, in order to obtain further information. If the reader is still unhappy, he or she may find it useful to consult the author of the relevant chapter. This is because the author is, by definition, likely to have a particular interest in that aspect of reptile medicine, and also because it is the editors' hope that this manual will be revised and reprinted in due course. At that stage feedback from practitioners (even if it is a complaint over certain omissions) will be of great value.

Reptiles, like all other animals, are susceptible to a wide range of diseases. These can be conveniently divided into infectious and non-infectious diseases. Throughout this manual the term disease is used in its broad sense of "any impairment of normal physiological function" and, therefore, not only diseases due to micro-organisms are covered, but also those attributable to or associated with such factors as adverse environment, incorrect feeding, toxins or genetic disorders.

A problem faced by the editors from the outset was how best to cover the various diseases. There are two standard ways of doing this:-

1. On the basis of the organ system affected - for example, respiratory system, cardiovascular system, integument etc.

2. On the basis of aetiology and pathogenesis - for example, bacterial diseases, viral diseases, traumatic diseases, toxic diseases etc.

In this manual the former system is used but it must be stressed that this approach has disadvantages. For example, diseases affecting the heart, eg. gout due to deposition of urates, may not reflect a primary cardiovascular disease but be secondary to another - in this case, renal failure. This and other examples emphasise how important it is that the reader makes full use of the Index and, by appropriate cross-referencing, refers to all the entries in the text.

The study of reptile diseases has developed greatly over the past twenty years and, as a result, captive animals are subject to a far higher standard of veterinary care. At the same time, as mentioned earlier, there is increasing involvement of the veterinary profession in work on free-living reptiles, with particular reference to factors that may contribute to their decline. It is hoped that this manual will play a part in enhancing veterinary knowledge in this new and exciting field.

TABLE 1

Families of lizards (Sauria).

GEKKONIDAE	Geckos	85 genera	800 species
PYGOPODIDAE	Snake lizards	8 genera	31 species
XANTHUSIIDAE	Desert night lizards	4 genera	16 species
DIBAMIDAE	Blind lizards	1 genus	4 species
IGUANIDAE	Iguanas	55 genera	650 species
CHAMAELEONIDAE	Chameleons	4 genera	85 species
AGAMIDAE	Agamas, flying dragons	53 genera	300 species
SCINCIDAE	Skinks	85 genera	1275 species
CORDYLIDAE	Girdle-tailed lizards	10 genera	50 species
LACERTIDAE	Wall and sand lizards	22 genera	200 species
TEIIDAE	Whiptails	40 genera	227 species
ANGUIDAE	Anguids, slow-worms	8 genera	75 species
XENOSAURIDAE	Crocodile lizards	2 genera	4 species
VARANIDAE	Monitor lizards	1 genus	31 species
HELODERMATIDAE	Beaded lizard, Gila monster (venomous species)	1 genus	2 species

TABLE 2

Families of snakes (Serpentes).

LEPTOTYPHLOPIDAE	Thread snakes	2 genera	78 species
TYPHLOPIDAE	Blind worm snakes	4 genera	180 species
ANOMALEPIDAE	Dawn blind snakes	4 genera	20 species
ACROCHORDIDAE	Asian wart snakes	2 genera	3 species
ANILIIDAE	Pipe snakes	3 genera	9 species
UROPELTIDAE	Shield-tailed snakes	8 genera	44 species
XENOPELTIDAE	Sunbeam snake	1 genus	1 species
BOIDAE	Boas, pythons	27 genera	88 species
COLUBRIDAE	Colubrids, boomslang	285 genera	1562 species
ELAPIDAE	Cobras, mambas	61 genera	236 species
VIPERIDAE	Vipers	11 genera	187 species

TABLE 3
Families of worm lizards (Amphisbaenia)

AMPHISBAENIDAE	European worm lizards	18 genera	130 species
RHINEURIDAE	Florida worm lizard	1 genus	1 species
TROGONOPHIDAE	Somali edge-snouted worm lizards	4 genera	6 species
BIPEDIDAE	Two-legged worm lizards	1 genus	3 species

TABLE 4
Families of side-necked chelonians (Pleurodira).

| PELOMEDUSIDAE | American side-necked turtles | 5 genera | 24 species |
| CHELIDAE | Snake-neck turtles | 9 genera | 37 species |

TABLE 5
Families of hidden-necked chelonians (Cryptodira).

KINOSTERNIDAE	Mud turtles	2 genera	20 species
STAUROTYPIDAE	Narrow-bridged musk turtles	2 genera	3 species
DERMATEMYIDAE	Central American river turtle	1 genus	1 species
CHELYDRIDAE	Snapping turtles	2 genera	2 species
PLATYSTERNIDAE	Big-headed turtle	1 genus	1 species
EMYDIDAE	Red-eared and freshwater terrapins	30 genera	85 species
TESTUDINIDAE	Land tortoises	10 genera	41 species
DERMOCHELYIDAE	Leather backed sea turtle	1 genus	1 species
CHELONIIDAE	Sea turtles	4 genera	6 species
TRIONYCHIDAE	Soft-shell turtles	6 genera	22 species
CARETTOCHELYIDAE	Big-nosed soft-shell turtle	1 genus	1 species

TABLE 6

Families of crocodiles (Crocodylia).

CROCODYLIDAE	Crocodiles	3 genera	14 species
ALLIGATORIDAE	Alligators and caimans	4 genera	7 species
GAVIALIDAE	Gharial	1 genus	1 species

REFERENCES AND FURTHER READING

BELLAIRS, A.d'A. (1970). *The Life of Reptiles.* Weidenfeld and Nicolson, London.

COOPER, J.E. and JACKSON, O.F. (1981). Eds. *Diseases of the Reptilia, Vols. 1 and 2.* Academic Press, London.

COOPER, J.E. and LAURIE, A.W. (1987). Investigation of deaths in marine iguanas (*Amblyrhynchus cristatus*) on Galapagos. *Journal of Comparative Pathology* **97**, 129.

COOPER, M.E. (1987). *An Introduction to Animal Law.* Academic Press, London.

COOPER, M.E. (1991). Legislation. In: *Manual of Exotic Pets.* New Edn. (Eds. P.H. Beynon and J.E. Cooper). BSAVA, Cheltenham.

FRYE, F.L. (1991). *Biomedical and Surgical Aspects of Captive Reptile Husbandry.* 2nd Edn. Krieger, Malabar.

GANS, C. (and often guest editors). (1969-1988). *Biology of the Reptilia, Vols. 1 – 14.* Academic Press, New York.

GANS, C. and BILLETT, F. (1985). Eds. *Biology of the Reptilia, Vol. 15.* John Wiley and Sons, New York.

GANS, C. and HUEY, R.B. (1988). Eds. *Biology of the Reptilia, Vol. 16.* Alan R. Liss, New York.

GREGORY, P.T. (1982). Reptilian hibernation. In: *Biology of the Reptilia, Vol.13. Physiology D: Physiological Ecology.* (Eds. C. Gans and F.H. Pough). Academic Press, New York.

HARE, T. and WOODWARD, J. (1989). *Illustrated Encyclopaedia of Wildlife, Vols. 26 – 29.* Orbis, London.

HOFF, G.L., FRYE, F.L. and JACOBSON, E.R. (1984). Eds. *Diseases of Amphibians and Reptiles.* Plenum Press, New York.

LAWRENCE, K. (1983). The use of antibiotics in reptiles: a review. *Journal of Small Animal Practice* **24**, 741.

MARCUS, L.C. (1981). *Veterinary Biology and Medicine of Captive Amphibians and Reptiles.* Lea and Febiger, Philadelphia.

REICHENBACH-KLINKE, H.H. and ELKAN, E. (1965). *The Principal Diseases of Lower Vertebrates.* Academic Press, London.

CHAPTER TWO

MANAGEMENT IN CAPTIVITY

Andrew A Cunningham BVMS MRCVS
Claudia Gili DMV

INTRODUCTION

Poor husbandry, usually through ignorance, is, either directly or indirectly, the major cause of disease in captive reptiles (Harper, 1987; Frye; 1991; Price, 1991). The veterinary care of reptiles must, therefore, be based on a knowledge of the basic biological and environmental requirements of the patient. This knowledge is of vital importance in both the prevention and cure of diseases in reptiles.

Reptiles, even within a genus, can vary greatly in their environmental, behavioural, nutritional and other needs. These needs have to be catered for if reptiles kept in captivity are to remain healthy and thrive. When a reptile is brought to the surgery, the owner will look to the veterinary surgeon for treatment and advice. As it is not practicable to detail the needs of every species of reptile, the aim of this chapter is to introduce the basics of reptile husbandry and management to the veterinary surgeon, providing general guidelines and an indication of where to find further information.

HOUSING

A large variety of reptiles are kept by private individuals and suitable accommodation must be provided to meet the requirements of the different species. Height may be an important consideration for reptiles which climb or live in trees, eg. desert spiny lizards (*Sceloporus magister*), anoles (*Anolis* spp.) and chameleons (*Chamaeleo* spp.), whereas a large floor area is more important for large or active terrestrial species, such as tortoises (*Testudo* spp.) and monitor lizards (*Varanus* spp.). Arboreal species require branches on which to climb. Aquatic and semi-aquatic species have special requirements which must be met; these include varying depths of water and a landing area for basking. Hiding areas must be provided, particularly for shy species, eg. the royal python (*Python regius*), or if two or more animals are to be kept in the same enclosure.

There are very few guidelines on space requirements for captive reptiles. Although body size can be used as a rough guide, the species must also be taken into consideration. Some snakes, eg. boa constrictors (*Boa constrictor constrictor*), may be kept adequately in a relatively small vivarium where the curled-up occupant touches two sides of its cage, although it must be allowed out for "exercise" several times a week. Most other reptiles will require a much larger cage relative to their body size. A useful general rule is that any enclosure must be spacious enough and furnished in such a way as to allow the occupant(s) room to move around and exhibit normal behaviour, eg. climbing, burrowing etc. (Avery, 1985; Bels, 1986; Warwick, 1989).

Furnishing can be either "naturalistic" with the provision of plants and a sand/peat/bark chippings substrate, or more "clinical" with a substrate of newspaper or tissue paper and a few easily-cleaned accessories. From a hygiene standpoint, a tendency towards the latter should be encouraged. However, this may not be suitable for all species (Avery, 1985). Pine snakes (*Pituophis* spp.) and sand boas (*Eryx* spp.), for instance, must be provided with a suitable substrate into which they can burrow.

TEMPERATURE

Many species of reptiles can tolerate a wide temperature range but within this range there is the optimum temperature at which their bodies function. This is known as the preferred body temperature (PBT) (Davies, 1981; Jacobson, 1988) and it varies between species (Davies, 1981; Marcus, 1981). Although a few species can raise their body temperature to some degree by endogenous means, eg. monitor lizards and pythons (Hutchison *et al*, 1966), the body temperature of a reptile is largely dictated by the environmental temperature.

Although ambient temperature influences the body temperature of reptiles, particularly the rate of body heat loss, many reptiles rely more on conduction (from surfaces warmed by the sun) and radiation to increase their body temperature. Temperature regulation is, therefore, dependent on behaviour, such as mouth gaping, burrowing, seeking shady areas or posturing (Peaker, 1969; Avery, 1985; Meek and Avery, 1988). Posturing may involve flattening the body against a warm surface, orientating the body so that a large surface area is exposed to the sun to absorb heat, or pointing the head in the direction of the heat source to minimise heat absorption. Some lizards can alter the pigmentation of their skin to assist thermoregulation: dark to absorb heat, light to reflect heat (Peaker, 1969; Frye, 1991).

In order to allow reptiles to regulate their body temperature in captivity, a temperature gradient is required within the vivarium. This is usually achieved by providing a "hot-spot" in one area of the cage and maintaining the air temperature within the natural range for the species being kept (see Table 1). The ambient temperature must never be allowed to exceed the PBT of the animal being kept; this should be carefully monitored, ideally with at least two maximum-minimum thermometers, one in the warmest area of the cage and one in the coolest. Heat sources (other than the "hot-spot") should be thermostatically controlled and the temperature under the "hot-spot" should be monitored to ensure that it does not exceed the maximum temperature that the animal can tolerate. A simple system used to maintain a temperate species, the viviparous lizard (*Lacerta vivipara*), in captivity has been described by Avery (1985).

Many different methods are available for providing supplementary heat in vivaria, but it is important that the heat and light sources can be operated independently. Underfloor heating is superior to most methods. The heat source, eg. a heating pad or heat cable, is used to heat one third to one half of the floor area in order to help maintain a temperature gradient. However, this type of heat source must not be used for burrowing species. These reptiles, eg. sand boas (*Eryx* spp.), burrow instinctively to escape excess heat and do not differentiate between heated and unheated substrate. Therefore, for these species the heat source should be overhead.

A "hot-spot" should be provided for basking: this allows an opportunity to provide an ultra-violet (UV) light source if this is required (see later). Basking has been reported to be of particular importance in some reptiles after feeding and, possibly, when gravid (Neitman, 1982). The heat source can be a dull infra-red emitter or an incandescent light bulb. It must be positioned carefully to prevent direct contact with the occupant(s); many snakes are burned by getting too close to a heat lamp. Electrically heated rocks are available for the provision of "hot-spots" and a UV source can be placed above these if necessary.

LIGHTING

Correct lighting is important for the well-being of reptiles. Light stimulates activity and appetite (Blatchford, 1987), provides essential UV wavelengths and often triggers certain physiological processes, eg. breeding or preparation for hibernation (see later). As with heating, the natural lifestyle and environment of the species must be considered when deciding what kind of lighting is required.

Both the photoperiod and the quality (wavelength) of light are important if stress is to be kept to a minimum, particularly if reptiles are to breed. It is generally believed that many reptile species, particularly lizards, require exposure to ultra-violet "B" (UVb) light (wavelengths between 280 – 315 nanometers) in order to synthesise vitamin D_3, which is essential for the utilisation of dietary calcium (see "Nutritional Diseases"). The provision of supplementary UV light to captive reptiles has been reported to bring about an improvement in growth rates and health (Cooper and Jackson, 1981). However, not all studies support these findings (Allen, 1988).

Table 1
Environmental and dietary requirments for some commonly kept species of reptile.

Species	Temp.* °C	Relative humidity %	Habitat	Diet
Boa constrictor (*Boa constrictor constrictor*)	25 – 30	50 – 70	arboreal	C
Sand boa (*Eryx* sp.)	25 – 30	20 – 30	burrowing	C
Indian python (*Python molurus*)	25 – 30	70 – 80	arboreal	C
Royal python (*Python regius*)	25 – 30	70 – 80	arboreal	C
Garter snake (*Thamnophis sirtalis*)	21 – 26	60 – 80	semi-aquatic	P/c
King snake (*Lampropeltis getulus*)	23 – 30	50 – 70	terrestrial	Op/c
Spur-thighed tortoise (*Testudo graeca*)	20 – 27	30 – 50	terrestrial	H/om
Common box-tortoise (*Terrapene carolina*)	24 – 29	60 – 80	semi-aquatic	C/f
Red-eared terrapin (*Trachemys scripta elegans*)	20 – 22	80 – 90	semi-aquatic	C
Common iguana (*Iguana iguana*)	26 – 30	60 – 80	arboreal	H/c
Leopard ground gecko (*Eublepharis macularius*)	25 – 30	20 – 30	terrestrial	I
Green anole (*Anolis carolinensis*)	23 – 29	70 – 80	arboreal	I/c
Jackson's chameleon (*Chamaeleo jacksoni*)	28 – 30	50 – 70	arboreal	I
Plumed basilisk (*Basiliscus plumifrons*)	23 – 30	70 – 80	arboreal	C/f
Water dragon (*Physignathus lesueuri* sp.)	24 – 30	80 – 90	arboreal/ semi-aquatic	I/om

C = carnivorous; F = frugivorous; H = herbivorous; I = insectiverous; Om = omnivorous; Op = ophiophagous; P = piscivorous.

Uppercase letters denote principal dietary requirements.

* Temperatures shown are ideal ambient daytime temperature gradients. These should be allowed to fall by approximately 5°C during the night. "Hot-spot" temperatures should generally be 5°C greater than the highest temperature shown.

Lighting should be provided via fluorescent tubes rather than incandescent light bulbs: the latter produce heat and, therefore, interfere with temperature control. Fluorescent tubes which have a "natural" light spectrum, eg. Truelite, Duro-Test International, have been found to be beneficial and should be used (Blatchford, 1987). The UV component of the light spectrum can be provided by a "natural" light source or by the provision of a specialised UV source ("Blacklight"), eg. Actinic 09, Philips Electrical, most of which also emit visible blue light. In order to ensure correct exposure of UV light, the manufacturers' instructions must be closely followed. Some appliances are not effective unless situated within a short distance of the reptile. Likewise, the lifespan of bulbs must be noted as the UV output will deteriorate with use (Blatchford, 1987).

HUMIDITY

Humidity is an often overlooked requirement when considering reptile husbandry. Examination of the animal's natural habitat will suggest the correct relative humidity levels for any individual species. Many species of pet reptiles will tolerate a relative humidity of around 50 – 70% (see Table 1), although the relative humidity must be kept low for desert species (Boyer, 1988). Conversely, rain-forest dwellers, such as the emerald tree boa (*Corallus caninus*), require a relative humidity of 70 – 80%, or higher. For these species, occasional spraying with tepid water can be beneficial and may stimulate feeding (Mehrtens, 1987). A misting system suitable for such species has been described by Blueman (1988). Incorrect relative humidity will often lead to diseases, eg. slough retention (humidity too low) or blister disease (humidity too high) (see "Integument").

FEEDING

A balanced diet is required for the maintenance and breeding of reptiles in captivity. Nutritional diseases are particularly common in reptiles. Many of these are caused by incorrect feeding and are easily preventable (see "Nutritional Diseases").

Compared with endotherms, reptiles have a lower metabolic rate and hence require a lower calorific intake per kilogram for the maintenance of bodyweight (Kirkwood, 1991). Dietary requirements are not only species dependent but are also affected by other factors, such as age, size, season of the year and nutritional and breeding status (Kirsche, 1979; Townsend, 1979). Environmental influences should be taken into consideration: temperature, humidity, photoperiod and light quality can all affect the frequency of feeding and the quantity of food taken (Bellairs, 1969; Peaker, 1969; Marcus, 1981; Neitman, 1982; Avery 1985). In general, the smaller the reptile the more frequently it should be fed, although even large herbivores and insectivores should be offered food daily. Guidelines on the required food intake of some commonly kept snakes are given by Coborn and Lawrence (1987a).

The diet should have an overall calcium (Ca) to phosphorus (P) ratio of 1.5:1 and, in captivity, reptiles may require supplementary vitamin D_3 (see "Nutritional Diseases"). Commercial mineral and vitamin preparations are readily available for this purpose, although in some products the Ca:P ratio is too low.

Food type and presentation are important; the egg-eating snake (*Dasypeltis* spp.), for example, will only eat whole bird eggs. Some snakes may refuse food until and unless it is presented in a certain way. Although some may think that reptiles, particularly snakes, will only accept live food, they can be "trained" to accept dead prey (Frye, 1991). The feeding of live vertebrate prey should be avoided, both for humane and legal reasons (Cooper, 1987). In addition, live prey can inflict injuries to a reptile (Rosskopf and Woerpel, 1981). Young reptiles may require their food to be chopped into bite-sized pieces but most will take small versions of the adult's prey, eg. newborn mice, small fry. Young captive-bred snakes may initially require assisted or forced feeding before they start taking food voluntarily (Campden-Main and Campden-Main, 1983; Coote, 1985). If fed in groups, reptiles should be monitored to ensure sufficient food intake for each animal, particularly when weaker, subordinate or younger animals are present. In some cases it may be necessary to separate such individuals from the rest of the group during feeding. This may also be necessary for potentially cannibalistic species, such as king snakes (*Lampropeltis* spp.).

Individual animals will exhibit food preferences and it is important to ensure that these do not preclude the animal from receiving a balanced diet: a young common iguana (*Iguana iguana*) with a fruit fetish will develop osteodystrophy unless the diet is supplemented. The eating of a wide variety of foods should be encouraged. New food items may initially be refused but a gradual introduction, often with a favoured food, and perseverance will usually result in the novel item being accepted.

Food should be consumed within 30 – 40 minutes of being offered; any uneaten food should be removed. Nervous, sick or newly acquired animals should be allowed more time to eat. For large lizards and chelonians, feeding is best carried out in a set area of the enclosure as these animals often spread their food around. Aquatic reptiles should be fed in water. They will quickly pollute their environment if housed and fed in the one tank, possibly resulting in bacterial or other disease problems (see "Integument").

Inanition and the opposite extreme, obesity, are two common problems of captive reptiles (see "Nutritional Diseases"). For this reason, the regular recording of food offered, food eaten and bodyweight should be encouraged.

Chelonians

Terrestrial species tend to be herbivorous but some, eg. the Mediterranean tortoises (*Testudo* spp.), will occasionally feed on animal products from which they derive extra protein and calcium (Bellairs, 1969; Kirsche, 1979; Reid, 1987). Terrapins and turtles are usually omnivorous or predominantly carnivorous, although some species are only carnivorous when young, becoming herbivorous as they mature (Mahmond and Klicka, 1979).

The predominantly herbivorous species require a high percentage of dietary fibre if digestive and metabolic dysfunctions are to be avoided. Sugar-free breakfast cereals are an excellent supplement, being rich in fibre and often a good source of calcium (Frye, 1991).

Crocodilians

Crocodilians are carnivorous and should be fed on a wide variety of prey (Webb *et al*, 1987). Although the diet for adult animals should be based on fish, young crocodilians can be fed invertebrates (eg. earthworms and shrimp), small mammals and birds as well as amphibians and fish (Bellairs, 1969). If whole fish are fed, vitamin and mineral supplements are unnecessary unless the food has been pre-frozen, when extra thiamine may be required (see "Nutritional Diseases"). Overfeeding of these voracious animals is easy; food intake should be restricted to approximately 8% of bodyweight per week (Joanen and McNease, 1979).

Lizards

Lizards range from herbivorous to insectivorous and from carnivorous to omnivorous. As with some chelonians, certain lizards change their food preferences as they mature and their nutritional requirements change; for example, young water dragons (*Physignathus* spp.) are insectivorous while the adults are omnivorous (Lawton, 1991). Carnivores, eg. monitors (*Varanus* spp.) and heloderms (*Heloderma* spp.), can be fed eggs, small mammals, eg. mice, and birds, either whole or parts thereof, depending on the size of the lizard.

Insectivorous lizards, eg. leopard ground geckos (*Eublepharis macularius*), are easily maintained on a supplemented diet of crickets or locusts, occasionally embellished with other invertebrates, eg. mealworms and flies (Thorogood and Whimster, 1979). Some insectivores, eg. the Tokay gecko (*Gekko gekko*) and the collared lizard (*Crotaphytus collaris*), may also accept small mice or lizards (Mattison, 1989), while others, eg. the green anole (*Anolis carolinensis*) and the basilisk (*Basiliscus* sp.), will also take fruit (Bloxam and Tonge, 1982). Chameleons (*Chamaeleo* spp.) are probably the most fastidious of the insectivores, requiring a large variety of food, including live winged insects. Flies, locusts, crickets and caterpillars are amongst preferred chameleon food items, although each may need to be offered repeatedly before being accepted.

Snakes

Snakes are carnivorous and swallow prey whole. The size of the food taken is dependent on the size of the reptile, although snakes do have flexible articulations between the bones of the oral cavity and skull, thus allowing prehension and swallowing of prey much larger than their head. The metabolic rate of snakes allows them to maintain their bodyweight with relatively small amounts of food (Kirkwood, 1991). Frequency of feeding is dependent on the species and the environmental conditions (see earlier): active species, eg. rat snakes (*Elaphe* spp.), will require food more frequently than relatively lethargic species, eg. gaboon viper (*Bitis gabonica*). Some species, eg. the boa constrictor (*Boa constrictor constrictor*) and royal python (*Python regius*), will not feed during the breeding season, which occurs during the cooler period of the year. Gravid snakes and snakes approaching ecdysis may also refuse to eat but should feed well post-parturition or post-sloughing.

Most commonly kept snakes can be maintained on a diet of whole rodents, rabbits and/or chicks, although some species, eg. grass snakes (*Natrix natrix*) and garter snakes (*Thamnophis* spp.), prefer fish and amphibians, or invertebrates, such as slugs and earthworms. Other species are even more specialised and may not be suitable as pets; for example, Jackson's tree snake (*Thrasops jacksoni*) feeds on lizards (Lawton, 1991) and the favoured food of king snakes (*Lampropeltis* spp.) are other snakes, ie. they are ophiophagous, although they will take small mammals and birds (Mehrtens, 1987).

WATER

Water should always be available, even though some reptiles will not drink directly from a bowl. Chameleons (*Chamaeleo* spp.), for example, will only take water from the foliage, cage walls or dripping from the roof as if during rainfall; therefore, their cages should be sprayed with a fine mist of water several times each day. Many species of reptiles may have a particular requirement for drinking water following egg laying or hibernation. Newly-hatched babies may not eat until the yolk sac has been fully absorbed but they drink quite frequently during this period.

Water is not only needed for drinking: many reptiles like to bathe and this may be a requirement for successful ecdysis (Lawton, 1991), particularly if the relative humidity of the cage is low (Kauffeld, 1969). For most pet species, the provision of a wide, shallow dish of water is adequate (Boyer, 1988).

HIBERNATION AND AESTIVATION

Many reptils, particularly those from temperate regions, undergo periods of dormancy as a protective response when environmental conditions become unfavourable. Aestivation is a period of stupor (see "Introduction") brought about by conditions which are too hot and/or too dry for normal activity to take place, while hibernation is generally a response to conditions being too cold. During periods of torpor the metabolic rate and physiological functions may be reduced (Davies, 1981). Most species of reptiles will aestivate during periods of drought or extreme heat. This ranges from the European adder (*Vipera berus*), which may only aestivate for a few days, to aquatic and semi-aquatic terrapins, eg. bog turtles (*Clemmys muhlenbergi*), which lie dormant in the mud of dried-up ponds during the summer months (Ernst and Barbour, 1972).

A series of factors may trigger the onset of hibernation. In Mediterranean tortoises (*Testudo* spp.) it would appear to be a combination of decreasing temperature and daylength (Gregory, 1982). Although some reptiles, eg. the European adder (*Vipera berus*), may hibernate for many months (Mehrtens, 1987), for most species periods of torpor tend to last for only a few days or weeks. Provided they are not exposed to the environmental stimuli which induce its onset, very few reptiles have an innate need to undergo hibernation (Gregory, 1982), although some species may require such a stimulus for breeding (Laszlo, 1984; Coote, 1985; Tryon, 1985).

The duration of dormancy is dependent on the local environment: for example, the duration of hibernation for Mediterranean tortoises in the wild depends on whether they are in a relatively cold or warm part of their natural habitat. In the UK, *Testudo* spp. may hibernate for up to six months. If they are not in prime condition prior to such a long period of dormancy, or if such periods are prolonged, reptiles can become weakened and predisposed to disease or even fail to survive. When in captivity, Mediterranean tortoises should be examined prior to hibernation to see if they are fit enough to survive the winter months (see "Examination and Diagnostic Techniques"). Coborn and Lawrence (1987b) described a method for limiting the hibernation of Mediterranean tortoises to four months in captivity. Juvenile reptiles, or those unfit to hibernate for other reasons, eg. illness, should be prevented from doing so by keeping them in a heated vivarium with an artificial daylength of 12 hours during the autumn and winter months (Kirsche, 1979; Frye, 1991). Feeding should be stopped several weeks prior to dormancy in order to allow emptying of the gastrointestinal tract and prevent fermentation, which may cause death or disease (Marcus, 1981).

Many tropical species do not hibernate but go through a period during the cooler months of the year when they cease to feed. This is often associated with the rainy season and is coincidental with the breeding season for many species. Therefore, if reptiles are to be properly maintained in captivity, the natural environment should be carefully researched.

HANDLING

Reptiles should be transported in a dark secure box or bag; the arrival at the surgery of a large boid wrapped around the neck of the owner should be strongly discouraged. When handling reptiles it is important to take precautions to avoid damage to either the animal or the handler, particularly when dealing with venomous snakes or lizards. These should only be handled if two or more people are present and the antivenin, if appropriate, is readily available. The use of antivenin is not always indicated following a bite from a venomous snake (Russell, 1980). In some cases allergic reactions to the antivenin, or its horse serum component, may be more problematical than the snake bite itself. Venomous reptiles are covered by the Dangerous Wild Animals Act (Cooper, 1978) (see "Appendix Four - Legal Aspects"), their possession requiring a local authority licence. Venomous snakes should only be examined by veterinary surgeons who are experienced in such techniques (Lawton, 1989).

Reptiles may carry zoonotic infections: for example, many species, particularly red-eared terrapins (*Trachemys scripta elegans*), often carry *Salmonella* spp. as commensal intestinal flora (Cooper, 1985; Frye, 1991). Although cases of salmonellosis contracted by humans from these animals have been reported many times in the USA, such incidents are rare in the UK (Lawton and Cooper, 1991). Other potential zoonoses from reptiles include other bacteria, eg. *Pseudomonas* spp., rickettsiae, eg. Q fever, and parasites, eg. pentastomids. Therefore, attention must be paid to personal hygiene when handling these animals.

Figure 1
Handling a docile tortoise (*Testudo* sp.).

Chelonians

Small tortoises and terrapins can be picked up by holding the whole shell between both hands slightly cranial to the hindlimbs (see Figure 1), by gripping both sides of the carapace just above the hindlimbs, or via the supracaudal shields. Box-terrapins (*Terrapene* spp.) have a hinged plastron and the animal's ability to close its shell should be borne in mind when handling this species. When held upside down, chelonians tend to relax and this may facilitate certain procedures (see "Examination and Diagnostic Techniques").

Handling the relatively docile Mediterranean tortoises (*Testudo* spp.) and red-eared terrapin (*Trachemys scripta elegans*) (see Figure 2) may be simple, but restraining some species of chelonians, eg. the snapping turtle (*Chelydra serpentina*) and the alligator snapping turtle (*Macroclemys temmincki*), can present problems. These species tend to be aggressive, particularly when out of water, and have a very powerful bite (Fowler, 1978; Jackson, 1991). Soft-shelled terrapins (*Trionyx* spp.) are also prone to bite and should be handled with care. Holding these animals on both sides of the carapace immediately above the hindlimbs is the safest method. Small specimens of these aggressive species can be examined more thoroughly by using the animals' forelimbs to restrain the head, as demonstrated by Lawton (1989). Chemical immobilisation should be employed before attempting close examination of the head or forelimbs of larger specimens (see "Examination and Diagnostic Techniques" and "Anaesthesia"). A cloth or gloves should be used when handling soft-shelled terrapins in order to ensure a secure hold (see Figure 3).

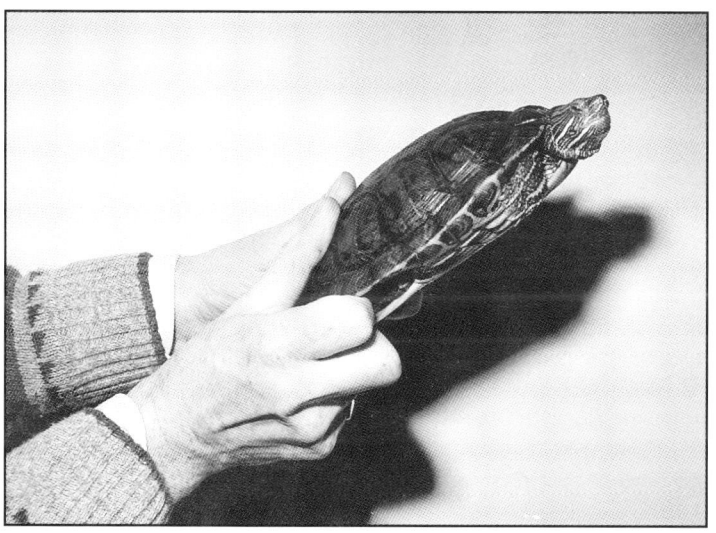

Figure 2
Handling a red-eared terrapin (*Trachemys scripta elegans*).

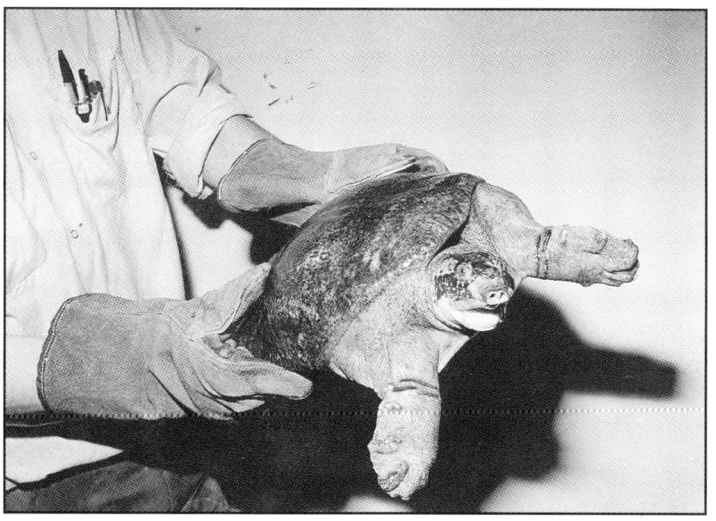

Figure 3
Handling a soft-shelled terrapin (*Trionyx* spp.).

Crocodilians

Apart from the obvious danger of being bitten, wounds can be inflicted by the claws, powerful tail and abrasive skin. Generally, the degree of risk is in direct proportion to the size of the animal but even young crocodilians can inflict serious injuries (Fowler, 1978).

Small specimens can easily be handled by grasping the animal behind the head and by the tail simultaneously. It may be advisable to pin the head to the ground or cover the eyes with a damp cloth before carrying out this procedure. For slightly larger specimens, a noose can be placed around the snout and the animal restrained by pinning it to the ground. This procedure should be carried out quickly to prevent damage to the neck or strangulation due to the animal struggling and the noose slipping (Fowler, 1978).

Particular care must be taken when handling large crocodilians. Several people are required and the tail should be immobilised with ropes or heavy nets. A noose can then be tossed around the jaws, making sure that the nostrils are free. A canvas muzzle, described by Ball (1974), has been devised for facilitating this procedure.

Lizards

As lizards can bite, scratch with their claws and lash out with their tails they should be held by the neck and pectoral girdle with one hand (which also controls the forelimbs), while the other hand supports the body near the pelvis and holds the hindlimbs (Coborn and Lawrence, 1987a) (see Figure 4). It may be necessary to pin down the head of some of the larger lizards, eg. monitors (*Varanus* spp.), before carrying out this procedure. The hindlimbs and tail of large lizards can be taped together to avoid scratching and gloves can be worn for protection. Care should be taken with tiny animals to avoid either an escape or too much pressure on the body (Ball, 1974). The skin of geckos is delicate and easily damaged and a soft cloth can be used to catch these animals (Ball, 1974; Coborn and Lawrence, 1987a). The head of a small lizard can be controlled between the index finger and thumb to prevent it biting. Lizards should never be caught or lifted by the tail; many species will shed their tail, autotomy being a natural defence mechanism against predators.

The gila monster (*Heloderma suspectum*) and the beaded lizard (*Heloderma horridum*) are the only venomous lizards and, although they are not particularly aggressive, the handler must take care to avoid their bite (Russell, 1980). It may be advisable to use a snake hook to control the animal's head before picking it up, holding the neck to restrain the head.

Figure 4
Restraining a savannah monitor lizard (*Varanus* sp.).

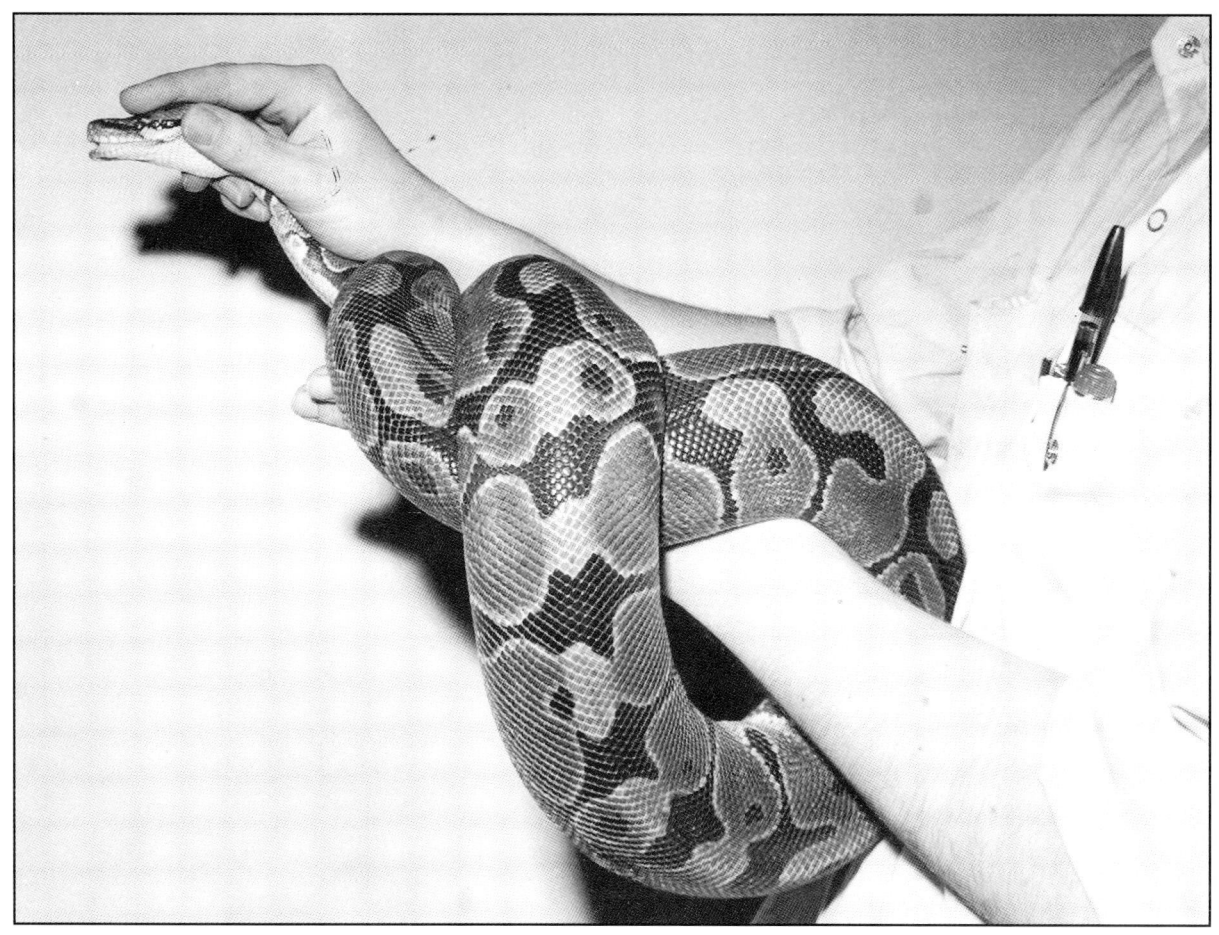

Figure 5
Controlling a royal python (*Python regius*).

Snakes

As with chelonians, aggressiveness in snakes is almost always species-related. The reticulated python (*Python reticulatus*), anacondas (*Eunectes* spp.) and brown water python (*Liasis fuscus*) are examples of the more irascible of the commoner species kept in captivity.

Non-venomous snakes can be controlled with a gentle, but firm, grasp with the thumb and second finger behind the occiput, while the index finger applies slight pressure to the top of the head. Reptiles have only one occipital condyle and rough handling can cause damage to the cervical spine. To avoid this, it is important to support the body of the snake. This can be done by allowing smaller animals to coil around the handler's arm so that the snake is gripping the handler rather than *vice versa* (Lawton, 1989) (see Figure 5). However, this technique should be avoided when handling larger snakes, ie. those more than 2–3 metres in length, the body of such animals being supported with the help of other people (Ball, 1974; Coborn and Lawrence, 1987a). Care should be taken not to grip the animal too tightly, as this may cause bruising which can result in debility or death (Lawton, 1989; 1991).

Equipment is now available from specialist dealers to facilitate the handling of snakes (Ball, 1974). The use of such implements, eg. snake hooks and snake tongs, is advisable when handling venomous species (even newly-hatched snakes can be dangerous). When examining these snakes, use should be made of restraining devices, eg. rigid plastic tubes or tight fitting boxes which can be modified as squeeze cages (see "Examination and Diagnostic Techniques"). However, a full clinical examination will normally require chemical immobilisation (see "Anaesthesia"). When dealing with snakes which can spit venom, eg. spitting cobras (*Naja nigricollis*) and ringhals (*Hemachatus hemachatus*), eye visors should be worn (Fowler, 1978).

SEXING

True sexual dimorphism is present in only a few species of reptiles: for example, mature male three-horned chameleons (*Chamaeleo jacksoni*) have facial horns but females do not (Frye, 1991); female European adders (*Vipera* spp.) will grow to 60 – 90cm in length whereas males rarely exceed 45cm (Mehrtens, 1987). Although external secondary sexual characteristics may become evident in many species during the breeding season, observation of mating, direct comparison with specimens of known sex, or a physical examination may be necessary if animals are to be sexed accurately.

Chelonians

The most reliable method of sexing many chelonians, eg. Testudinidae, is by examining the tail. The tail of the male is often longer and wider at the base than the female, and the cloaca of the male more caudal then the female's when compared with the rear edge of the plastron (see Figure 6). In some species the males have longer front claws than the females, eg. the red-eared terrapin (*Trachemys scripta elegans*) and some marine turtles. The iris of the mature male common box-tortoises (*Terrapene carolina*) is frequently red, while in females it is usually brown. Males are often smaller and lighter then females, eg. map turtles (*Graptemys* spp.), but the opposite holds true for some species, eg. gopher tortoises (*Xerobates* spp.).

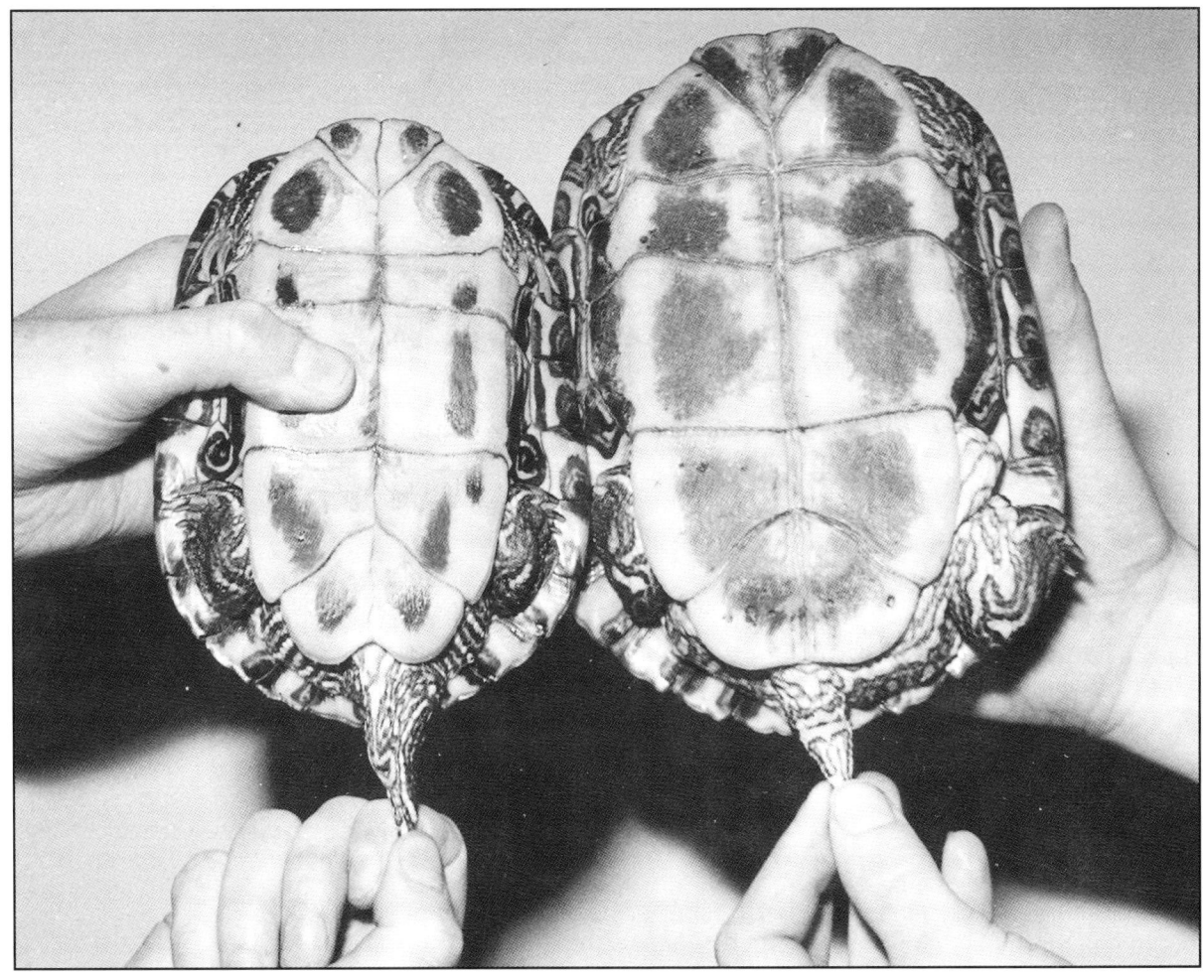

Figure 6
Sexing red-eared terrapins (*Trachemys scripta elegans*). Note: the male (on the left) has a longer tail and the distance between the end of the plastron and the cloaca is greater than that of the female (on the right). The male is smaller and has longer claws on his fore-feet.

Male tortoises are often cited as having a concave plastron in contrast to the flat plastron of females (Frye, 1991). Although this is true for some genera, eg. *Xerobates* spp., it is not true for Mediterranean tortoises (*Testudo* spp.) (Jackson and Lawrence, 1985; Coborn and Lawrence, 1987b).

Crocodilians

Sexing crocodilians is by cloacal palpation. A lubricated, gloved finger is inserted into the cloaca and directed cranially. In males, a small pointed structure, representing the penis, can be felt (Brazaitis, 1968; Jackson, 1985). This structure is absent in the female. Light pressure on either side of the cloaca can be used to demonstrate the penis, which will then be extruded.

Lizards

Males tend to be larger and stronger then females and they are often more colourful during the breeding season. Males tend to have more body appendages then females, eg. horns (*Chamaeleo* spp.), helmets (*Basiliscus* spp.), heel spurs (*Chamaeleo* spp.) or larger gular folds (*Iguana iguana*). The pre-anal, anal and femoral pores are often more developed and numerous in the male; this is particularly so in the Lacertidae. Males have two hemipenes situated slightly distal to the cloaca; these may be seen as a bulge in the tail. They can be gently extruded for sexing or carefully probed (Wagner, 1988).

Snakes

Although the tail is usually longer (determined by counting the paired post-cloacal scales) (see "Examination and Diagnostic Techniques") and has a wider base (due to the housing of the hemipenes) in males (Lawton, 1991), in some species there is an overlap between the sexes. Boas and pythons have small spurs on either side of the cloaca. These are often larger in the male. The most reliable way of sexing snakes is by probing of the hemipenis or vaginal sac. This procedure can result in severe damage to the snake if incorrectly performed. The probe should be of the correct size, be blunt-ended and well lubricated. Purpose-made probes are commercially available, but a Jackson's cat catheter (Arnolds) can be used successfully in many cases. The probe is placed caudally into the vent. In males the probe can be gently inserted to a depth of approximately 8 – 16 sub-caudal scales, while in females the probe can be inserted to a depth of only 2 – 6 sub-caudal scales (Laszlo, 1984; Lawton, 1991).

BREEDING

In order to breed in captivity, reptiles must attain breeding condition, ie. have adequate fat reserves and have developed active gonads, and must be exposed to environmental conditions conducive to triggering breeding behaviour (Coote, 1985).

Many different parameters affect the onset of breeding in reptiles, eg. temperature patterns, photoperiod, relative humidity and availability of food. Such parameters vary in importance between species, depending on whether the reptile is a spring breeder, autumn breeder or non-seasonal breeder. The male:female ratio may be important. In some cases the presence of extra males will only result in fighting and territorial displays, eg. leopard ground gecko (*Eublepharis macularius*) (Allen, 1987); in others it may be a necessary stimulus for mating to take place, eg. Sinaloan milk snake (*Lampropeltis triangulum sinaloae*) and gila monster (*Heloderma suspectum*) (Risley, personal communication). Separation and reintroduction of the sexes may also trigger breeding activity. For some species of snakes repeated spraying with chilled water may stimulate breeding (Laszlo, 1984).

Courtship rituals vary between the species but generally consist of three phases: the male chasing the female, the male positioning himself with respect to the female and copulation. During courtship and mating the male may be aggressive towards the female. This is particularly so for some lizards and many chelonians. The male leopard ground gecko, for example, licks and bites the female around the legs and neck prior to intromission (Thorogood and Whimster, 1979). The Mediterranean tortoises, amongst others, violently butt their intended mates with their shells and bite at their legs. Previously silent reptiles may vocalise during mating, eg. the spur-thighed

Table 2
Table of reproductive parameters for some commonly kept species of reptile.

Species	Method of reproduction	Period of gestation/ incubation (days)	Incubation Temperature °C	Relative humidity %
Boa constrictor (*Boa constrictor constrictor*)	V	120 – 240		
Sand boa (*Eryx* sp.)	V	120 – 180		
Indian python (*Python molurus*)	Ov	56 – 65	30 – 32	80 – 100
Royal python (*Python regius*)	Ov	90	30 – 32	80 – 100
Garter snake (*Thamnophis sirtalis*)	V	90 – 110		
King snake (*Lampropeltis getulus*)	Ov	50 – 60	25 – 30	80 – 100
Spur-thighed tortoise (*Testudo graeca*)	Ov	60	28 – 32	90 – 100
Common box-tortoise (*Terrapene carolina*)	Ov	50 – 90	25 – 30	80 – 100
Red-eared terrapin (*Trachemys scripta elegans*)	Ov	59 – 93	25 – 30	80 – 100
Common iguana (*iguana iguana*)	Ov	73	30	75
Leopard ground gecko (*Eublepharis macularius*)	Ov	55 – 60	26 – 33	80 – 100
Green anole (*Anolis carolinensis*)	Ov	60 – 90	28 – 30	80 – 100
Jackson's chameleon (*Chamaeleo jacksoni*)	V	90 – 180		
Plumed basilisk (*Basiliscus plumifrons*)	Ov	60 – 64	27	
Water dragon (*Physignathus lesueuri*)	Ov	90		

V = viviparous; Ov = oviparous;

tortoise (*Testudo graeca*) and the red-footed tortoise (*Chelonoidis carbonaria*) (Davis, 1979). The male red-eared terrapin (*Trachemys scripta elegans*) uses his long front claws during the courtship display. This species, like most terrapins and turtles, copulates in water. The duration of each phase is very variable between species and individuals with copulating alone ranging from a few seconds, eg. Australian skinks (*Tiliqua* spp.), to over an hour, eg. some chelonians. Some female reptiles, eg. common box-tortoises (*Terrapene carolina*), can store viable sperm for months, or years, after mating (Porter, 1972).

A few species are parthenogenetic, eg. the whip-tailed lizard (*Cnemidophorus* spp.) and the Caucasian rock lizard (*Lacerta saxicola*) (Marcus, 1981). The Central America night lizard (*Lepidophyma flavimaculata*) may also exhibit parthenogenesis (Holmback, 1984).

Reptiles are either oviparous, eg. crocodilians, chelonians and pythons, or viviparous, eg. boas and garter snakes (*Thamnophis* spp.) (Mehrtens, 1987). Most Families of lizards contain oviparous and viviparous species, although one type usually predominates within each Family. The term "ovoviviparous" used to be applied to most viviparous reptiles; however, such a distinction is now considered to be imprecise and its use is no longer advocated (Shine, 1985; Yaron, 1985). If closely confined in captivity, offspring may be taken as food by the adults; therefore, it is advisable to separate them as soon as possible (Wagner, 1979; Townson, 1985). This will allow improved monitoring of health, feeding habits and growth rates.

The duration of incubation is both species and temperature dependent. In general, the length of time from fertilisation to live-birth or hatching is about three months (Hubert, 1985). Average incubation periods for some live-bearing snakes commonly kept in captivity are given in Table 2.

A suitable environment must be provided for oviparous reptiles if egg laying is to be successful and eggbinding is to be prevented (Jackson, 1991; Lawton, 1991) (see "Urogenital System"). Oviparous chameleons (*Chamaeleo* spp.) and many chelonians and crocodilians, for example, require a deep sand or peat substrate in which to build a nest chamber (Davis, 1979; Bustard, 1989). Some crocodilians, eg. the broad-nosed spectacled caiman (*Caiman latirostris*), are mound-nesting and require suitable vegetation and other material with which to build their nest (Widholzer *et al*, 1986).

All crocodilians and most chelonians produce eggs with a calcified shell (calcareous eggs) (Bellairs, 1969). In contrast, the eggs of snakes, lizards and many turtles are not calcareous and have rubbery, flexible shells (soft-shelled eggs) (Bellairs, 1969). The eggs of most geckos are sticky and rubbery when first laid, but the shells quickly dry and become firm with an outer calcareous layer (Bellairs, 1969; Thorogood and Whimster, 1979).

The eggs of oviparous species should be removed for artificial incubation once they are laid. Indian pythons (*Python molurus*), however, can often be left undisturbed with their clutch as they will coil around, regulating the incubation temperature with heat generated by muscle contractions (Hutchison *et al*, 1966; Mehrtens, 1987). The eggs of some geckos are firmly attached to the walls or floor of the cage (Thorogood and Whimster, 1979). In such cases a layer of paper can be placed in the cage, or a plastic tube can be provided for egg laying. The tube can then be placed in an incubator.

The eggs of most species can be incubated in plastic bags containing a suitable water-absorbent substrate, eg. vermiculite, which has been pre-soaked in water. The shells of reptiles' eggs are permeable to water. In contrast to birds' eggs, reptiles' eggs absorb water from the atmosphere and gain weight during incubation (Thorogood and Whimster, 1979). Therefore, relative humidity is a major factor in the success, or otherwise, of hatching. The ratio of vermiculite to water (by weight) for most species should be 1:1, providing a relative humidity of 80 – 100%; a ratio of 1:0.75 is best for desert species. Eggs should be marked with a pencil prior to collection and half-buried in the vermiculite, paying particular attention to positioning them exactly as they were laid, using the mark already made. The bags should then be sealed, labelled and placed in a thermostatically controlled incubator. Incubation temperature and duration varies between species (see Table 2). Unlike birds' eggs, reptiles' eggs should never be turned but the plastic bags should be ventilated once or twice a week to restore any water loss and avoid hypoxia which can result in developmental abnormalities. Incorrect environmental conditions during part, or all, of the incubation period can result in abnormal embryonic development (Bellairs, 1981). Temperature and humidity are of particular importance in this respect (Bellairs, 1981). The development of the eggs can be monitored by regular weighing and candling. Mould often grows rapidly over the shells of eggs which are infertile or which die during incubation (Ross, 1980). In-contact healthy eggs are usually not affected (Ross, 1980) and it is probable that reptiles' eggs have an innate resistance to fungal infection (Austwick and Keymer, 1981).

Sex determination of the offspring of snakes and most lizards is, as in endotherms, genetically determined. However, for chelonians, crocodilians and a few lizards, eg. the leopard ground gecko (*Eublepharis macularius*), sex determination is environmentally dependent with incubation temperature being the most important factor (Hubert, 1985). Generally, higher incubation temperatures result in a predominance of females in chelonians and a predominance of males in lizards and crocodilians.

Newly-hatched reptiles will not usually feed until the yolk sac has been absorbed and they have undergone their first moult (Thorogood and Whimster, 1979; Townsend, 1979). The latter usually occurs within the first 2–3 weeks but can take place within a few minutes of birth/hatching (Brodsky, 1969). If they continue to refuse to eat, forced feeding may be necessary (see earlier). Water should be available for drinking and bathing during this period (see earlier).

Acknowledgements

The authors would like to thank Mr. D. Ball, retired Assistant Curator of Reptiles, and Mr. D. Risley, Head Keeper, Reptile House, London Zoo, for their help with this chapter. Terry Dennett of the Institute of Zoology, took the photographs.

REFERENCES

ALLEN, M.E. (1988). The effect of three light treatments on growth in the green iguana (*Iguana iguana*). *Proceedings of the Joint Conference of the American Association of Zoo Veterinarians and the American Association of Wildlife Veterinarians.*

ALLEN, R. (1987). Captive care and breeding of the leopard gecko, *Eublepharis macularius*. In: *Reptiles: Proceedings of the 1986 UK Herpetological Societies Symposium on Captive Breeding.* (Ed. J. Coote). British Herpetological Society, London.

AUSTWICK, P.K.C. and KEYMER, I.F. (1981). Fungi and actinomyceters. In: *Diseases of the Reptilia, Vol.1.* (Eds. J.E. Cooper and O.F. Jackson). Academic Press, London.

AVERY, R.A. (1985). Thermoregulatory behaviour of reptiles in the field and in captivity. In: *Reptiles: Breeding, Behaviour and Veterinary Aspects.* (Eds. S. Townson and K. Lawrence). British Herpetological Society, London.

BALL, D. (1974). Handling and restraint of reptiles. *International Zoo Yearbook* **14,** 138.

BELLAIRS, A. d'A. (1969). *The Life of Reptiles, Vol. 1 and 2.* Weidenfeld and Nicolson, London.

BELLAIRS, A. d'A. (1981). Congenital and developmental diseases. In: *Diseases of the Reptilia, Vol. 2.* (Eds. J.E. Cooper and O.F. Jackson). Academic Press, London.

BELS, V.L. (1986). A behavioural approach to the maintenance of anolis lizards in captivity. In: *Exotic Animals in the Eighties. Proceedings of the 25th Anniversary Symposium of the British Veterinary Zoological Society.* (Eds. P.W. Scott and A.G. Greenwood). British Veterinary Zoological Society, London.

BLATCHFORD, D. (1987). Environmental lighting. In: *Reptiles: Proceedings of the 1986 UK Herpetological Societies Symposium on Captive Breeding.* (Ed. J. Coote). British Herpetological Society, London.

BLOXAM, Q. and TONGE, S. (1982). Breeding and maintenance of the plumed basilisk (*Basiliscus plumifrons*) at the Jersey Wildlife Preservation Trust. In: *Proceedings of the 5th Annual Reptile Symposium on Captive Propagation and Husbandry.* The Reptile Symposium on Captive Propagation and Husbandry Zoological Consortium, Thurmont.

BLUEMAN, T. (1988). Some husbandry techniques developed at Brookfield Zoo's reptile house. In: *Proceedings of the 11th International Herpetological Symposium on Captive Propagation and Husbandry.* (Ed. K.H. Petersen). International Herpetological Symposium on Captive Propagation and Husbandry Zoological Consortium, Thurmont.

BOYER, D.M. (1988). Watering techniques for captive reptiles and amphibians. In: *Proceedings of the 10th International Herpetological Symposium on Captive Propagation and Husbandry.* (Ed. K.H. Peterson). International Herpetological Symposium on Captive Propagation and Husbandry Zoological Consortium, Thurmont.

BRAZAITIS, P. (1968). The determination of sex in living crocodilians. *British Journal of Herpetology* **4**, 54.

BRODSKY, O. (1969). Breeding the great house gecko (*Gekko gecko*) at Prague Zoo. *International Zoo Yearbook* **9**, 37.

BUSTARD, R. (1989). Keeping and breeding oviparous chameleons. *The British Herpetological Society Bulletin* **27**, 18.

CAMPDEN-MAIN, S.M. and CAMPDEN-MAIN, L.K. (1983). A guide to the care of neonate ophidia. In: *Proceedings of the 6th Annual Reptile Symposium on Captive Propagation and Husbandry.* (Ed. D.L. Marcellini). The Reptile Symposium on Captive Propagation and Husbandry Zoological Consortium, Thurmont.

COBORN, J. and LAWRENCE, K. (1987a). Snakes and lizards. In: *The UFAW Handbook on the Care and Management of Laboratory Animals.* 6th Edn. (Ed. T. Poole). Longman Scientific and Technical, Harlow.

COBORN, J. and LAWRENCE, K. (1987b). Tortoises and terrapins. In: *The UFAW Handbook on the Care and Management of Laboratory Animals.* 6th Edn. (Ed. T. Poole). Longman Scientific and Technical, Harlow.

COOPER, J.E. (1981). Bacteria. In: *Diseases of the Reptilia, Vol. 1.* (Eds. J.E. Cooper and O.F. Jackson). Academic Press, London.

COOPER, J.E. (1985). The significance of bacterial isolates from reptiles. In: *Reptiles: Breeding, Behaviour and Veterinary Aspects.* (Eds. S. Townson and K. Lawrence). British Herpetological Society, London.

COOPER, J.E. and JACKSON, O.F. (1981). Miscellaneous diseases. In: *Diseases of the Reptilia, Vol.2.* (Eds. J.E. Cooper and O.F. Jackson). Academic Press, London.

COOPER, M.E. (1978). The Dangerous Wild Animals Act 1976. *Veterinary Record* **102**, 475.

COOPER, M.E. (1987). *An Introduction to Animal Law.* Academic Press, London.

COOTE, J. (1985). Breeding colubrid snakes, mainly *Lampropeltis.* In: *Reptiles: Breeding, Behaviour and Veterinary Aspects.* (Eds. S. Townson and K. Lawrence). British Herpetological Society, London.

DAVIES, P.M.C. (1981). Anatomy and physiology. In: *Diseases of the Reptilia, Vol. 1.* (Eds. J.E. Cooper and O.F. Jackson). Academic Press, London.

DAVIS, S. (1979). Husbandry and breeding of the red-footed tortoise (*Geochelone carbonaria*) at the National Zoological Park, Washington. *International Zoo Yearbook* **19**, 50.

ERNST, C.H. and BARBOUR, R.W. (1972). *Turtles of the United States.* The University Press of Kentucky, Lexington.

FOWLER, M.E. (1978). *Restraint and Handling of Wild and Domestic Animals.* Iowa State University Press, Ames.

FRYE, F.L. (1991). *Biomedical and Surgical Aspects of Captive Reptile Husbandry.* 2nd Edn. Kreiger, Malabar.

GREGORY, P.T. (1982). Reptilian hibernation. In: *Biology of the Reptilia, Vol. 13: Physiology D.* (Eds. C. Gans and F.H. Pough). Academic Press, London.

HARPER, R. (1987). The veterinary aspects of captive reptile husbandry. *Reptiles: Proceedings of the 1986 UK Herpetological Societies Symposium on Captive Breeding.* (Ed. J. Coote). British Herpetological Society, London.

HOLMBACK, E. (1984). Parthenogenesis in the Central American night lizard (*Lepidophyma flavimaculatum*) at San Antonio Zoo. *International Zoo Yearbook* **23**, 157.

HUBERT, J. (1985). Embryology of the Squamata. In: *Biology of the Reptilia, Vol. 15: Development B.* (Eds. C. Gans and F. Billett). John Wiley and Sons, New York.

HUTCHISON, V.H., DOWLING, H.G. and VINEGAR, A. (1966). Thermoregulation in a brooding female Indian python (*Python molurus bivittatus*). *Science* **151**, 694.

JACKSON, O.F. (1985). Crocodilians. In: *Manual of Exotic Pets.* Revised Edn. (Eds. J.E. Cooper, M.F. Hutchison, O.F. Jackson, and R.J. Maurice). BSAVA, Cheltenham.

JACKSON, O.F. (1991). Chelonians. In: *Manual of Exotic Pets.* New Edn. (Eds. P.H. Beynon and J.E. Cooper). BSAVA, Cheltenham.

JACKSON, O.F. and LAWRENCE, K. (1985). Chelonians. In: *Manual of Exotic Pets.* Revised Edn. (Eds. J.E. Cooper, M.F. Hutchison, O.F. Jackson and R.J. Maurice). BSAVA, Cheltenham.

JACOBSON, E.R. (1988). Evaluation of the reptile patient. In: *Exotic Animals.* (Eds. E.R. Jacobson, and G.V. Kollias). Churchill Livingstone, New York.

JOANEN, T. and McNEASE, L. (1979). Culture of the American alligator. *International Zoo Yearbook* **19**, 61.

KAUFFELD, C. (1969). The effect of altitude, ultra-violet light and humidity on captive reptiles. *International Zoo Yearbook* **9**, 8.

KIRKWOOD, J.K. (1991). Energy requirements for maintenance and growth of wild mammals, birds and reptiles in captivity. *Journal of Nutrition* **121**, supplement: S29.

KIRSCHE, W. (1979). The housing and regular breeding of Mediterranean tortoises (*Testudo* spp.) in captivity. *International Zoo Yearbook* **19**, 42.

LASZLO, J. (1984). Further notes on reproductive patterns of amphibians and reptiles in relation to captive breeding. *International Zoo Yearbook* **23**, 166.

LAWTON, M.P.C. (1989). *Examination and Handling of Reptiles.* UVCE Video. Unit for Veterinary Continuing Education, London.

LAWTON, M.P.C. (1991). Lizards and snakes. In: *Manual of Exotic Pets.* New Edn. (Eds. P.H. Beynon and J.E. Cooper). BSAVA, Cheltenham.

LAWTON, M.P.C. and COOPER, J.E. (1991). Salmonella in terrapins. *Veterinary Record* **129**, 127.

MAHMOND, I.Y. and KLICKA, J. (1979). Feeding, drinking and excretion. In: *Turtles: Perspectives and Research.* (Eds. M. Harless and H. Morlock). John Wiley and Sons, New York.

MARCUS, L.C. (1981). *Veterinary Biology and Medicine of Captive Amphibians and Reptiles.* Lea and Febiger, Philadelphia.

MATTISON, C. (1989). *Lizards of the World.* Blandford Press, London.

MEEK, R. and AVERY, R.A. (1988). Mini-review: thermoregulation in chelonians. *Herpetological Journal* **1**, 253.

MEHRTENS, J.M. (1987). *Living Snakes of the World.* Sterling Publishing, New York.

NEITMAN, K. (1982). The role of thermoclines and temperature variance in reptile husbandry. In: *Proceedings of the 5th Annual Reptile Symposium on Captive Propagation and Husbandry.* The Reptile Symposium on Captive Propagation and Husbandry Zoological Consortium, Thurmont.

PEAKER, M. (1969). Some aspects of the thermal requirements of reptiles in captivity. *International Zoo Yearbook* **9**, 3.

PORTER, K.R. (1972). *Herpetology.* W.B. Saunders, Philadelphia.

PRICE, H. (1991). Exotic pet husbandry. *Veterinary Practice* **23 (1)**, 1.

REID, D. (1987). Rearing juvenile spur-thighed tortoises (*Testudo graeca*). In: *Proceedings of the 1986 UK Herpetological Societies Symposium on Captive Breeding.* (Ed. J. Coote). British Herpetological Society, London.

ROSS. R. (1980). The breeding of pythons (Sub-family *Pythononae*) in captivity. In: *Reproductive Biology and Diseases of Captive Reptiles.* (Eds. J.B. Murphy and J.T. Collins). Society for the Study of Amphibians and Reptiles, Ohio.

ROSSKOPF, W.J. and WOERPEL, R.W. (1981). Rat bite injury in a pet snake. *Modern Veterinary Practice* **62**, 871.

RUSSELL, F.E. (1980). *Snake Venom Poisoning.* J.B. Lippincott, Philadelphia.

SHINE, R. (1985). The evolution of viviparity in reptiles: an ecological analysis. In: *Biology of the Reptilia, Vol. 15 : Development B.* (Eds. C. Gans and F. Billett). John Wiley and Sons, New York.

THOROGOOD, J. and WHIMSTER, I.W. (1979). The maintenance and breeding of the leopard gecko (*Eublepharis macularius*). *International Zoo Yearbook* **19**, 74.

TOWNSEND, C.R. (1979). Establishment and maintenance of colonies of parthenogenetic whiptail lizards (*Cnemidophorus* spp.). *International Zoo Yearbook* **19**, 80.

TOWNSON, S. (1985). The captive reproduction and growth of the yellow anaconda (*Eunectes notaeus*). In: *Reptiles: Breeding, Behaviour and Veterinary Aspects.* (Eds. S. Townson and K. Lawrence). British Herpetological Society, London.

TRYON, B.W. (1985). Snake hibernation and breeding: in and out of the zoo. In: *Reptiles: Breeding, Behaviour and Veterinary Aspects.* (Eds. S. Townson and K. Lawrence). British Herpetological Society, London.

WAGNER, E. (1979). Breeding king snakes (*Lampropeltis* spp.). *International Zoo Yearbook* **19**, 98.

WAGNER, E. (1988). Sexing Australian skinks. In: *Proceedings of the 11th International Herpetological Symposium on Captive Propagation and Husbandry.* (Ed. M.J. Rosenberg). International Herpetological Symposium on Captive Propagation and Husbandry Zoological Consortium, Thurmont.

WARWICK, C. (1989). The welfare of reptiles in captivity. In: *First World Congress of Herpetology Abstracts.* (Eds. T. Halliday, J. Baker and L. Hosie). Open University, UK.

WEBB, G., MANOLIS, S. and WHITEHEAD, P. (1987). *Wildlife Management: Crocodiles and Alligators.* Surrey Betty and Sons, Chipping Norton.

WIDHOLZER, F.L., BORNE, B. and TESCHE, T. (1986). Breeding the broad-nosed caiman (*Caiman latirostris*) in captivity. *International Zoo Yearbook* **24/25**, 226.

YARON, Z. (1985). Reptilian placentation and gestation: structure, function and endocrine control. In: *Biology of the Reptilia, Vol. 15: Development B.* (Eds. C. Gans and F. Billett). John Wiley and Sons, New York.

CHAPTER THREE

EXAMINATION AND DIAGNOSTIC TECHNIQUES

Oliphant F Jackson PhD MRCVS
Martin P C Lawton BVetMed CertVOphthal FRCVS

INTRODUCTION

The basic principles of clinical examination and use of diagnostic techniques are the same for the whole of the Class Reptilia. However, such is the diversity in size, physical appearance, ease of handling and environmental requirements, that for convenience this chapter will be divided into chelonians, snakes and lizards.

Before embarking on a clinical examination of a reptilian patient, a full case history must be taken. It is also advantageous, wherever possible, to examine the reptile in its environment. Careful questioning of the owner will allow one to ascertain if the correct environmental conditions (temperature, humidity, photoperiod) are being provided. In an anorexic patient one must establish if the correct type of food is being offered, or if the reptile is being handled too frequently.

Good herpetologists will keep records of the behaviour and physiological processes of the reptiles in their care; all reptile owners should be encouraged to keep such records (Jackson, 1987). These are not only useful for monitoring seasonal changes in breeding activity, shedding and feeding, but often indicate the first signs of disease. The keeping of records showing when the reptile has fed, defaecated, shed, urinated and been lethargic or active, together with regular recordings of bodyweight, is strongly advised.

All reptiles presented to a veterinary surgeon should be routinely weighed and measured (see later). Measurement of the rostrum to cloaca length should always be made (see "Appendix Two – Clinical Examination Sheets").

Handling of reptiles is covered very briefly in this chapter and is confined to hints on dealing with the difficult patient. Further details on handling of reptiles can be found elsewhere in this manual (see "Management in Captivity") and in Lawton (1991a).

CHELONIANS

Handling

Handling and restraining tortoises of the genus *Testudo* presents few problems. The occasional tortoise will, however, be head-shy and keep its head withdrawn deeply within the shell. It is possible to withdraw the head by applying whelping or sponge-holding forceps over the cranium and under the mandible. The head can then be withdrawn slowly but steadily. Once the head is sufficiently retracted, the normal method of holding the head with finger and thumb behind the occipital condyles can be used.

Limbs can usually be handled without difficulty providing one waits patiently for one to be withdrawn. As the volume within the shell is restricted, it is possible to force the forelimbs and the head out slightly by pressing both hindlimbs deeply into the inguinal fossae; sometimes, the hindlimbs can be forced slightly into the open by pressing both forelimbs and head into the cranial opening.

Handling box-tortoises (*Terrapene* spp.) requires care as closure of either hinge on a finger can be painful. Gentle pressure on the posterior hinge often causes the anterior hinge to open slightly; placing a finger and thumb between the carapace and plastron and applying gentle pressure to open the hinge will allow a limb to be withdrawn, which in itself will keep the hinge open (Lawton, 1991a).

Terrapins and turtles should be held at the rear of the carapace, ideally with one's fingers in the inguinal fossae. The snapping turtle (*Chelydra serpentina*) has a long neck which can reach an unsuspecting hand and inflict a severe bite. Soft-shelled turtles (*Trionyx* spp.) also have long necks and a tendency to bite; with this genus it may be advisable to wear gloves or allow them to bite on to a towel, as once they have bitten on to something which remains between their jaws, they can be examined more easily and safely.

Clinical examination

A full clinical examination is not as impossible as it might at first appear. Clinical examination of all reptiles should be methodical, starting with the head and working one's way down the body, ending with the tail. Patience is often required in order to carry out a full examination and, in some aggressive or larger species, sedation is advised (see "Anaesthesia").

Prior to carrying out a clinical examination, it is important to measure the total bodyweight and carapace length of the chelonian. In certain species (*Testudo* spp.) calculations can be made based on bodyweight and carapace length which give an indication of the health status of the individual and allow the clinician to assess whether the patient is a satisfactory weight for its length, is underweight or is even overweight (Jackson's ratio) (see Figure 1) (Jackson, 1980). It is also important to assess whether they are alert or depressed, whether or not they are able to lift themselves off the table top, and whether or not they are withdrawn within their shell during the whole time of the examination.

The nostrils should be clear with no sign of mucoid or purulent discharge. Any discharge should be examined further, samples being taken for microscopy and bacteriological examination (see "Respiratory System"). The eyelids should be free to open, with no distension or discolouration. The eyes should be clear and bright. It is often normal, especially in *Testudo* spp., for there to be an overflow of tears down the side of the face. Any swelling, discolouration or pain should be

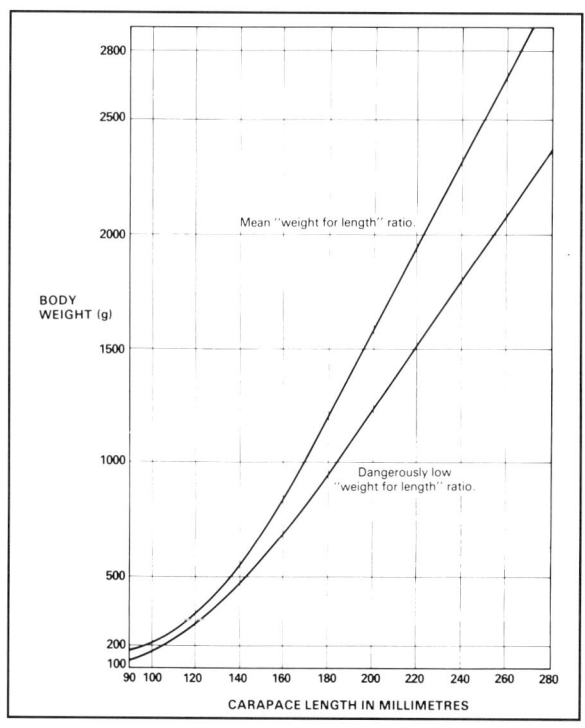

Figure 1
Growth of healthy *Testudo graeca* and *T. hermanni* in the UK, plotting mass against length.

investigated further (see "Ophthalmology"). The tympanic membranes should be identified in all chelonians; these are situated on the lateral side of the head, just above the ramus of the mandible. This area should be flat or slightly concave; any bulging is usually associated with otitis media (ear abscesses) which will require surgical treatment (see "Surgery"). The beaks of chelonians should be evenly apposed. Any overgrowth of either the maxillary or mandibular beak should be clipped and/or filed into shape. The mouth should always be opened in order to examine the colour of the tongue and mucous membranes, and to make sure that there is no discharge or foreign bodies around the internal nares. The mouth and surface of the tongue should also be checked for excess saliva or any abnormal or necrotic material (see "Gastrointestinal System"). Pus should not be present at the opening of the Eustachian tubes in the pharynx; the presence of pus may be associated with otitis media.

Next, the neck and limbs should be examined. Obvious swellings must be differentiated into soft (possibly oedematous or fatty) or hard (abscesses, fibrous material, steatitis or tumours) swellings. If necessary, needle aspirations or biopsies may be required (see "Laboratory Investigations"). The skin should be carefully assessed for its colouration and condition; any abnormality should be investigated further (see "Integument"). Specific examination is required for parasites, scarring or signs of dysecdysis or excessive sloughing. As in mammals, the degree of dehydration may be assessed by the elasticity of the skin.

Limbs should be examined for any abnormalities or swelling. Any enlargement or swelling should be investigated radiographically to assess whether or not there is a fracture, soft tissue injury or osteolysis due to bacterial infection. The feet and the plastron should be examined for signs of excessive wear: this can be associated with being kept on unsuitable substrate, eg. concrete, or attempts at digging a nest in females, associated with eggbinding.

The shell should be examined, looking for abnormalities which may be associated with past or present nutritional disturbances (see "Nutritional Diseases"). Note should be made of any sign of softness or haemorrhages under the shields, especially in species where the shell is normally firm. Softness and haemorrhages can be associated with nutritional problems, bacterial infections or trauma.

The cloaca should be examined for signs of discharges, prolapses, excessive wear or swellings. If necessary, the cloaca may be investigated using a speculum or endoscope, or by digital examination. Any swelling or discharge should be investigated further (see "Urogenital System"). If faeces or urine are passed by the chelonian during the clinical examination, it is important that these are examined promptly under a microscope for signs of parasitic infection.

The ophthalmoscope and stethoscope are useful aids in the clinical examination of chelonians. Radiographic examination is often helpful (see "Radiological and Related Investigations"). Haematological and biochemical evaluations may be required to reach a final diagnosis (see "Laboratory Investigations"). The use of diagnostic endoscopy is often overlooked as a valuable tool for examining within the shell of chelonians. Endoscopy can be useful for assessing liver and kidney function as well as the reproductive state of the chelonian (Brearley *et al*, 1991).

Sexing

Many male chelonians have a longer tail than the female. This is easy to assess when both sexes are compared, but with practice it soon becomes possible to identify the sex of individual chelonians. Other sexual differences are recognised: for example, the male American box-tortoise (*Terrapene carolina*) has red eyes and the female has brown eyes. Further information can be found in "Management in Captivity".

SNAKES

Handling

Many snakes are transported to the surgery in a bag and are usually coiled up in the bottom. When dealing with non-venomous species one should pick up the whole snake still coiled; this gives an assessment of the muscular tone of the snake. How the snake is coiled also has a bearing on the reptile's state of health; sick snakes are less likely to remain coiled when being picked up and often become limp. Soon the snake will uncoil and start moving over one's hands and arms and the tongue will be flicked in and out to pick up scent droplets. The snake should be allowed to move over one's hands; in a healthy snake there is a feeling of strength as it glides over the hand. Constricting and arboreal snakes often throw a tail coil round the wrist in order to give themselves some support. If a snake has a coil of the caudal part of its body round the wrist, one should allow the head and body to hang down. A healthy snake will be strong enough to bring its head right up to the wrist as it would if it were hanging from a branch (see "Neurological Diseases").

For aggressive species it is advisable to control the head from outside the bag before placing one's hand inside. The head should be gripped from behind the occiput. For giant species, eg. Burmese pythons (*Python molurus*), it is sensible to have another pair of hands to hold the snake and prevent the risk of accidents. Dealing with venomous species is not advised for those with limited experience in handling snakes; this is described elsewhere (Lawton, 1991a).

Clinical examination

The owner should be asked to produce his or her snake records, which should include details of the quantity and type of food eaten as well as the date on which the snake last fed. The records should also include details of the weight and body length of the snake, when it sloughed and when it last passed faeces and urates.

Although at present there is no "Jackson's ratio" for snakes, it is nevertheless important that measurements are taken both of the bodyweight and of the rostrum to cloaca length. This is useful in assessing one member of a colony, or even for legal cases, as well as for future reference. It is also essential when administering an anaesthetic or medication to know the weight of the snake (see "Therapeutics").

The nostrils and infra-orbital pits should be clear with no signs of retained skin or discharges. Any discharge or sneezing should be investigated fully (see "Respiratory System").

The eyes of snakes should be clear with no wrinkling or abnormalities of the spectacle. The use of magnification is recommended for examining the rim of the spectacle at the junction with the facial scales for evidence of the snake mite (*Ophionyssus natricis*), as well as for looking under the scales of the body. If a snake is about to shed, the spectacle may appear bluish; this should be differentiated from opacities related to a buildup of lacrimal fluids in the sub-spectacular space, or even infection (see "Ophthalmology").

Prior to opening the mouth, the tongue should have been seen, as mentioned earlier, to flick in and out of the labial notch. The oral cavity should be examined to assess the colour of the mucous membranes. There should be no signs of hypersalivation, oedema, petechiation or necrotic material (see "Gastrointestinal System"). If there is evidence of fluid accumulation in the mouth, it is important to differentiate between excessive salivation associated with stomatitis and fluid discharging through the glottis associated with respiratory infection (see "Respiratory System").

The body of a snake should be rounded without any obvious bony prominences. In thin or emaciated snakes there is often sinking of the muscles around the vertebrae and ribs, giving a triangular appearance to the body. As in mammals, the degree of elasticity of the skin is an indicator of dehydration and this can be assessed by pinching a fold of skin and noting the rapidity with which it returns to its normal position. The scales should be in good condition, with no evidence of scale loss, haemorrhages, retained skin or blisters. Any abnormality of the skin should be investigated further (see "Integument"). The ventral scales, in particular, should be examined as these are often missed by owners when examining their own snakes.

There should be no palpable swellings in a healthy snake. Should a swelling be palpated within the coelomic cavity, knowledge of the organ positions (see Table 1) will be helpful in assessing its possible significance. Palpation of the coelomic cavity can be undertaken to a limited degree by digital examination under the ventral scales between the rib ends. In all snakes the length from rostrum to cloaca can be conveniently divided into three parts. In the cranial third can be found the oesophagus, trachea and heart; in the middle third the lungs, liver, stomach and cranial portion of the air sac; and in the caudal third the pylorus, duodenum, intestines, kidneys, gonads (in both sexes), uterine horns in females, vasa deferentia in males, fat body and cloaca. Table 1 can be used to find the position of an organ by measuring from the rostrum to the cloaca and from the rostrum to the organ in question.

Table 1

Average organ positions in boas and pythons.

Organ	Position expressed as % of total length from rostrum to cloaca
Heart	22 – 33
Lungs	33 – 45
Air sac	45 – 65
Liver	38 – 56
Stomach	46 – 67
Intestines	68 – 81
Right kidney	69 – 77
Left kidney	74 – 82
Colon and cloaca	81 – 100

The cloaca should be examined for oedema, erythema, discharge or hard swellings. As in chelonians, the use of a speculum or endoscope or digital examination is advantageous.

If the snake defaecates, a fresh faecal smear should be examined under a microscope (see "Laboratory Investigations") for nematode eggs (both ascarid and trichostrongyloid), live larvae and protozoa, eg. flagellates, ciliates and *Entamoeba invadens*.

Further investigations may be useful, especially haematology and biochemistry, which allow assessment of haematogenous parasites as well as liver and kidney function. Radiography and, particularly, ultrasonography are very useful in snakes (see "Radiological and Related Investigations"). Both flexible and rigid endoscopy can be used, especially in the examination of regurgitating snakes. In small and hatchling snakes transillumination is a very useful technique; a bright light source is placed against the body or within the oesophagus to transilluminate the internal organs.

Sexing

Accurate sexing of snakes requires experience. Often a skilled herpetologist can be of great assistance, especially when dealing with snakes that show sexual dimorphism - for example, the European adder (*Vipera berus*) where the males and females may be distinguished by their colour; males are reddish-brown and the females are more brownish-black. In snakes where the sex is not readily distinguished there are three main ways in which the sex may be determined; this is described by Coote (1985) and Lawton (1991b). Only one method is non-invasive and it is this method that should be advocated. The length of the tail is always less in females than in males, because in males the inverted twin hemipenes are housed caudal to the cloaca in two penile pockets. Therefore, the sub-caudal scale counts in females are usually less than for males of the same species (see Table 2). These scale counts can be carried out by the owner. It is helpful to advise owners that they are able to carry out more accurate counts, particularly in hatchling snakes, by making the counts on freshly sloughed skins and marking each scale with a felt tip pen (see Figure 2). Caudal scale counts for many more species of snakes are given in Fitzsimons (1961).

The two invasive methods of sexing snakes are probing and fluid injection (the latter is strongly discouraged). Probing involves passing an aseptic, lubricated metal probe or cannula into the inverted hemipenis (see Figure 3), the entry to which is just inside the caudal edge of the cloaca. In males, which have hemipenile openings, the probe will reach 6 – 8 sub-caudal scale lengths (sometimes more). In female snakes, which have two scent sacs (similar to anal sacs) at this site, the probe only reaches 2 – 4 sub-caudal scale lengths.

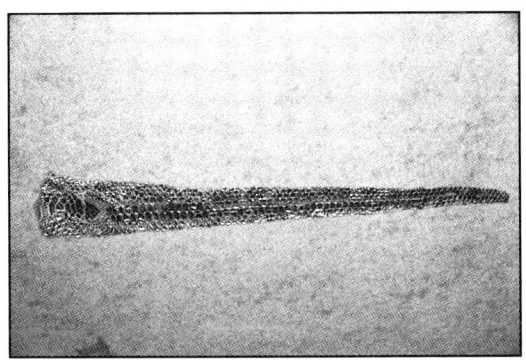

Figure 2
Paired caudal scale count
on a sloughed skin.

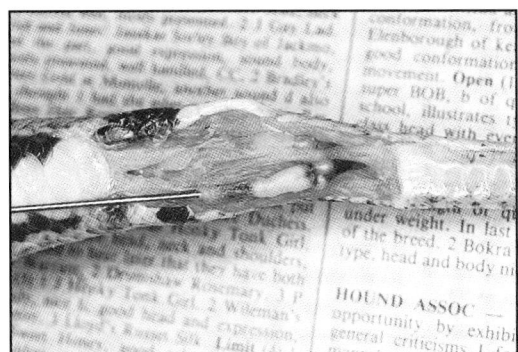

Figure 3
A dissected specimen showing
probing of the male hemipenis.

Table 2

Sub-caudal scale counts of some commonly kept species of snakes
(Jackson, unpublished data).

Species	Male	Female
Boa (*Boa constrictor constrictor*)	60	50
Black rat snake (*Elaphe obsoleta*)	93 – 98	83
Corn snake (*Elaphe guttata*)	67 – 86	41 – 62
Grass snake (*Natrix natrix*)	68 – 72	52 – 56
African house snake (*Boaedon fuliginosus*)	56 – 68	45 – 56
African rock python (*Python sebae*)	approx. 81	approx. 70
Indian python (*Python molurus*)	approx. 70	approx. 60
Royal python (*Python regius*)	approx. 40	27 – 36

LIZARDS

Handling

Lizards of all sizes should be transported to the surgery in a linen reptile bag, or in a pillow case (which should have a secure tie around the neck of the open end) or even in their vivarium. Smaller lizards and geckos are best controlled by placing the reptile bag flat on the examination table so that the outline of the lizard can be clearly seen, allowing the top of the neck and the front part of the thorax to be grasped, pinning the animal, still inside the bag, down on to the table. The tie around the opening of the bag can then be undone and the top of the bag turned inside out over the hand that is controlling the reptile's body. The lizard can then be transferred to the other hand. The safest hold for small lizards is round the pectoral girdle keeping the forelimbs back down the side of the thorax. One should be particularly careful not to hold any lizard by its tail. Many lizards can perform autotomy - the ability to shed part of the tail. The tail may be left twisting and turning in the hand while the lizard escapes and runs for cover. Even though the tail will regrow, the value of the lizard to the herpetologist may be reduced and the veterinary surgeon may be extremely embarrassed.

Clinical examination

All lizards should be weighed and the rostrum to cloaca length measured. In species which store fat in the base of the tail, it may be helpful to measure the diameter at this site.

The nostrils and eyes should be clear with no discharges. If there are any discharges, these need to be investigated further.

The tip of the mandible and maxilla should be examined for damage, especially in water dragons (*Physignathus* spp.) where repeated jumping at the side of their tank may result in rostral erosions. Tongue depressors or forceps may be useful for keeping the mouth open for continued examination. Mucous membranes should be clear and there should be no signs of discharge, oedema, petechiation or necrotic tissue in the mouth. While the mouth is open one should examine the medial aspect of the temporo-mandibular joints for white opacities, which may be indicative of uric acid tophi formation (see "Urogenital System"). The jaw should be carefully examined for any sign of softness and/or swellings (see Figure 4) which may be associated with infection or nutritional secondary osteodystrophy (see "Nutritional Diseases").

The skin should be bright without evidence of scarring, retained skin, blisters, haemorrhages or erosions. Any skin lesions should be investigated further (see "Integument" and "Laboratory Investigations"). External parasites should be eliminated as a possible complicating factor.

All the limbs must be examined carefully for signs of swelling or deformity: this may be associated with either abscesses or secondary nutritional osteodystrophy. If any abnormality or swelling is found, radiographs should be taken (see "Radiology and Related Investigations"). An opportunity should be taken to examine the lizard whilst it is moving; in particular one should make sure that there is no twitching or muscular fasciculation: this may be associated with hypocalcaemic tetany

Figure 4
Bilateral swelling of the mandible
in an *Iguana iguana* caused by
osteodystrophia fibrosa.

(see "Neurological Diseases"). The digits and the tip of the tail should be examined for evidence of retained skin; this may produce a band of material which could result in a tourniquet effect and distal necrosis.

The cloaca should be free of any sign of swelling, erythema or discharges. The use of a speculum or endoscope or digital examination may be indicated.

Radiographs should be taken as a routine to rule out nutritional osteodystrophy which is a common problem in lizards. Routine blood sampling to measure the calcium:phosphorus ratio may be indicated. Where the abdomen is extended or a mass is palpated, ultrasound, endoscopy or even exploratory laparotomy (see "Surgery") should be considered as part of any further investigation.

Any urine or faeces passed during the clinical examination should be examined promptly under a microscope for internal parasites.

Sexing

In many species of lizards the differences between the sexes is subtle while in other species the differences are more obvious (Howard, 1964). The following points should be taken into consideration when sexing lizards:-

1. Male lizards are often more colourful than females.
2. Males are often more aggressive towards each other.
3. Display forms, such as head bobbing, dewlap extension, raising the body off the ground on extended limbs, tail raising and tail curling, are usually only made by males.
4. During the breeding season males sometimes show an increase in the size of the hemipenile pockets caudal to the cloaca. This is seen as a bulge and can easily be palpated.
5. The number of visible femoral pores on the ventral surface of the thighs is diagnostic in Iguanidae, ie. usually more than 12 in males and fewer than 12 in females. When males are adult the femoral pores become more prominent.

In some lizards it is possible to probe the hemipenile pockets as has been described for snakes. However, this method is not recommended by the authors unless one has done a considerable number *post mortem.*

REFERENCES

BREARLEY, M.J., COOPER, J.E. and SULLIVAN, M. (1991). *A Colour Atlas of Small Animal Endoscopy.* Wolfe, London.

COOTE, J. (1985). Breeding colubrid snakes, mainly *Lampropeltis.* In: *Reptiles: Breeding, Behaviour and Veterinary Aspects.* (Eds. S. Townson and K. Lawrence). British Herpetological Society, London.

FITZSIMONS, V.F.M. (1961). *Snakes of Southern Africa.* Purness, Cape Town.

HOWARD, C.J. (1964). Points to look for when sexing lizards. *Herptile* **9 (2)**, 44.

JACKSON, O.F. (1980). Weight and measurement data on tortoises (*Testudo graeca* and *T. hermanni*) and their relationship to health. *Journal of Small Animal Practice* **21**, 409.

JACKSON, O.F. (1987). Record keeping. In: *Handbook on the Maintenance of Reptiles in Captivity.* (Ed. K.R.G. Welch). Krieger, Malabar.

JACKSON, O.F. (1991). Chelonians. In: *Manual of Exotic Pets.* New Edn. (Eds. P.H. Beynon and J.E. Cooper). BSAVA, Cheltenham.

LAWTON, M.P.C. (1991a). Reptiles. In: *Practical Animal Handling.* (Eds. R. Anderson and A.T.B. Edney). Pergamon Press, Oxford.

LAWTON, M.P.C. (1991b). Lizards and snakes. In: *Manual of Exotic Pets.* New Edn. (Eds. P.H. Beynon and J.E. Cooper). BSAVA, Cheltenham.

CHAPTER FOUR

POST-MORTEM EXAMINATION

John E Cooper BVSc CertLAS DTVM FRCVS MRCPath FIBiol

INTRODUCTION

Post-mortem examination is an important part of work with reptiles (Cooper and Jackson, 1981; Hoff *et al*, 1984; Frye, 1991). It is an integral part of diagnosis but must be coupled with an adequate case history, assessment of environment/management and analysis of any clinical data (see Figure 1).

Specimens for *post-mortem* examination can be sent to a diagnostic laboratory or the practising veterinary surgeon may like to carry out the gross examination him/herself and submit samples for further investigation. The latter has much to commend it. Even a gross *post-mortem* examination, without supporting laboratory tests, may be sufficient to make, confirm or refute a diagnosis. In addition, *post-mortem* examination can yield valuable information on (for example) the reproductive system or body condition and thus be of assistance to those who are attempting to keep and breed reptiles in captivity. The veterinary surgeon is likely to be presented with a fairly fresh carcase and this is another reason for doing the examination him/herself rather than despatching it elsewhere.

In this chapter *post-mortem* examination is described and discussed, together with those laboratory techniques which are specific to *post-mortem* material. Laboratory investigations that are more properly part of clinical work, eg. haematology, blood biochemistry, biopsies, microbiology, parasitology and serology, are covered in the chapter on "Laboratory Investigations".

Figure 1
Diagnosis of disease in reptiles.

Case history   Assessment of   Clinical   *Post-mortem*
                environment/   investigation   investigation
                management

→ PLUS ←

Laboratory tests
↓
Diagnosis
↓
Appropriate action

SUBMISSION OF MATERIAL

A carcase should be fresh or only chilled prior to examination. Freezing can cause tissue damage, making interpretation of histological sections difficult. Other adverse effects of attempting to preserve a carcase in this way may include rapid spread and multiplication of bacteria during thawing, the development of artefactual changes and a deceptively haemorrhagic or congested appearance to the viscera. On the other hand freezing is an excellent way of preserving carcases and tissues for subsequent toxicological or (usually) virological investigation. Fixing a carcase in buffered formalin or ethanol will help to preserve organs (so long as sufficient fixative is used and, for larger specimens, the body cavity is opened) but will render microbiology useless. It can also change the gross appearance of tissues, eg. uric acid deposits on or in tissues may be dissolved. The use of formalin may be contra-indicated because the tissues may not be suitable for DNA studies; in this case ethanol fixation or freezing are preferable. For electron microscopy glutaraldehyde or other special fixatives are to be preferred but formalin-fixed tissues can be processed if necessary.

These points must be borne in mind when specimens are to be examined *post mortem*, especially if there is likely to be some delay in carrying out the necropsy or material needs to be kept for subsequent re-examination. As a general rule, a carcase can be kept for up to 96 hours at normal refrigerator temperature before alternative methods of preservation have to be considered.

If necropsy material is to be sent in the post it must be well wrapped, with the carcase in at least one sealed plastic bag, and the package properly padded and protected. Post Office Regulations and any relevant legislative requirements must be followed (see later). The history should accompany the carcase but be wrapped separately to avoid contamination. It is often preferable to send the material by courier or datapost.

RECEPTION OF MATERIAL

The practice or laboratory which receives material for *post-mortem* examination must have a protocol for dealing with carcases. This will help to avoid mistakes, eg. the accidental translocation of material, and should minimise the risk of spread of infection to staff, visitors or animals. Specimens must be given a reference number and wrappings disposed of hygienically.

PERFORMING THE *POST-MORTEM* EXAMINATION

A full *post-mortem* examination with supporting tests can be both time-consuming and expensive. An extensive investigation is not always necessary, however, and busy practitioners may find it easier to divide cases into two categories: a) those that they can do themselves, with laboratory support as necessary, and b) those that should be referred elsewhere for special investigation, eg. when a zoonotic hazard is suspected or there may be legal action. Carcases in category b) should be despatched promptly, paying due heed to the points above. Prior contact with the laboratory may be prudent.

However basic an examination, suitable facilities and equipment are necessary. Particular attention must be paid to the health and safety of those who are involved in the examination or may come into contact inadvertently. Essential requirements are a well ventilated room with washing facilities and adequate drainage, a table or bench area, refrigerator, appropriate instruments, protective clothing, incinerator, macerator or other means of disposal, disinfectants, a steriliser or autoclave, balance or scales, a ruler or measurer, bottles for storing specimens, fixatives and ethanol or other wetting agent.

The risk of zoonoses will be minimised if basic rules are followed and there is a clearly established division between "clean" and "dirty". Those involved in the examination must practise good hygiene at all times: in particular, it is important not to contaminate papers or the outsides of bottles or swabs. Carcases and tissues for disposal must be carefully double-wrapped and incinerated.

The gross *post-mortem* examination can be conveniently divided into three stages:–

1. External examination and sampling.

2. Opening the carcase and examining internal organs.

3. Opening, dissecting or removing organs for examination and sampling.

Space does not permit detailed discussion of *post-mortem* techniques. This is essentially a practical subject which is best perfected by experience. A recommended approach for busy practitioners is as follows:–

1. Obtain and read all relevant history.

2. Examine the reptile externally (including mouth and cloaca) - record any parasites, lesions or abnormalities. Weigh the carcase and carry out routine measurements (rostrum – cloaca, cloaca – tail tip). A whole-body radiograph may be useful if skeletal changes are to be investigated or a radio-opaque foreign body is suspected (see Figure 5).

3. Open the reptile (see Figures 2 – 4) using a saw if necessary. Examine the internal organs and record any lesions or abnormalities. Note whether the stomach contains food and the presence or absence of fat bodies.

4. Fix tissues in buffered formalin - lung, liver and kidney should always be taken plus any abnormality. These samples will not necessarily be processed but can be stored and used later if required. Touch preparations of internal organs, suitably stained, can provide valuable information and may obviate the need for histopathology or microbiology.

5. Open all portions of intestinal tract and look with naked eye for parasites. Make two wet preparations in saline of **lower** intestine and search for parasites under the microscope.

6. Take samples for bacteriology if indicated. If there is to be any delay in processing these, or if they are to be sent elsewhere, use swabs in transport medium. Although aerobic culture is often adequate the role of anaerobes in reptile disease should not be overlooked (Stewart, 1989). Touch preparations and/or histopathology may help demonstrate anaerobes and fastidious aerobes.

7. Save carcase (chilled) until preliminary results are received or (frozen) if to be kept for subsequent investigations, eg. virology, toxicology.

The importance of recording findings cannot be overemphasised. The person carrying out the examination should take written notes or record findings on a (voice-activated) tape recorder. If two people are involved, one can remain "clean" and take notes while the other does the dissecting. Drawings can be made or photographs taken as necessary.

A suggested format for a basic *post-mortem* record sheet is given in "Appendix Two": this can easily be modified to suit individual circumstances.

It must be remembered that eggs and embryos may also need to be examined: a slightly different approach will be required and there may be merit in using a dissecting microscope or binocular loop and ophthalmic instruments for small specimens.

LABORATORY TECHNIQUES

Laboratory techniques are an important adjunct to *post-mortem* examination (Cooper, 1986). The range of investigations that may be used to supplement the necropsy is listed below:–

haematology	clinical chemistry/biochemistry
parasitology	microbiology
histology	cytology
electron microscopy	radiography
toxicology	chemical analysis
examination of stomach/gut contents.	

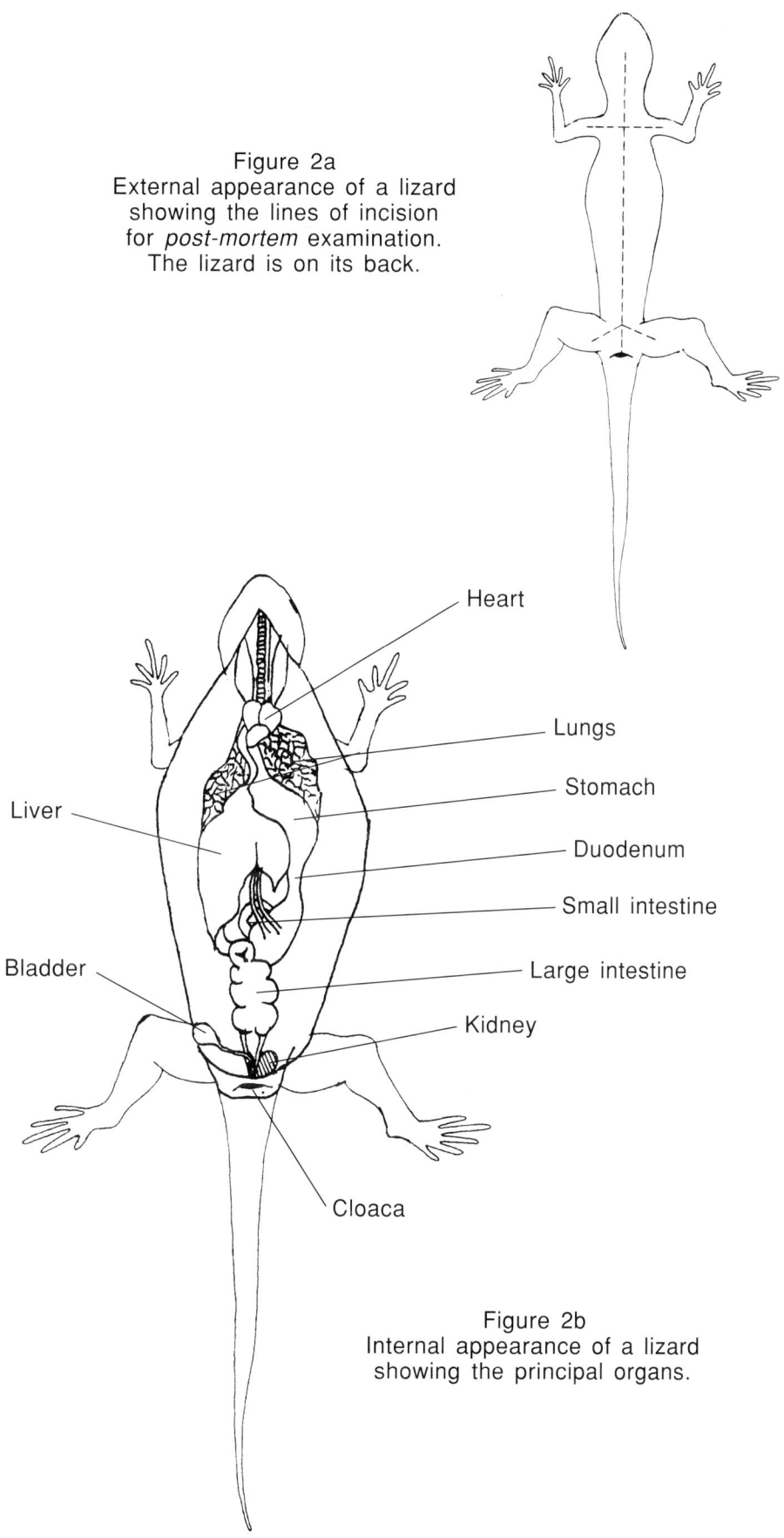

Figure 2a
External appearance of a lizard showing the lines of incision for *post-mortem* examination. The lizard is on its back.

Figure 2b
Internal appearance of a lizard showing the principal organs.

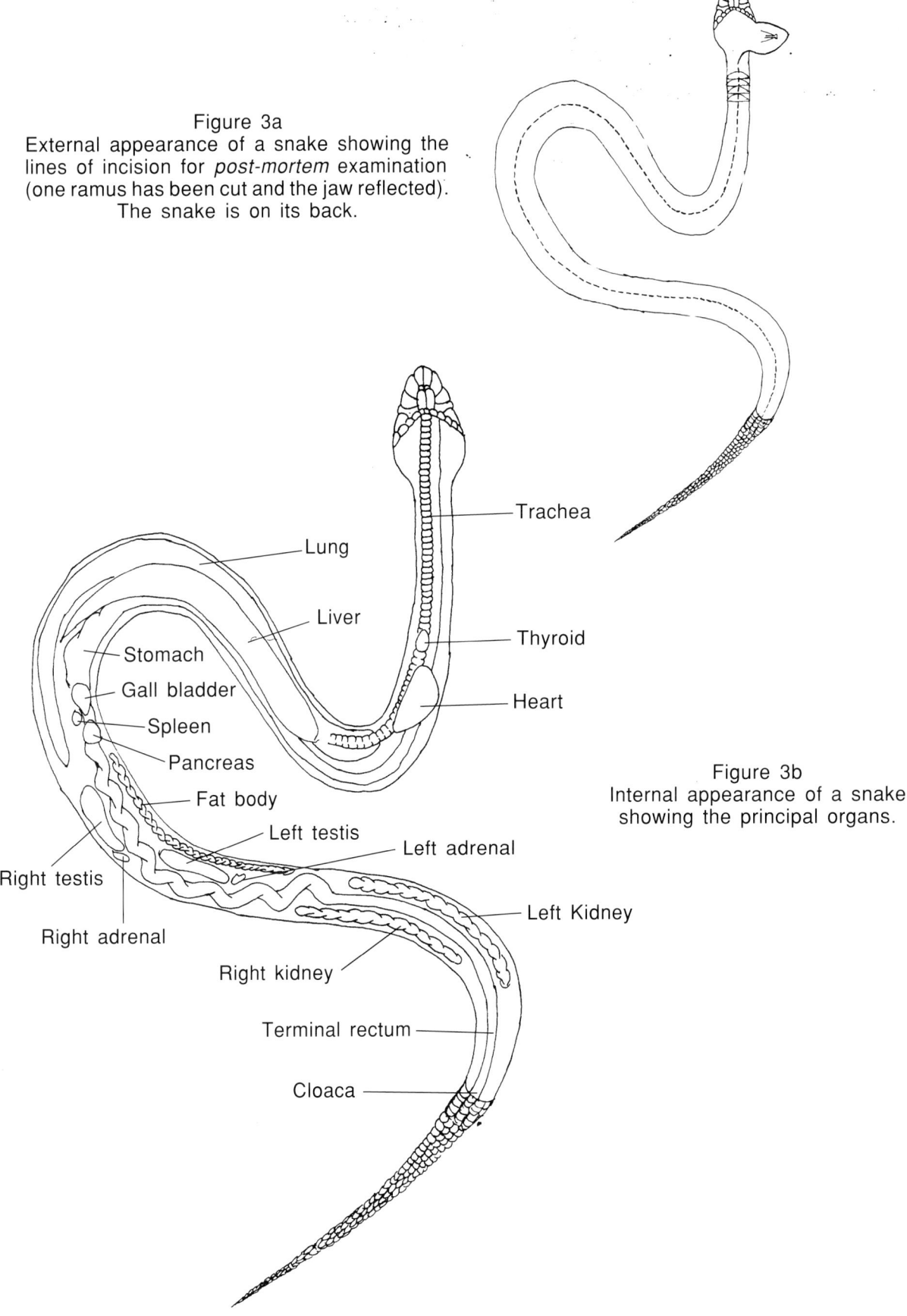

Figure 3a
External appearance of a snake showing the lines of incision for *post-mortem* examination (one ramus has been cut and the jaw reflected). The snake is on its back.

Figure 3b
Internal appearance of a snake showing the principal organs.

Figure 4a
External appearance of a chelonian showing the lines of incision for *post-mortem* examination. The chelonian is on its back.

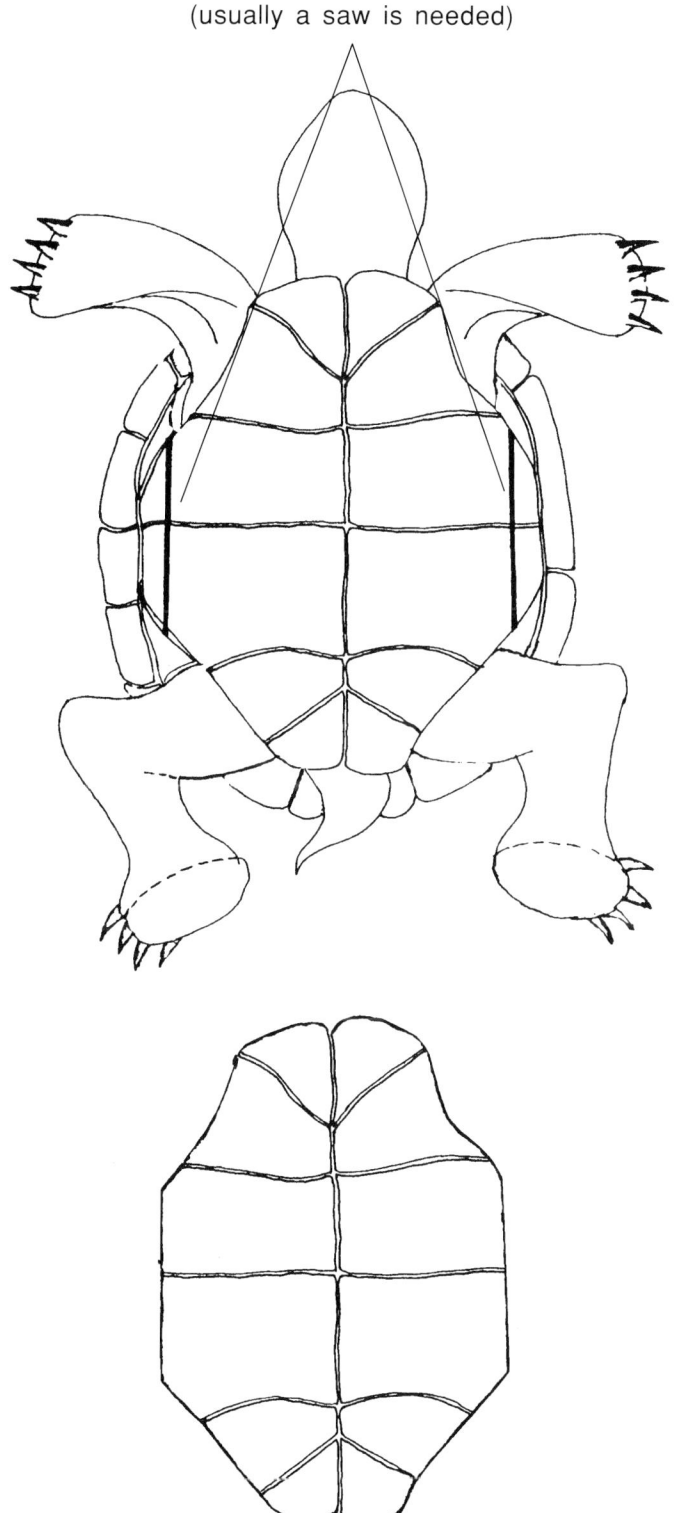

The piece of plastron following removal

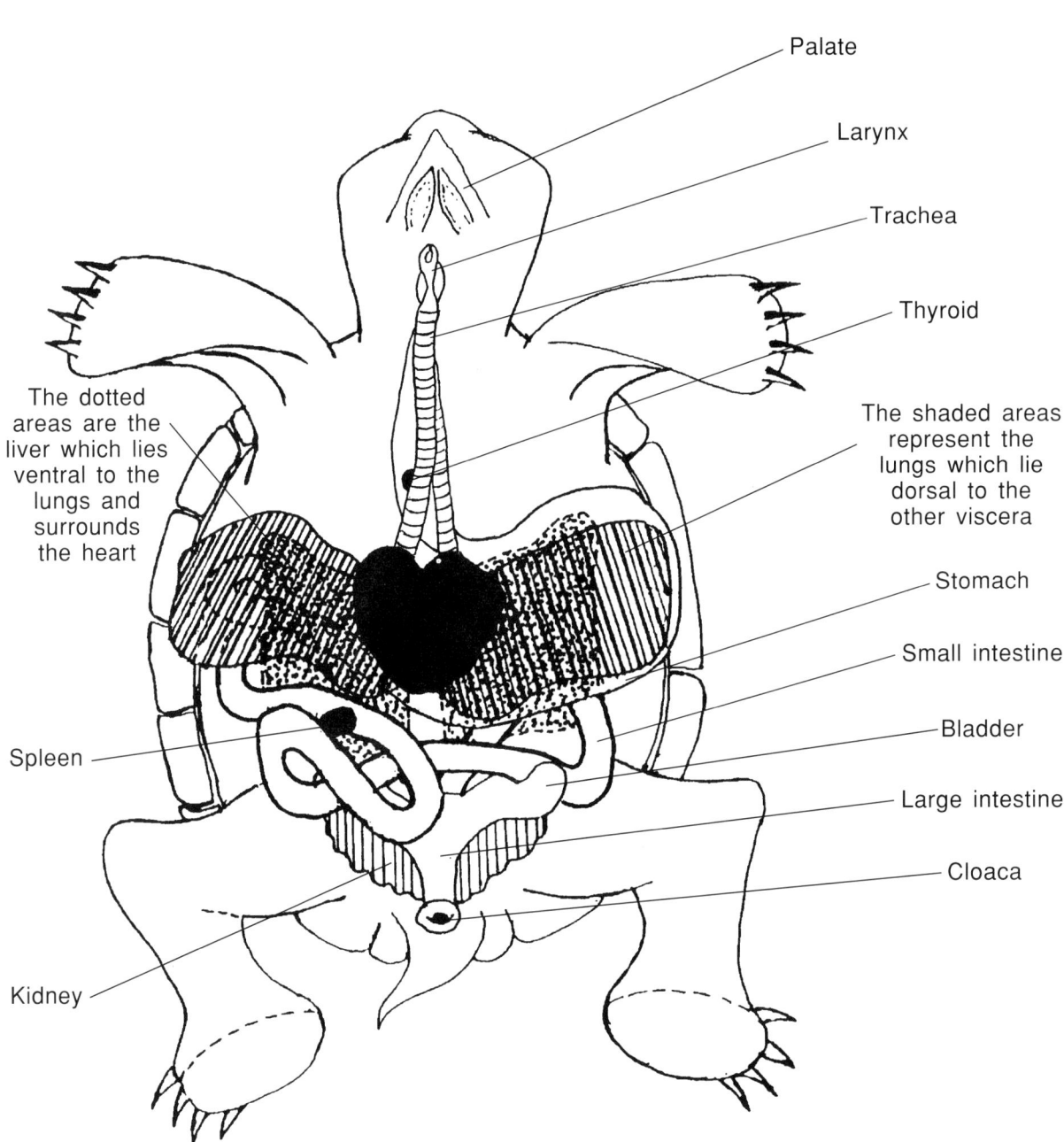

Figure 4b
Internal appearance of a chelonian showing the principal organs.
The buccal cavity and structures in the neck have also been exposed

Figure 5
A *post-mortem* radiograph can provide valuable information.
This Round Island gecko (*Phalsuma guentheri*) shows nutritional osteodystrophy,
relatively small calcium deposits in its storage areas (neck)
and evidence of a regenerated tail.

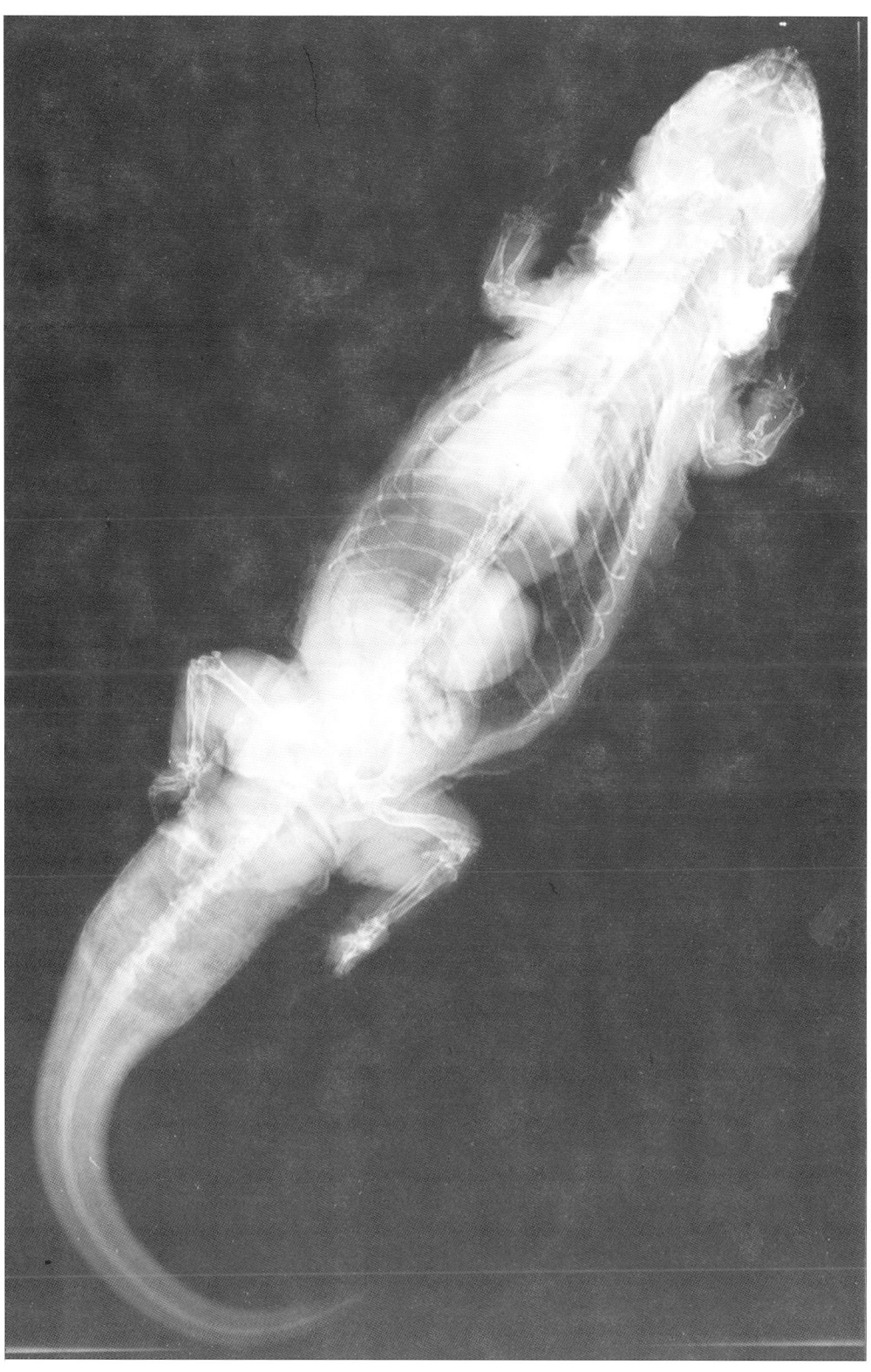

Techniques listed which are also used as part of clinical investigation are covered in more detail elsewhere (see "Laboratory Investigations").

As a general rule it is always preferable to take material for laboratory investigation, even if it is not subsequently used: tissue in formalin, for example, can be retained and sections cut if the need arises. It may also be wise to store (freeze) the carcase - not only may further tests be carried out at a later date but also the carcase may be required or requested by the owner, a museum or as an exhibit in a court case. A number of reference collections of threatened species have been established (Cooper and Jones, 1986) and, whenever possible, material should be deposited in these. Environmental monitoring usually includes laboratory tests, eg. water analysis, and is mentioned briefly later (see Colony Problems).

Samples for laboratory investigation must be carefully taken, labelled, stored and despatched. Again, Post Office Regulations must be followed and, if material is being sent to, or from, another country, the relevant animal health and conservation legislation must be observed (Cooper, 1987).

INTERPRETATION AND SIGNIFICANCE OF FINDINGS

A *post-mortem* examination may reveal signs of disease or injury but a proper diagnosis must be based on careful analysis of a combination of inter-related factors (see Figure 1). Sometimes the cause of death cannot be ascertained with certainty. Often more than one factor is involved.

In some cases a diagnosis as such is not required – the reptile may have been killed on humane grounds or, alternatively, the immediate cause of death is apparent – but the *post-mortem* examination is a useful way of providing background data on that animal and, to a certain extent, others in the same environment. In this case the *post-mortem* examination should be considered part of "health monitoring" which, coupled with clinical screening (live reptiles), is an increasingly important part of preventive medicine. Health monitoring is also advisable when dealing with threatened species in the wild. The general rule should be that whenever a reptile dies or has to be killed, a *post-mortem* examination is performed and, if personnel and facilities permit, certain (standard) supporting tests are carried out. In this case careful interpretation of findings is necessary since the aim is to collate and store data on the reptile's health status. Computerised records are desirable and will facilitate retrieval and analysis of information.

COLONY PROBLEMS

Although as a general rule individual reptiles will be presented in the surgery, "colony problems" can occur, especially when large numbers of animals are kept together – for example, in zoos, laboratories, captive-breeding centres or private collections. In such cases the reasons for consultation will vary: disease and/or deaths are the most likely but more subtle signs of "ill health" may have been noted, eg. poor weight gains, impaired reproductive performance.

Colony problems are covered here because *post-mortem* examination and associated laboratory tests are often an important, sometimes vital, part of such investigations. A similar approach may be needed when investigating morbidity and mortality in free-living reptiles.

The following is recommended:–

1. Where possible visit the premises. Spend time with the owner and get to know him/her as well as the animals and facilities.

2. Prepare and complete (where necessary with the assistance of the client) a check sheet/ questionnaire. Full use should be made of the owner's records and these should be analysed in as many ways as possible, eg. by tabulating the data and preparing graphs and histograms.

3. Consider using a) a tape-recorder and b) a camera to record observations and incidents.

4. While on the premises carry out a thorough inspection. Check the species that are kept together. Be prepared to take samples of live/dead animals for investigation. Culling of animals may be advisable or necessary to provide fresh material. Have a systematic approach.

5. Check temperature and relative humidity. Remember the value of environmental monitoring, eg. use of bacteriological "settle plates", water sampling, smoke tests, as well as clinical and pathological investigation of animals. Carefully scrutinise incubators and other items of equipment used for breeding.

6. Acquaint yourself fully with the "normal" by reading appropriate literature or visiting other similar premises.

7. Do not hesitate to consult colleagues or "specialists".

8. Analyse all data and findings before attempting to make a diagnosis.

CONCLUSIONS

Post-mortem examination and associated laboratory tests play an important part in the diagnosis of disease and monitoring of health of captive reptiles. To prove of maximum value they need to be carried out systematically and detailed records must be kept.

Acknowledgement

I am grateful to Miss Clare Knowler for drawing Figures 2 – 4 and for reading and commenting on this manuscript.

REFERENCES

COOPER, J.E. (1986). The role of pathology in the investigation of diseases of reptiles. *Acta Zoologica et Pathologica Antverpiensia* **70**, 15.

COOPER, J.E. and JACKSON, O.F. (1981). Eds. *Diseases of the Reptilia.* Academic Press, London.

COOPER, J.E. and JONES, C.G. (1986). A Reference Collection of Endangered Mascarene Specimens. *The Linnean* **2**, 327.

COOPER, M.E. (1987). *An Introduction to Animal Law.* Academic Press, London.

FYRE, F.L. (1991). *Biomedical and Surgical Aspects of Captive Reptile Husbandry.* 2nd Edn. Krieger, Malabar.

HOFF, G.L., FRYE, F.L. and Jacobson, E.R. (1984). Eds. *Diseases of Amphibians and Reptiles.* Plenum Press, New York.

STEWART, J.S. (1989). Anaerobic bacterial infections in reptiles. *Proceedings of the American Association of Zoo Veterinarians.* Queensboro, North Carolina.

CHAPTER FIVE

LABORATORY INVESTIGATIONS

Elliott R Jacobson DVM PhD DipACZM

INTRODUCTION

As in birds and mammals, evaluating the health status and diagnosing disease problems in captive reptiles often requires the collection of appropriate biological samples for laboratory investigations. While some samples can be collected using manual restraint alone, others will require the use of sedation, or local or general anaesthesia. Anaesthesia has been addressed elsewhere (see "Anaesthesia") and will not be discussed further here.

The information gleaned from any of the samples collected will only be as good as the techniques used in collecting, handling and processing these samples. The ultimate usefulness of the samples will depend upon the clinician's ability to interpret the results and use them for developing a medical management programme.

In this chapter methods for obtaining and processing various biomedical samples for diagnosing disease problems in reptiles will be discussed. The emphasis is on samples from clinical cases. *Post-mortem* techniques are discussed elsewhere (see "*Post-Mortem* Examination"). The most commonly collected samples include:-

1. Blood for haematological, biochemical and serological evaluations.

2. Biopsies for cytology, histopathology, electron microscopy, microbial isolation attempts and heavy metal, toxicological, mineral and vitamin analyses.

3. Scrapings, washings and exudates for microbiological and cytological evaluations.

4. Fluids and faeces for negative-staining electron microscopy.

5. Faecal samples for parasitological investigations.

6. Urine and spinal fluid for analysis.

BLOOD EVALUATIONS

COLLECTION

The total amount of blood which can be safely withdrawn from a reptile depends upon the reptile's size and health status. The total blood volume of reptiles varies between species but, as a generalisation, is approximately 5 – 8% of total bodyweight (Lillywhite and Smits, 1984; Smits and Kozubowski, 1985). Thus, a 100g snake has an estimated blood volume of 5 – 8ml. Since clinically healthy reptiles can lose 10% of their blood volume over a short period without any detrimental consequences, 0.7ml of blood can be withdrawn safely from a snake weighing 100g. Much larger percentages of blood can be removed over an extended period of time (Lillywhite *et al*, 1983). However, this practice is limited to experimental animals under controlled laboratory conditions.

Chelonians (turtles and tortoises)

Several sites can be used to obtain blood from chelonians, each having advantages and disadvantages. Sites include the heart, jugular vein, brachial vein, ventral coccygeal vein, orbital sinus and trimmed toe-nails (Gandal, 1958; McDonald, 1976; Taylor and Jacobson, 1981; Rosskopf, 1982; Stephens and Creekmore, 1983; Avery and Vitt, 1984; Samour et al, 1984; Nagy and Medica, 1986).

Cardiac sampling, although not recommended, has been used. In young chelonians, before the shell has calcified, a needle can be passed through the plastron into the heart. Older tortoises with calcified shells require either drilling a hole through the plastron over the heart, or using a spinal needle for percutaneous sampling through soft tissues in the axillary region at the base of the forelimbs. In all situations a sterile technique is necessary since contamination of the pericardial sac with bacteria and other potential pathogens can lead to pericarditis and death. A sterile drill bit should be used to create a hole. The hole should be sealed with an appropriate sealant, such as bone wax (Johnson and Johnson, Somerville, USA), and a methacrylate resin (Cyanoveneer, Ellman International, Hewlett, New Jersey, USA).

In turtles and tortoises orbital sinus sampling can be used for collecting small volumes of blood in capillary tubes (Nagy and Medica, 1986). However, in order to prevent damage to periocular tissues and possible trauma to the cornea, care must be taken when using this technique. The end of the capillary tube is placed into the lateral canthus of the orbit and, using a gentle twisting motion, blood can be collected. A further problem with this technique is that dilution of the blood sample with extravascular fluids and secretions may alter the composition of plasma and affect the volume percentages of cellular components. Blood samples may also be obtained from the scapular vein, brachial vein and brachial artery of chelonians (Rosskopf, 1982; Avery and Vitt, 1984). However, vessels associated with limbs can rarely be visualised through the skin and sampling is usually blind. In addition, since lymphatics are well developed in chelonian forelimbs (Ottaviani and Tazzi, 1977), obtaining blood samples from these vessels may result in haemodilution with lymph. At times pure lymph only may be obtained.

The author has found that the only peripheral blood vessels which can be consistently visualised in many small and moderate sized tortoises are the jugular vein and carotid artery (Jacobson et al, 1992). The major problem encountered when sampling from these vessels is that manual extension and restraint of the head of the tortoise beyond the margins of the plastron are required, which at times may be difficult or impossible. One method is to push in or lightly touch the hindlimbs, which usually causes the tortoise to extend its head from the shell and allows the sampler to restrain the tortoise's head. The head is held with one hand and, while sitting, the sampler positions the tortoise between his/her knees, with the tortoise's head pointing towards the sampler's body. The jugular vein and carotid artery are well developed on both right and left sides of the neck. Once the head is extended, the jugular vein can often be seen as a bulge through the cervical skin, coursing caudally from the level of the tympanic membrane to the base of the neck. The carotid artery is deeper and more difficult to visualise and is located ventral and parallel to the jugular vein. Once either vessel is identified, the skin over the puncture site should be cleaned with 70% ethanol and a 23G or 25G butterfly catheter used for obtaining the sample. With the cap removed, blood will flow down the tube once the needle is inserted into the vessel. The technique described above can be used in Mediterranean tortoises (*Testudo* spp.) but is not always successful: in these species sampling from the dorsal tail vein (Samour et al, 1984) is the method of choice.

Crocodilians (crocodiles, alligators, gharial)

Blood samples can be obtained from the supravertebral vessel located caudal to the occiput and immediately dorsal to the spinal cord (Olson et al, 1975). The skin behind the occiput is cleaned with an organic iodine solution and 70% ethanol. A 3.75cm, 22G or 23G needle is inserted through the skin in the midline directly behind the occiput and is slowly advanced in a perpendicular direction. As the needle is advanced, gentle pressure is placed on the plunger. If the needle is passed too deep, the spinal cord will be pithed. Other sites of blood collection which are commonly used include the heart (via cardiocentesis) and the ventral coccygeal vein (Jacobson, 1984). The heart is located in the ventral midline, approximately 11 scale rows behind the forelimbs. When collecting blood from the coccygeal vein, the crocodilian is placed in dorsal recumbency and the needle is inserted through the skin towards the caudal vertebrae.

Lizards

Blood samples can be obtained from several sites. In large lizards, blood is easily obtained from the ventral tail vein (Esra *et al*, 1975). Toe-nails can be clipped and blood collected in a microcapillary tube (Samour *et al*, 1984). Microcapillary tubes can be used to obtain blood samples from the orbital sinus (LaPointe and Jacobson, 1974) in a similar fashion to that described under chelonians.

Snakes

Blood samples can be obtained from a variety of sites, including the palatine veins, ventral tail vein and via cardiocentesis (Olson *et al*, 1975; Samour *et al*, 1984). The author prefers heart puncture to other methods and, so long as the heart is not excessively traumatised with multiple attempts at sampling, considers the procedure to be safe and effective. However, this does require experience. This method should be limited to snakes over 300g (Jackson, 1981). Essentially, the heart is located either directly by seeing it beating through the ventral scales, or by palpation. The heart is relatively moveable within the coelomic cavity and is easy to move manually several scale rows both cranially and caudally. Once the heart is located it is stabilised by placing a thumb at its apex and forefinger at its base. A 23G or 25G needle attached to a 3 – 6ml syringe is advanced under a ventral scale, starting at the apex and aiming toward the base. With gentle suction, a sample can be obtained. Sometimes, a clear fluid is withdrawn; this is pericardial fluid. In such cases, the needle should be withdrawn, a new syringe and needle secured and the procedure repeated. Sometimes, in obtaining a sample from the heart the clinician will note that blood can be withdrawn with each beat. The use of cardiocentesis is less widely accepted in the UK and most clinicians will prefer to use the ventral tail vein or palatine vessels (see "Anaesthesia").

HAEMATOLOGIC EVALUATIONS

Complete blood counts (CBCs) should be routinely performed on ill reptiles. The author prefers using microcontainer tubes containing lithium heparin (Fisher Scientific, Orlando, Florida, USA). Immediately following collection of the sample, 0.6ml of blood is added to a tube and the tube is inverted several times to prevent clotting. Other anticoagulants which can be used for CBCs include sodium heparin and ammonium heparin. Since potassium ethylenediaminetetraecetic acid (EDTA) results in haemolysis of chelonian red blood cells, this anticoagulant is not recommended (Jacobson, 1987).

When calculating CBCs, analyses should include:-

1. Red blood cell counts.
2. White blood cell counts.
3. Differential white blood cell counts.
4. Packed cell volumes (PCV).
5. Haemoglobin concentrations.

Red blood cell counts are determined using an automated Coulter counter. White blood cell counts can be determined either manually using a haemocytometer (Schermer, 1967) or as an estimated count from a blood film, similar to that used in birds (Campbell and Coles, 1986). PCVs are determined following centrifugation of a sample in a microhaematocrit tube. Haemoglobin values are calculated using a haemoglobinometer. Although total protein values are often determined using a refractometer, the author's experience suggests that the accuracy of this method with reptile blood is questionable. Proper methods for determining total protein values of reptile blood will be discussed under biochemical evaluations. Normal haematological values can be found elsewhere in this manual (see "Appendix Three – Haematological and Biochemical Data") and in Jacobson (1988).

BIOCHEMICAL EVALUATIONS

Biochemical evaluations of blood generally involve analysis of plasma or serum samples for the following inorganic and organic constituents: sodium, potassium, chloride, calcium, phosphorus, glucose, urea, uric acid, creatinine, cholesterol, aspartate, aminotransferase activity (AST, formerly GOT), alanine aminotransferase activity (ALT, formerly GPT), alkaline phosphatase activity (ALP) and total protein. The author prefers to use plasma rather than serum since a greater volume of plasma can be collected per unit volume of blood compared with serum. Also, it is more common for serum to clot than plasma. Serum removed from the blood sample following centrifugation may suddenly clot into a gelatinous mass. Although the causes of this phenomenon are unknown, clotting is more common in glass tubes and, possibly, the electric charge on the glass is an initiating factor.

Plasma samples are collected from blood placed into two microcontainer lithium heparin tubes. The tubes should be centrifuged immediately following collection and the plasma removed and submitted for evaluation. Since plasma potassium values will increase over the time plasma is in contact with blood, it should be removed and frozen immediately following centrifugation (Jacobson *et al*, 1992). In field work on blood biochemical values of free-ranging reptiles, plasma can be placed in cryotubes and frozen with liquid nitrogen. If samples need to be sent to a laboratory, they should be transported frozen on dry ice.

To date, no information appears to have been published on the accuracy of different methodologies used in determining various plasma biochemical analytes in reptiles. A variety of automated machines have been developed for use in determination of plasma/serum biochemical profiles of humans. The veterinary profession also uses these machines for determination of blood biochemical values of both domestic and non-domestic animals. Although it is quite simple to submit a sample and have it analysed on an automated machine, it is difficult to determine the accuracy of these values. In studies the author has conducted on replicate plasma samples submitted from the same sample of blood and analysed on different machines, flame photometry appears to be more accurate for determination of sodium and potassium values than machines using ion-exchange electrodes.

With automated machines, procedures for determining total protein values are often based on the Biuret method (Kingsley, 1972) and the author has found results by this method significantly different from values for the same samples determined by refractometry. The Biuret method appears to be more accurate and should be the preferred method. Albumin values are generally determined using various dyes such as bromcresol green (BCG) and globulin values by subtracting albumin from total protein. In studies on blood proteins of reptile plasma the author has found significant differences when comparing albumin and globulin values determined by electrophoresis with those determined using BCG. Based on such experience, it is likely that albumin values determined by dye techniques are inaccurate and methods using these chemicals should be avoided. Serum/plasma protein electrophoresis is more costly to perform but is more accurate.

The most important points to remember when submitting plasma/serum samples for biochemical evaluations are:-

1. Try to use the same blood collection techniques at all times.

2. Handle the blood in a constant fashion. Use the same anticoagulant and try to add the same volume of blood to the collection tube.

3. Centrifuge the blood immediately following collection and remove the plasma immediately following centrifugation. The warmer the ambient temperature, the quicker potassium will move out of red blood cells into surrounding fluid, resulting in falsely elevated values.

4. Freeze the sample following collection, preferably on dry ice, in liquid nitrogen or in an ultra-cold freezer at -70°C.

5. The sample should be transported frozen to the laboratory, preferably on dry ice.

6. Try and use the same clinical pathology laboratory using the same machine. Make sure the samples are handled similarly prior to analysis. For instance, small plasma volumes which have to be diluted in order to reach a minimum volume necessary for the machine will have a dilution error superimposed upon other technique errors.

Normal biochemical values of reptile blood can be found elsewhere in this manual (see "Appendix Three – Haematological and Biochemical Data") and in Jacobson, 1988.

SEROLOGY

Very few serological tests have been developed or are available for diagnosing disease problems in captive reptiles. Serological studies have been performed on free-ranging reptiles and reptiles used in experimental studies in order to determine the presence of antibodies to various togaviruses (Jacobson, 1986a). However, infections with these viruses do not appear to be clinically important in reptiles. A haemagglutination inhibition test has been developed for determining the presence of antibody to viperid paramyxovirus (Jacobson and Gaskin, in press) and has been used for monitoring exposure to this pathogen in zoological and private collections of snakes. However, the only laboratory currently performing this test is that of Dr. Jack Gaskin, Department of Infectious Diseases, University of Florida, Gainesville, Florida, USA. Serum banks from healthy and ill reptiles need to be established and maintained for retrospective studies as more serological tests and laboratories performing these tests become available.

BIOPSIES

The collection of biopsies is often necessary to diagnose disease problems in reptiles. These samples can provide invaluable information necessary to monitor, manage and treat the patient.

In collecting biopsy specimens multiple samples should be obtained for:-

1. Histopathology.

2. Electron microscopy.

3. Cytology.

4. Microbiology.

Biopsies are commonly obtained from the integumentary and digestive systems and the author will focus on these. Similar techniques can be used for collecting samples from other systems.

INTEGUMENTARY SYSTEM

Reptiles are often presented with skin disease (see "Integument") and a great variety of infectious and non-infectious diseases can result in pathological changes in the integument. The clinician needs to determine if the changes are due to local infection or are a manifestation of systemic illness. Many systemic diseases will result in pathological changes in the integumentary system.

The key to evaluating skin lesions of reptiles is collecting and properly evaluating a good biopsy specimen for both histopathology and microbial culture. Microbial culture by itself is often misleading because secondary infection of skin lesions is common.

Chelonians

Of all the reptiles, chelonians present the greatest challenge for biopsy, especially when lesions involve the shell. The reptile shell is a very hard, biological structure that makes biopsy somewhat difficult. Under anaesthesia (see "Anaesthesia") a rotary power saw or bone trephine can be used to cut a wedge out of the shell. Ideally, the biopsy should include normal tissue along with the diseased component. A piece should be fixed in neutral buffered 10% formalin for histopathological evaluation and a portion (with the most superficial contaminated portion removed) submitted for microbial culture. For initial attempts at isolation, the author often uses a broth such as tryptic soy broth. The defect created in the shell should be filled with calcium hydroxide dental paste and covered with a methacrylate resin. This technique is routinely used in repair of the chelonian shell.

For biopsy of soft tissues, a 2% xylocaine block is satisfactory. This can be infiltrated around the biopsy site and the skin cleaned with 70% ethanol and allowed to dry. If the sample is to be cultured, sterile saline is used instead of ethanol. If there is epidermal involvement, a biopsy punch can be used for collecting the sample. Following punch biopsy, the skin may require a single suture for closure. Monofilament nylon is routinely used. If a subcutaneous mass is present, fine-needle aspiration can be performed. This is a rapid method, resulting in minimal trauma to the patient. A 22G needle is inserted into the mass and, using a 6 – 12ml syringe, full negative pressure is developed by quickly pulling back on the plunger. While maintaining negative pressure on the syringe, the needle is moved throughout the mass in multiple planes. After several passes through the mass, the plunger is released and the needle removed. Pressure should not be applied to the plunger while removing the needle from the mass since this will cause the sample to be aspirated into the syringe barrel. The preparation of the specimen will be discussed later.

Crocodilians and lizards

A full-thickness biopsy may be difficult in those areas of the crocodilian integument which bear osteoderms. Small crocodilians and most lizards can be manually restrained, whereas large crocodilians and large monitors must be chemically immobilised. The area around the biopsy site should be infiltrated with 2% xylocaine and a full-thickness skin incision taken with a biopsy punch. As with chelonians, a minimum of two biopsies should be taken, one for histopathology and one for microbiology. For microbial culture, the lesions can be ground in a sterile tissue grinder and samples applied to appropriate media. This appears to be particularly important for isolation of fungi from reptile skin lesions: the author has had more success in isolating fungi when the skin is ground prior to attempts at isolation.

Snakes

Snakes are ideally suited for skin biopsy. Harmless species can be manually restrained and poisonous species can be guided into a plexiglass tube for restraint or anaesthesia. Affected scales can be removed with a scalpel blade, or a sterilised one-hole paper punch can be used for biopsies of individual scales. In such cases the area around the lesion should be infiltrated with 2% xylocaine. In certain skin diseases, such as vesiculating skin lesions, larger samples may be needed. Similarly, for sampling subcutaneous masses, 2% xylocaine can be infiltrated subcutaneously around the mass. Once removed, the mass should be split into several portions for various diagnostic tests (see later).

DIGESTIVE SYSTEM

Numerous reptiles are submitted with clinical signs of regurgitation, anorexia or passing loose stools (faeces) that appear to be abnormal. A regurgitation syndrome is commonly seen in wild emerald tree boas (*Corallus caninus*) imported into the USA and the cause of this syndrome has not been completely elucidated.

For problems of the upper gastrointestinal tract, endoscopic examination of the oesophageal and gastric mucosa and collection of biopsy specimens for histopathology and microbial culture are routine. For those clinicians without such equipment, a sterile nasogastric tube can be placed via the oesophagus into the stomach. Stomach washings can be obtained for preparation of wet mounts, cytological evaluation and microbial culture. Radiographic evaluation using normal films followed by contrast techniques can be used in cases of regurgitation (see "Radiological and Related Investigations").

For persistent cases of regurgitation in which upper gastrointestinal involvement has been ruled out, the next approach would be examination of the lower gastrointestinal tract. The reptile should be anaesthetised and a vaginal speculum used to dilate the cloaca and locate the colonic opening. A flexible scope will allow the clinician to observe the colonic mucosa directly and biopsy samples can be secured for histopathology and microbial culture.

PATHOLOGICAL EVALUATIONS

Biopsies can be evaluated by light and electron microscopical techniques. For histopathology, neutral buffered 10% formalin (NBF) is the most commonly used fixative (see "*Post-Mortem Examination*"). The NBF volume to tissue ratio should be at least 10:1. For best penetration, the tissue should not exceed 6mm in thickness. A fixative which the author routinely uses for light and electron microscopy is Trump's solution (McDowell and Trump, 1976) which is a combination of 4% formaldehyde and 1% glutaraldehyde. This solution needs to be refrigerated prior to use and is stable at refrigeration temperatures for several months.

Following light microscopical examination of a paraffin–embedded sample, the pathologist can determine whether or not to embark on an electron microscopical evaluation. Electron microscopy is costly and labour intensive and is only used for selected specimens. Electron microscopy can be performed on paraffin–embedded tissue samples. A small piece of the paraffin-embedded tissue can be removed from the block and processed for electron microscopy. In diagnostic reptile medicine, electron microscopy is invaluable in determining the nature of various intracytoplasmic and intranuclear inclusions and the presence of viruses, as well as unusual bacterial organisms, such as *Mycoplasma* spp. and *Chlamydia* spp.

CYTODIAGNOSTICS

Examination of touch impressions and wet mounts of various lesions is extremely helpful in diagnosing problems in reptiles. Because of the ease and rapidity of processing these samples, much information can be gathered in a short time. For external lesions, samples can be collected with relative ease. Depending upon the size and nature of the patient, manual restraint alone may be all that is needed. In larger, more fractious reptiles, or in those patients with internal masses, sedation or anaesthesia will be required.

Collection and preparation

The collection method used will depend upon the nature and consistency of the lesion being sampled. Prior to collection of the sample, the distribution, colour, morphological description, size and odour (if present) of the lesion should be noted. For cutaneous lesions, scrapings can be collected from keratinised structures, such as shell lesions of chelonians, or hyperkeratotic surfaces. A sample can be suspended in a small quantity of saline and a coverslip placed over the sample. Next, a "squash-prep" can be made by applying gentle pressure with a fingertip (in a twisting motion) to the coverslip. This will help in detecting pathogens contained within keratinised material. Potassium hydroxide can be used to digest keratinised material and improve examination for pathogens.

Lesions containing fluid or purulent material can easily be sampled with a needle. If the lesion is cutaneous, it should be cleaned with a small amount of 70% ethanol and allowed to dry prior to sampling. As discussed earlier, fine-needle aspiration biopsy specimens can also be collected from firm masses.

Aspirates should be examined first as a wet mount. This will give the clinician an idea of what additional preparations should be made. Some organisms, such as protozoa and nematode larvae, are better appreciated in a wet mount than in a stained preparation. Aspirates with high cellularity can be used to prepare direct smears using the conventional wedge method or the technique used for preparing blood films on coverslips. Aspirates containing tenacious material with thick cellular fragments should be prepared using a squash technique. After the sample is placed on a microscope slide, another slide is used to flatten the material and both slides are quickly pulled apart.

Snakes and those lizards with spectacles (see "Ophthalmology") will often develop accumulations of clear or purulent fluid in the sub-spectacular space (Millichamp *et al*, 1983). The spectacle should be cleaned with sterile saline, a fine-gauge needle (25G – 27G) inserted under the spectacle and the aspirate collected. Since the spectacle is an extremely vascular structure, it is not uncommon to collect blood cells in the aspirate. In reptiles with intraerythrocytic parasites, such as *Hepatozoon* spp., red cells may occasionally rupture releasing these parasites into the surrounding fluid. The clinician should be aware of this and if protozoan parasites are seen in the aspirate, their possible origin from red blood cells should be considered

Lung washes should be routinely collected from reptiles with respiratory disease (see "Respiratory System"). While samples can be collected from some reptiles using manual restraint, other patients will have to be sedated or anaesthetised. The jaws of the reptile are held apart and a sterile catheter is guided through the glottis into the lung field. The location of the lung field will vary between each major group of reptiles, and with snakes will differ even between members of the same Family. The clinician needs to know the location of the lung(s) before collecting a lung wash. With a syringe attached to the catheter, sterile saline (1ml for a 200g snake) can be introduced into the lung field and aspirated several times. Material collected can be used for both cytological evaluation and microbial culture.

If the reptile is large enough, samples can be collected via bronchoscopy. Bronchoscopy has an added advantage of permitting the lower respiratory tract to be examined directly. The author prefers the patient to be anaesthetised: the flexible fibreoptic bronchoscope is passed through the endotracheal tube. A T-tube connected to the endotracheal tube allows the technique to be performed while the patient remains connected to the gas anaesthesia machine. Using this technique, the tracheobronchial system can be examined methodically and sampling procedures, such as lavage, culture, brushing and transbronchial biopsy, can be performed on specific areas of the respiratory tract. For collecting samples for culture, the plugged telescope catheter brush system is the method of choice (Schaer et al, 1989).

Aspirates and lung washes of low cellularity can be concentrated using a number of techniques. Using the conventional wedge method, cells can be marginated by lifting the spreader slide from the smear slide just prior to reaching the end of the smear. Cells can also be concentrated by simple centrifugation in a plastic-capped tube followed by examination of the pelleted sediment. If samples are sent to a clinical pathology laboratory, cells can be concentrated on a microscopic slide using a cytocentrifuge.

The collection of biopsy specimens has already been discussed and, in addition to histopathology, biopsy specimens can be evaluated cytologically. The cut surface of a specimen can either be scraped with a sterile scalpel blade or touched several times with a microscopic slide to make impression smears. If the surface of the specimen contains a moderate amount of blood or clots of blood, this can be removed by gently cleaning the surface with sterile saline via a syringe and needle or gently rolling a saline-moistened cotton wool swab across the surface. If blotting paper or a medical wipe is used there is less chance of fibres adhering. A swab can also be used to collect cellular samples for examination on a microscope slide.

Staining

Depending upon the suspected disease or presence of pathogenic organisms, a variety of staining techniques can be used in evaluating smears. The method of fixation of the smear to the slide will depend upon the staining procedure used.

The most common method used for initial evaluation of smears is the Wright's-Giemsa stain. Prior to staining, the smear is fixed in absolute methanol for approximately 10 seconds. Quick staining techniques are commercially available and allow staining of smears in a few seconds. Other stains which are commonly used in evaluating smears of lesions from reptiles include Gram's stain for bacteria, acid-fast stain for mycobacteria and new methylene blue for fungi.

NEGATIVE-STAINING ELECTRON MICROSCOPY

Negative staining in conjunction with transmission electron microscopy (TEM) is a useful and rapid method of examining clinical specimens. The principle of negative staining is that there is no reaction between the stain and the specimen. The most commonly used negative stains are uranyl acetate (0.5 – 1.0%) and potassium phosphotungstate (PTA) (0.5 – 3.0%). In cases where a viral agent is suspected this technique may be used in a variety of specimens. Depending on the nature of the tissue, different ways of processing the sample are required to detect viral particles. Fluid from vesicles can be obtained with a sterile pipette and may be placed directly on a Formvar-coated 200 mesh copper grid, while large amounts of fluid (serum, urine, liquor) require centrifugation for clarification. In these cases the supernatant after low speed centrifugation (1,500g), or the diluted pellet after high speed centrifugation (15,000g), is placed on the grid for staining. Faecal material requires suspension and concentration and should be placed in distilled water or phosphate buffered saline (PBS). A very useful method is to mix faecal material with PBS to give a 20% suspension in 5ml. After centrifugation, a drop of the supernatant is placed on a grid and negatively stained. Alternatively, one drop of a 1:1 mixture of supernatant and stain is placed directly on a grid for examination. If needed, bacitracin as a wetting agent may be added. If the grids are to be stored and cannot be examined immediately, the stain of choice is uranyl acetate because it will not have adverse effects on the samples.

Tissues, such as liver, kidney, spleen etc, require grinding to obtain a homogenate which will be processed in the same way as the above samples. In general, the method used for processing specimens depends on the concentration of viruses suspected in the sample.

MICROBIOLOGY

Swab specimens, aspirates and biopsy specimens can be collected and submitted for microbial isolation attempts, including those for:-

1. Viruses.

2. Aerobic bacteria.

3. Anaerobic bacteria.

4. Special bacterial organisms, such as *Chlamydia* spp. and *Mycoplasma* spp..

5. Fungi.

Proper collection technique is a prerequisite for successful recovery of micro-organisms responsible for an infectious disease. The recovery of contaminants may result in improper or even harmful treatment. While isolation of aerobic organisms is relatively inexpensive and fairly rapid, isolation of the other groups of pathogens listed earlier requires special techniques and is far more costly. Viruses, *Chlamydia* spp. and *Mycoplasma* spp. require special media and conditions for isolation and most human and veterinary diagnostic laboratories have little experience with these pathogens in reptiles. The clinician will need to establish a special rapport with either a university research laboratory or diagnostic laboratory in order to culture these organisms. This author rarely cultures for viruses, *Mycoplasma* spp., *Chlamydia* spp. or fungi unless histopathology or cytology supports or suggests their presence. Fluids, aspirates and biopsy specimens can be frozen on dry ice or in an ultra-freezer at -70°C until a decision is made as to the specific isolations that will be attempted. Most diagnostic laboratories have ultra-freezers for storage of biological samples.

The author has rarely attempted viral isolation on *ante-mortem* cases. For the most part, viral isolation attempts are made following histopathological evaluation of a necropsy case. Where a viral disease is suspected, representative portions of major organs, such as liver, spleen, kidney and heart, are frozen for possible future isolation. In the author's experience, most reptile viruses are best isolated in reptile cell cultures, and in the USA several reptile cell lines are commercially available (American Type Culture, Bethesda, Maryland, USA). In some cases cell cultures from the species being evaluated need to be established. If a laboratory is available to attempt viral isolation, samples should be submitted either:-

1. As fresh tissue samples in a sterile container on wet ice.

2. As minced pieces in a transport medium, such as Eagle's minimal essential medium.

3. As fresh frozen samples transported on dry ice or in liquid nitrogen.

In an *ante-mortem* case, if a viral disease is suspected, aspirates, washings or biopsy specimens can be handled and transported similarly.

For isolation of aerobic bacteria, a variety of sterile swabs with associated transport media are commercially available. The size of the swab may be important. Once the sample is collected, the swab should be returned to its protective holder, placed in a plastic zip-lock bag and sent to the laboratory, on ice, as rapidly as possible. Some private practitioners have established their own microbiology laboratories and, in such cases, swab specimens can be immediately streaked on to appropriate agar plates.

Samples taken in the field from reptiles can be placed in a cryotube containing tryptic soy broth and frozen in a tank containing liquid nitrogen. Once frozen, the cryotubes can be transported on dry ice to the laboratory for culture.

It is only recently that anaerobic bacteria have been appreciated as potential pathogens in reptiles (Stewart, 1990). Samples submitted for anaerobic culture require special collection techniques to improve the success of recovery. For obligate anaerobes, anaerobic swab culturettes are available and for biopsy specimens the samples should be delivered to the clinical pathology laboratory in an anaerobic pack system. These are available through various microbiology and scientific supply companies. It is important that the specimen is placed in the anaerobic transport system as soon after collection as possible.

Samples for blood culture should be collected from reptiles suspected of being septicaemic. Prior to collection the skin should be thoroughly cleaned with three applications of both an organic iodine solution and 70% ethanol. As large a volume of blood as is safe to collect should be obtained in a sterile syringe and quickly added to the blood culture bottle containing the appropriate culture medium. The top of the blood culture bottle needs to be cleaned several times with 70% ethanol prior to insertion of the needle. Sterile technique must be used at all steps in the collection process. Both aerobic and anaerobic bottles need to be used. Blood culture bottles are commercially available from most microbiology or medical supply companies.

Fastidious microbial organisms, such as *Chlamydia* spp. and *Mycoplasma* spp., are only recently being appreciated as causes of illness and death of reptiles. While *Chlamydia* spp. have been seen in tissue sections of puff adders (*Bitis arietans*) dying of pericarditis and hepatitis (Jacobson *et al*, 1989) and in monocytes and spleen of a flap-necked chameleon (*Chamaeleo dilepis*) in Tanzania (Jacobson and Telford, 1990), as yet this organism does not appear to have been isolated from a reptile. If *Chlamydia* spp. are suspected, biopsy specimens should be submitted in an appropriate transport medium, eg. one routinely used for samples collected from birds and cats.

It is only recently that *Mycoplasma* spp. have been appreciated as a potential pathogen in reptiles. The only described mycoplasma from a reptile is *M. testudinis* which was isolated from the cloaca of a Greek tortoise (*Testudo graeca*) (Hill, 1985). Multiple isolates of *Mycoplasma* spp. have been obtained from nasal exudate of desert tortoises (*Xerobates agassizi*) and other tortoises (*Geochelone* spp.) with chronic upper respiratory tract disease (Jacobson *et al*, 1992). This organism appears to be an important pathogen in this complex disease. A special enrichment medium, SP_4 medium, is used both as a broth and agar in culturing for this organism in chelonians (Jacobson *et al*, 1992). A tom cat catheter attached to a syringe containing tryptic soy broth is inserted through the external nares and an aspirate obtained. The aspirate can be either placed in a cryotube containing tryptic soy broth, frozen on dry ice and shipped to a diagnostic laboratory, or placed immediately in SP_4 medium, if available. In the USA, SP_4 medium is commercially available (Remel, Lenexa, Kansas, USA).

Fungal diseases are commonly encountered in reptiles, particularly in cutaneous or subcutaneous lesions in lizards and snakes (Jacobson, 1980; Austwick and Keymer, 1981; Migaki *et al*, 1984). Biopsies of suspected mycotic disease are preferable to swabs or scrapings. Biopsy specimens can be minced or, if the sample is heavily keratinised, gently ground in a sterile tissue grinder containing sterile saline and antibiotics, such as gentamicin or amikacin. Minced or ground samples can be cultured on an appropriate mycotic medium, eg. Sabouraud's agar, containing antibiotics to suppress bacterial growth. Fungi will also grow readily in tissue culture media, such as Eagle's minimal essential medium. New, previously unreported, fungal organisms are constantly being cultured from reptiles (Austwick and Keymer, 1981) and for identification require a mycologist familiar with the particular Family of fungus isolated. It may take well over one year to have an isolate definitively identified. Often a new species will need to be described.

FAECAL EXAMINATIONS AND COLONIC WASHES

Of all the biological samples collected from reptiles, faecal samples are generally the easiest to obtain. If a faecal sample is not available and the reptile has not defaecated recently, a colonic wash can be obtained. For most reptiles this is fairly easy to do using manual restraint. Some tortoises will have to be anaesthetised to collect a sample. A lubricated French catheter filled with sterile saline and attached to a syringe containing saline, is passed into the colon via the cloaca. The catheter should slide in fairly easy and should not be forced. The cloaca and colon are relatively thin-walled and if the clinician is not careful the tip of the catheter can penetrate the walls of these structures resulting in rupture and subsequent peritonitis. Once the catheter has been inserted, several millilitres of saline are flushed into the colon, followed by aspiration. This can be repeated several times. Once a sample has been collected the following can be examined:-

1. A direct wet mount for protozoa and helminth ova.

2. Sediment, following centrifugation, for protozoa and trematode ova.

3. A flotation specimen using similar techniques to those in domestic animals for identification of nematode ova.

Faecal samples can be evaluated similarly.

A few papers with photomicrographs of gastrointestinal parasites and helminth ova of reptiles are available (Barnard, 1986; Jacobson, 1986b) and should be used as guides for proper identification. The clinician needs to become familiar with parasites and parasite ova of prey species fed to carnivorous reptiles since these may also be encountered in the faeces. For instance, eggs of lice found on mice are commonly found in the faeces of snakes being fed rodents. The clinician does not need to treat a reptile because of the presence of non-infectious prey parasites which are simply passing through the intestinal tract.

In the author's opinion, Gram staining of faeces is of limited value. However, culturing can provide useful information.

URINE COLLECTION

Few normal data are available for urine and spinal fluid of reptiles. These fluids are not easy to obtain and without published data, interpretations are difficult.

In reptiles, such as crocodilians and snakes, which do not have bladders, collection of a pure urine sample is almost impossible unless the ureters can be catheterised. In these species urine enters the terminal colon via the cloaca and, when the animals "urinate", urine is mixed with faeces. The urinary papillae can be visualised, under anaesthesia, in the cloaca. The paired urinary openings are located on the dorsal aspect of the cloaca near the orifices to the reproductive tract. While the author has collected urine samples by this method, the procedure is not easy and may have more academic than practical value.

Urine samples are fairly easy to collect from chelonians, particularly tortoises. Often, when a tortoise is examined and manipulated, it will spontaneously urinate. The clinician needs to be ready with a pot to collect these samples. Another method commonly used is cystocentesis. The skin around the right hindlimb is cleaned with several alternating applications of povidone-iodine solution and 70% ethanol. The tortoise is then placed in a vertical position with the lateral side of the left margins of the shell up and the right hindlimb pulled back away from the skin. Using a 22G or 23G needle a percutaneous sample can be collected from the bladder. Because tortoises have large bladders, cystocentesis is fairly easy to perform. However, since the bladder wall is extremely thin, a small amount of urine will leak into the coelomic cavity following removal of the needle. Some normal data are available for tortoise urine (Dantzler and Schmidt-Nielsen, 1966; Nagy and Medica, 1986). In the wild, tortoises go through rather complex changes in urine composition dependent upon the amount of available water and food and the season of the year (Nagy and Medica, 1986).

SPINAL FLUID COLLECTION

Diseases of the central nervous system are commonly seen in snakes (see "Neurological Diseases"). Unfortunately, the author has found it extremely difficult to collect spinal fluid from snakes, because of their vertebral anatomy. Spinal fluid samples are difficult to collect from lizards for the same reason. However, samples can be collected from crocodilians and chelonians; the site in these reptiles is in the midline, immediately beyond the occiput of the skull. However, a large supravertebral vessel is present in this location and the needle must be passed through this vessel to be able to collect a spinal fluid sample. Contamination of the sample with blood will make evaluation difficult and, because of this, a "clean" spinal fluid sample is difficult to obtain.

Acknowledgement

The author would like to thank Dr. Juergen Schumacher for information provided on negative-staining techniques.

REFERENCES

AUSTWICK, P.K.C. and KEYMER, I.F. (1981). Fungi and actinomycetes. In: *Diseases of the Reptilia, Vol. 1.* (Eds. J.E. Cooper and O.F. Jackson). Academic Press, London.

AVERY, H.W. and VITT, L.J. (1984). How to get blood from a turtle. *Copeia 209.*

BARNARD, S.M. (1986). An annotated outline of commonly occurring reptilian parasites. In: *Maintenance and Reproduction of Reptiles in Captivity.* (Eds. V.L. Bels and P.V. Den Sande). *Acta Zoologica et Pathologica Antvenpiensia* **2**, 39.

CAMPBELL, T.N. and COLES, E.H. (1986). Avian clinical pathology. In: *Veterinary Clinical Pathology.* (Ed. E.H. Coles). W.B. Saunders, Philadelphia.

DANTZLER, W.H. and SCHMIDT-NIELSEN, B. (1986). Excretion in fresh-water turtle (*Pseudemys scripta*) and desert tortoise (*Gopherus agassizi*). *American Journal of Physiology* **210**, 198.

ESRA, G.N., BENIRSCHKE, K. and GRINER, L.A. (1975). Blood collecting techniques in lizards. *Journal of the American Veterinary Medical Association* **167**, 555.

GANDAL, C.P. (1958). Cardiac punctures in anesthetised turtles. *Zoologica* **43**, 93.

HILL, A.C. (1985). *Mycoplasma testudinis*, a new species isolated from a tortoise. *International Journal of Systematic Bacteriology* **35**, 489.

JACKSON, O.F. (1981). Clinical aspects of diagnosis and treatment. In: *Diseases of the Reptilia, Vol. 2.* (Eds. J.E. Cooper and O.F. Jackson). Academic Press, London.

JACOBSON, E.R. (1980). Necrotising mycotic dermatitis in snakes: clinical and pathological features. *Journal of the American Veterinary Medical Association* **177**, 838.

JACOBSON, E.R. (1984). Immobilisation, blood sampling, necropsy techniques and diseases of crocodilians: a review. *Journal of Zoo Animal Medicine* **15**, 38.

JACOBSON, E.R. (1986a). Viruses and viral associated diseases of reptiles. In: *Maintenance and Reproduction of Reptiles in Captivity.* (Eds. V.L. Bels and P.V. Den Sande). *Acta Zoologica et Pathologica Antverpiensia* **2**, 73.

JACOBSON, E.R. (1986b). Parasitic diseases of reptiles. In: *Zoo and Wild Animal Medicine.* 2nd Edn. (Ed. M.E. Fowler). W.B. Saunders, Philadelphia.

JACOBSON, E.R. (1987). Reptiles. In: *Veterinary Clinics of North America: Small Animal Practice, Vol. 17.* (Ed. J. Harkness). W.B. Saunders, Philadelphia.

JACOBSON, E.R. (1988). Evaluation of the reptile patient. In: *Exotic Animals.* (Eds. E.R. Jacobson and G.V. Kollias). Churchill Livingstone, New York.

JACOBSON, E.R. and TELFORD, S.R. (1990). Chlamydial and poxvirus infections of circulating monocytes of a flap-necked chameleon (*Chamaeleo dilepsis*). *Journal of Wildlife Disease* **26**, 572.

JACOBSON, E.R., GASKIN, J.M. and MANSELL, J. (1989). Chlamydial infection in puff adders (*Bitis arietans*). *Journal of Zoo and Wildlife Medicine* **20**, 364.

JACOBSON, E.R., SCHUMACHER, J. and GREEN, M.E. (1992). Techniques for sampling and handling blood for hematological and plasma biochemical determinations in the desert tortoise, *Xerobates agassizi. Copeia* January 1992.

JACOBSON, E.R. and GASKIN, J.M. Paramyxovirus infection of viperid snakes. In: *The Biology of Pit Vipers.* (Ed. J. A. Campbell). In Press.

KINGSLEY, G.R. (1972). Procedure for serum protein determinations. In: *Standard Methods of Clinical Chemistry, Vol. 7.* (Ed. G.R. Cooper). Academic Press, New York.

LAPOINTE, J.L. and JACOBSON, E.R. (1974). Hyperglycemic effect of neurohypophysial hormones in the lizard, *Klauberina reversiana. General and Comparative Endocrinology* **22**, 135.

LILLYWHITE, H.B. and SMITS, A.N. (1984). Lability of blood volume in snakes and its relation to activity and hypertension. *Journal of Experimental Biology* **110**, 267.

LILLYWHITE, H.B., ACKERMAN, R.A. and PALACIOS, L. (1983). Cardiorespiratory responses of snakes to experimental hemorrhage. *Journal of Comparative Physiology* **152**, 59.

McDONALD, H.S. (1976). Methods for the physiological study of reptiles. In: *Biology of the Reptilia, Vol. 5: Physiology A.* (Eds. C. Gans and W.R. Dawson). Academic Press, New York.

McDOWELL, E.M. and TRUMP, B.F. (1976). Historical fixatives suitable for diagnostic light and electron microscopy. *Archive of Pathology and Laboratory Medicine* **100**, 405.

MIGAKI, G., JACOBSON, E.R. and CASEY, H.W. (1984). Fungal diseases in reptiles. In: *Diseases of Amphibians and Reptiles.* (Eds. G. Hoff, F.L. Frye, and E.R. Jacobson). Plenum Press, New York.

MILLICHAMP, N., JACOBSON, E.R. and WOLF, E.D. (1983). Diseases of the eye and ocular adnexa in reptiles. *Journal of the American Veterinary Medical Association* **183**, 1205.

NAGY, K. and MEDICA, P.A. (1986). Physiological ecology of desert tortoises in southern Nevada. *Herpetologica* **42**, 73.

OLSON, G.A., HESSLER, J.R. and FAITH, R.E. (1975). Techniques for blood collection and intravascular infusions of reptiles. *Laboratory Animal Science* **25**, 783.

OTTAVIANI, G. and TAZZI, A. (1977). The lymphatic system. In: *Biology of the Reptilia, Vol. 6: Morphology E.* (Eds. C. Gans and T.S. Parsons). Academic Press, New York.

ROSSKOPF, W.J. (1982). Normal hemogram and blood chemistry values for Californian desert tortoises. *Veterinary Medicine/Small Animal Clinician* **77**, 85.

SAMOUR, H.J., RISLEY, D., MARCH, T., SAVAGE, B., NIEVA, O. and JONES, D.M. (1984). Blood sampling techniques in reptiles. *Veterinary Record* **114**, 472.

SCHAER, M., ACKERMAN, N. and KING, R.R. (1989). Clinical approach to the patient with respiratory disease. In: *Textbook of Veterinary Internal Medicine, Vol. I.* 3rd. Edn. (Ed. S.J. Ettinger). W.B. Saunders, Philadelphia.

SCHERMER, S. (1967). *The Blood Morphology of Laboratory Animals.* 3rd Edn. F.A. Davis, Philadelphia.

SMITS, A.W. and KOZUBOWSKI, M.M. (1985). Partitioning of body fluids and cardiovascular responses to circulatory hypovolemia in the turtle, *Pseudemys scripta elegans. Journal of Experimental Biology* **116**, 237.

STEPHENS, G.A. and CREEKMORE, J.S. (1983). Blood collection by cardiac puncture in conscious turtles. *Copeia* 522.

STEWART, J.S. (1990). Anaerobic bacterial infections in reptiles. *Journal of Zoo and Wildlife Medicine* **21**, 180.

TAYLOR, R.W. and JACOBSON, E.R. (1981). Hematology and serum chemistry of the gopher tortoise, *Gopherus polyphemus. Comparative Biochemistry and Physiology* **72A**, 425.

CHAPTER SIX

RADIOLOGICAL AND RELATED INVESTIGATIONS

Oliphant F Jackson PhD MRCVS
Anthony W Sainsbury BVetMed CertLAS MRCVS

RADIOGRAPHY

Radiography is a useful aid to diagnosis in all species of reptiles because they are very difficult to palpate (Holt, 1978; Jackson, 1981; Jackson and Fasal, 1981). Snakes have ribs that extend from the third cervical vertebra to the cloaca and protect the viscera. Many captive reptiles have an incorrect diet which leads to nutritional osteodystrophy: as a result many lizards have fragile bones. Chelonians have a "bony box" which protects the whole body except for the head, neck, limbs and tail and even these can often be withdrawn into the shell. A full description of the principles of radiology and radiography can be found in the BSAVA *Manual of Radiography and Radiology in Small Animal Practice* (Lee, 1989) and of radiology applied to exotic species in Rübel *et al* (1991).

With careful adjustments to kV and mAS it is possible to gain much valuable information from radiographs of reptiles. With the availability of computerised (axial) tomography (CT) in most National Health Service hospitals in the UK, even greater information can be obtained (see later), but CT in reptiles is still in its infancy.

Certain X-ray films - for example, LoDose film in a LoDose Cronex screen (Du Pont) - give the greatest detail in small lizards (particularly the limbs) and hatchling snakes. However, LoDose films are not universally available in veterinary surgeries and many X-ray machines used in practice cannot deliver less than 50kV. Kodirex non-screen or dental films are also recommended for use in small reptiles and for extremities.

When considering exposure factors one should keep the mAS as short as possible in order to avoid movement blur. The focal film distance should be at least 91cm when using LoDose film but with other types of film, the focal distance can be as close as 70cm. The mAS setting should be adjusted depending upon the film-intensifying screen that is being used. Good film detail can be enhanced by the use of high resolution, slow-speed, intensifying screens.

Specific details of the techniques required for taking radiographs of snakes, lizards and chelonians will be covered separately in this chapter.

When taking lateral views of reptiles it is always good practice to have the head pointing to the left and to use a horizontal beam, especially in chelonians (see later).

When interpreting reptile radiographs it is important that the clinician should follow a protocol as for any other animal in order to reach a correct diagnosis. The criteria to be considered are:-

1. Organ position, shape, size, density and homogeneity.
2. The comparative size of the various organs.
3. The state of nutrition of the reptile with respect to the gradation of density between the skeleton, the muscle mass and the soft tissues and the gastrointestinal organs and their contents.

In experienced hands ultrasound can assist by distinguishing between soft tissue organs more readily than is possible with plain radiographs, but ultrasound is unable to image through bone. Ultrasonography is covered in more detail later in this chapter.

RADIOGRAPHING SNAKES

Taking a radiograph of the skeleton of a coiled up snake on a 18 x 24cm plate is of little diagnostic value; the result is unlikely to be a true dorsoventral view of any part of the snake (see Figure 1). It is also extremely difficult to assess the position of any lesion on subsequent examination. Because all snakes have long, fusiform bodies with organs of similar shape, it is advisable to mark the snake at specific distances from the rostrum - for example, by attaching lead markers or letters to the snake every 10 – 20cm with micropore tape or zinc oxide plaster. In many cases the radiograph need only show areas which are suspected of having lesions rather than the whole snake. Various other abnormal bone artefacts can be seen using good radiographic techniques, eg. healed rib fractures.

For intestinal investigations it is useful to perform barium meal studies in order to assess the patency, size and emptying time of the gastrointestinal tract. This should only be done after plain radiographs have been taken and assessed. For a 2kg snake, 10ml of barium sulphate suspension given by oesophageal tube and followed immediately by 90ml of air, gives a very effective double contrast radiograph. The folds of the oesophagus, the two types of gastric rugae, the pyloric sphincter and the duodenal villi are readily identifiable 15 minutes after giving the barium (see Figure 2). Administration of 5mg metoclopramide either orally (Maxolon syrup, SmithKline Beecham) or by injection (Emequell, SmithKline Beecham) will reduce the passage time of the barium meal through the intestines. If metoclopramide is administered, barium will be seen at the cloaca after about 20 hours, whereas, if metoclopramide is not given, it may take up to seven days for the barium to reach the large bowel and cloaca, and even longer if the snake has been fed recently.

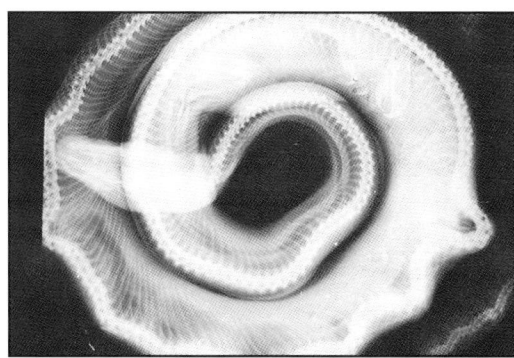

Figure 1
A whole body radiograph of a coiled reticulated python (*Python reticulatus*). Even though this shows kyphotic lesions of the spine, it is not possible to assess fully these lesions due to the poor positioning.

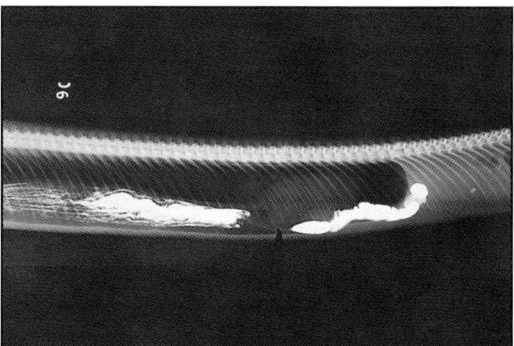

Figure 2
Barium meal study in a boa (*Boa constrictor constrictor*). The stomach is 90cm from the rostrum. The pylorus is arrowed. The abdominal air sac is dorsal to the stomach and duodenum

RADIOGRAPHING LIZARDS

The most common finding when radiographing Sauria kept under sub-optimal management conditions is poor skeletal density. All the members of this Sub-order of reptiles require high levels of calcium for good skeletal development. They probably also require ultra-violet light for the production of vitamin D_3, for the deposition of minerals into the bone and for absorption of calcium from the gastrointestinal tract (see "Nutritional Diseases").

In radiographs where the skeleton is of good density the difference in the muscular density and bony density is similar to that observed in mammalian radiographs. The majority of captive lizards of any size in the UK show signs of nutritional osteodystrophy (see Figure 3).

When taking radiographs of cases of suspected nutritional osteodystrophy the kV may have to be reduced, otherwise the result will be overexposure of the plate with little bone outline visible.

Figure 3
Radiograph of an iguana (*Iguana iguana*) with nutritional osteodystrophy, seen as poor definition between muscles and bones, thin cortices and apparent disappearance of the bone in some areas.

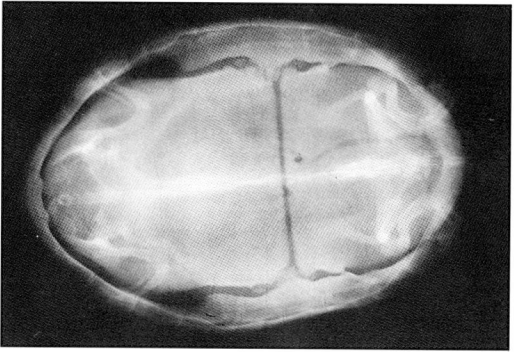

Figure 4
A dorsoventral radiograph of a 1.5kg American box-tortoise (*Terrapene* sp.) clearly showing the hinge between the two parts of the plastron, and the hinge between the pillars of the carapace and the plastron.

RADIOGRAPHING CHELONIANS

Positioning

Dorsoventral (DV) view. A healthy chelonian can move very rapidly over short distances when warm. However, sick chelonians are less active and it is usually possible to position the patient under the central beam of the X-ray machine, adjust the collimator, retire to the control panel and make the exposure without any patient movement. Movement of the head and forelimbs occurs during respiration but, as there is a pause between inspiration and expiration, it is possible to take exposures during this time.

Figure 4 is of a large box-tortoise (*Terrapene* sp.); one should note the large hinge on the plastron transversely across the body and between the pillars of the carapace. These uncalcified hinges also occur as normal physiological hinges, but with only minimal movement, on the plastron of *Testudo* spp. of either sex, and a more normal hinge is found on the caudal carapace of African hinge-backed tortoises (*Kinixyx* spp.).

Nutritional osteodystrophy is commonly found in captive terrapins fed an incorrect diet (see Figure 5).

Lateral view. No chelonian will lie on its side voluntarily nor should it be placed in such a position. The lungs of chelonians lie dorsal to the viscera and are separated from the viscera by a very thin membranous diaphragm. Because of this, tilting of chelonians on to their edge will cause distortion of the diaphragm and lungs due to the weight of the viscera, and an abnormal radiograph will be produced which cannot be diagnostically interpreted.

Lateral projections should, therefore, be taken with a horizontal beam. The chelonian should be placed in a normal position on a foam rubber cushion with the cassette supported perpendicular to the beam and close to the tortoise to ensure that the object-film distance is as short as possible. The X-ray beam should be centred on the 6th or 7th marginal shields at right angles to the vertebral column. The correct focal distance must be used.

Craniocaudal view. The craniocaudal view is particularly useful when comparing and contrasting the two lungs. A similar method to that described for the lateral projection is used but with the patient facing towards the X-ray beam. The central horizontal beam is centred on the nuchal shield (see Figure 6).

Head and/or limbs. Radiographing the head, neck and limbs requires a different approach. To facilitate the placing of the limbs on the cassette without needing to tie them down, it is preferable to anaesthetise the patient. The tortoise can then be placed on its back. The corner of a non-screen or dental film can be positioned between the marginal shields of the carapace and the limb, head or neck. The beam must be collimated to the smallest area possible in order to obtain the most detailed results and to reduce scatter. By tilting the patient using sponge rubber pads or sandbags it is possible to obtain two different views of the affected area. Until experience in reading chelonian limb radiographs is acquired, readers are advised to take the same views of the normal opposite limb. On only one occasion have the authors found a tortoise with similar lesions in both hind or forelimbs. Figure 7 shows a lesion involving the limb of a tortoise.

Contrast radiography. Both radiopaque and radiolucent contrast techniques can be employed in chelonians (Holt, 1978; Lee, 1978). It is possible to introduce both into the gastrointestinal tract by passing a stomach tube. Administering 2ml of barium sulphate followed by 18ml of air will produce a good double contrast radiograph in animals of over 1kg bodyweight (see Figure 8).

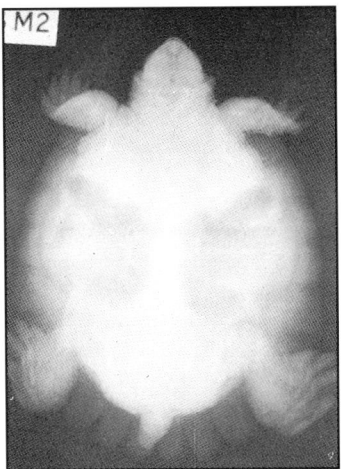

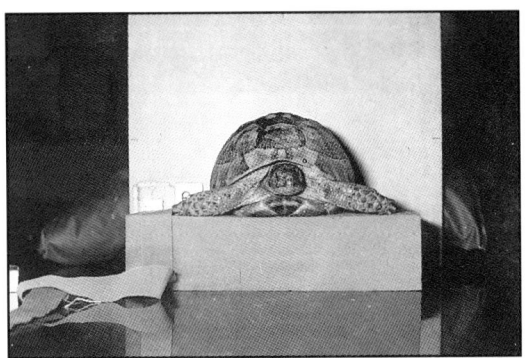

Figure 6
Positioning for a craniocaudal view of a spur-thighed tortoise (*Testudo graeca*).

Figure 5
A red-eared terrapin (*Trachemys scripta elegans*) with severe nutritional osteodystrophy. Note the apparent lack of bones in the limbs and the clear definition of left and right lungs.

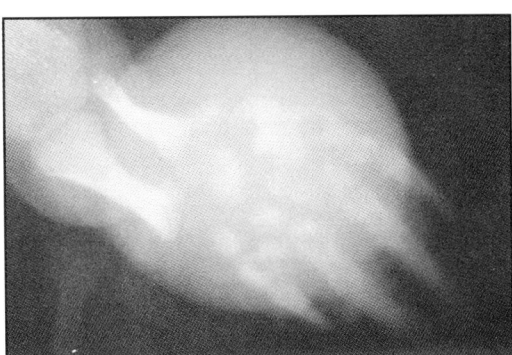

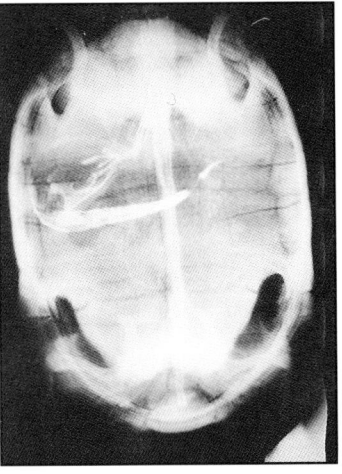

Figure 7
Radiograph of the distal right hindlimb of a *Testudo* sp. Severe bacterial osteolysis has resulted in destruction of the tarsal bones and the distal tibia and fibula.

Figure 8
Double contrast barium meal study defining the stomach and pylorus in a spur-thighed tortoise (*Testudo graeca*).

COMPUTERISED TOMOGRAPHY (CT) AND MAGNETIC RESONANCE IMAGING (MRI)

The equipment for CT examination is available in most National Health Service or private hospitals and is useful for examining organs inside bone-limited hollow cavities; hence its particular usefulness in chelonians. However, MRI, formerly called nuclear magnetic resonance (NMR), is at present restricted to research establishments: the Royal College of Surgeons of England has performed MRI examination on reptiles (Cooper, personal communication). As these facilities become more readily accessible, veterinary surgeons must become aware of how they can assist in the correct evaluation of abnormalities within the coelomic cavity and which organ systems require their usage. At present CT scanning and MRI are expensive procedures but they hold out exciting prospects for the future in the diagnosis of disease in reptiles.

Computerised tomography (CT)

Conventional X-ray films display only a small proportion of the data theoretically produced when gamma rays pass through living tissues. A new method of forming images from X-rays was developed and introduced into clinical use in 1972 (Ambrose and Hounsfield, 1973). CT uses multidirectional scanning of the patient to produce cross-sectional images. The X-rays fall not on a film but on detectors which record the attentuation values of the X-ray beam emerging from the patient. The attentuation values are directly related to the tissue density, as a larger proportion of the X-ray beam will be absorbed by denser tissues. An X-ray source and an array of detectors rotate around the patient during each scan.

The CT image is displayed on a computer video monitor and consists of a matrix of dots, called pixels. Each pixel displays the tissue density at one point in the cross-section; the outstanding feature of CT is that very small differences in tissue density can be visualised. Not only can fat be distinguished from other soft tissue but gradations in density within the soft tissue can also be recognised. Variations in tissue density are expressed in terms of CT numbers. This scale, composed of Hounsfield units (HU), relates the density values to that of water. Positive values are recorded for structures such as bone and soft tissues which have higher attenuations of X-rays than water, whereas negative values are obtained for fat and air. The scale ranges from +1000 to -1000, the CT number for air being -1000HU while that for fat is -60 to -100HU.

The different CT numbers are displayed as a grey scale image which forms the visual display. Once the image is displayed on the monitor it can be transferred to magnetic tape, floppy disc or hard disc for permanent storage.

A windowing system is used where the CT number range is spread to cover the full grey scale. The window level and width are manipulated to provide optimum viewing conditions for the area of interest, the window width being varied according to the tissue being analysed. For example, the lungs are viewed at a window level of from -600 to -700HU. At this level the soft tissue surrounding any infection appears white and the pulmonary vessels are seen as white structures traversing the lungs. Increasing the window level to +300HU demonstrates the architecture of the bone and, at this level, soft tissue detail is seen less clearly. Hard copies of any particular areas of interest can be taken using a multiformat camera or video printer so that photographic or thermal film and polaroid or bromide prints can be obtained (see Figure 9).

The patient is scanned while moving slowly through the gantry on a platform. The X-ray tube and detectors remain stationary. The image produced is analogous to a conventional dorsoventral radiograph and is used to define the beginning and end of the area to be scanned.

Scan times vary from 1 – 10 seconds. Fast scanning times reduce movement artefacts and allow a shorter overall scanning time. However, longer scanning times may be desirable for areas encased in bone, eg. within the shell of chelonians, where higher radiation doses result in better image quality. Slice thickness varies according to the area being examined, 2mm sections being the smallest.

Magnetic resonance imaging (MRI)

The physics of MRI is complicated. MRI avoids the use of ionising radiation and so far has not been associated with any significant hazards. MRI employs radio frequency radiation in the presence

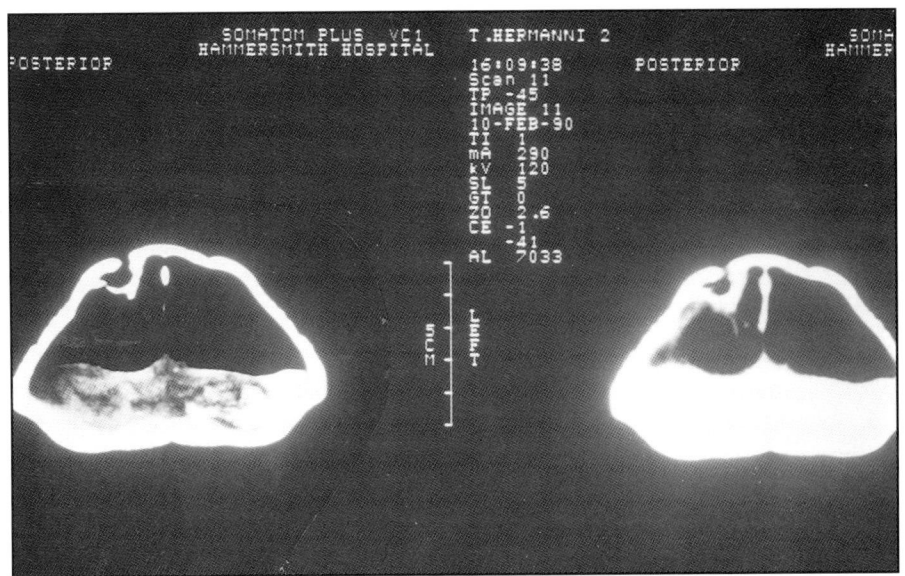

Figure 9
CT scan of a Hermann's tortoise (*Testudo hermanni*) showing a depressed fracture of the carapace. The median connective tissue between right and left lung carrying the pulmonary vessels is clearly visible.

of a carefully controlled magnetic field in order to produce high quality cross-sectional images in any plane. It portrays the distribution density of hydrogen nuclei (protons) and parameters relating to their motion in cellular water and lipids. Like CT, the display of soft tissue detail is excellent and the fact that bone is barely visualised makes this the technique of choice when investigating central nervous system disturbances in reptiles (Cooper, unpublished data).

The patient is placed centrally within a large electromagnet. The images produced depend upon numerous factors, including the proton density in the tissue, and, therefore, give different information to that provided by CT. An average examination may take from 45 – 60 minutes, ie. much longer than CT. The patient has to remain still during the procedure so this technique is not as easy in reptiles as it is in man, unless they are anaesthetised.

MRI scanners present the data as an analogue display of the pixel values of the signal in an anatomical cross-section. Figures 10 and 11 show the ova within the coelomic cavity of an adult female *Testudo graeca* in transverse and horizontal sections. The window width and level at which the slices are taken can be manipulated to produce optimal images.

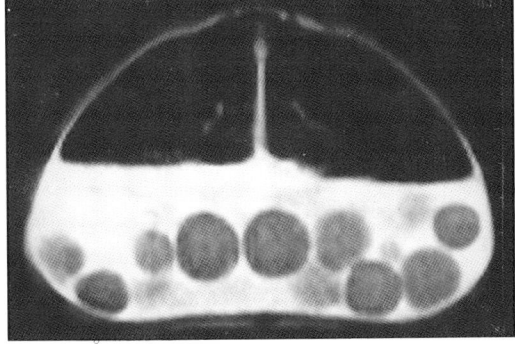

Figure 10
MRI transverse section through an adult spur-thighed tortoise (*Testudo graeca*) showing ova spread out across the midline, behind the liver.

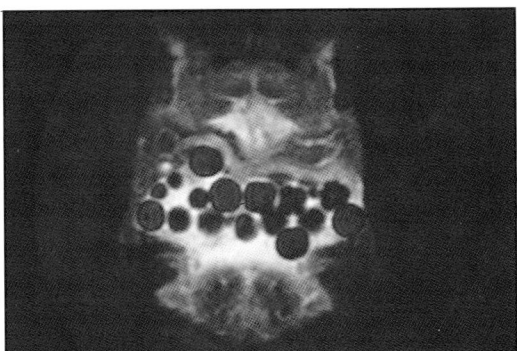

Figure 11
MRI horizontal coronal section across the same *T. graeca* as in Figure 10, showing the general distribution of ova in the coelomic cavity.

ULTRASONOGRAPHY

Ultrasonography has the same potential applications in reptiles as it has in mammals: the monitoring of reproductive function, the diagnosis of disease and assisting with other diagnostic techniques, such as ultrasound guided biopsy (Lamb et al, 1988). Little has been written on the use of ultrasound in reptiles (Rostal et al, 1990; Penninck et al, 1991; Sainsbury and Gili, 1991) but, as in mammals, it has the advantages of being a rapid and simple technique and the ability to distinguish between soft tissues is good. Because safety precautions are not necessary and an animal can be scanned whilst moving, ultrasound has advantages over radiography. Its non-invasive nature is important when compared with laparoscopy.

Herring and Bjornton (1985) can be consulted for a description of the physical properties of ultrasound, while Meire and Farrant (1982) and Barr (1991) give general information on ultrasonography, including the interpretation of images. Only real-time scanning, in which the ultrasound image is continuously updated to allow movement to be seen, will be dealt with in this chapter. The terminology used is that recommended by the American Institute of Ultrasound in Medicine (1990).

Scanning technique

It is rare for sedation to be required to scan reptiles, although several handlers may be needed to restrain lively animals. Lizards and snakes are usually scanned with the transducer directed through the ventral body wall with the animal held in a prone position. In view of their extensive bony carapace and plastron, through which the ultrasound beam will not penetrate, many chelonians are difficult to scan. Their internal organs can be approached with the transducer in any available space between the plastron, carapace and limbs: mediastinal, axillary, inguinal and paracloacal approaches are possible (Penninck et al, 1991). Therefore, in chelonians the size of the transducer in relation to the size of the animal is the main limiting factor for successful imaging. However, in soft-shelled turtles (*Trionyx* spp.) a picture can be obtained through any part of the plastron and carapace.

The scales of reptiles do not unduly impede ultrasound transmission. An aqueous coupling gel (Aquasonic Ultrasound Transmission Gel, Parker Laboratories; or K-Y Jelly, Johnson and Johnson) is required to achieve good contact between animal and transducer.

The ribs of snakes occur throughout the length of the body and nearly meet in the midline. They are robust and bony (rather than cartilaginous) (Davies, 1981); therefore, the ultrasound image can be broken with a regular array of shadows (see Figure 12). The ribs of lizards extend further caudally than in the majority of mammals and encompass parts of the coelom as well as the thorax (Hoffstetter and Gasc, 1969) and may impede some views.

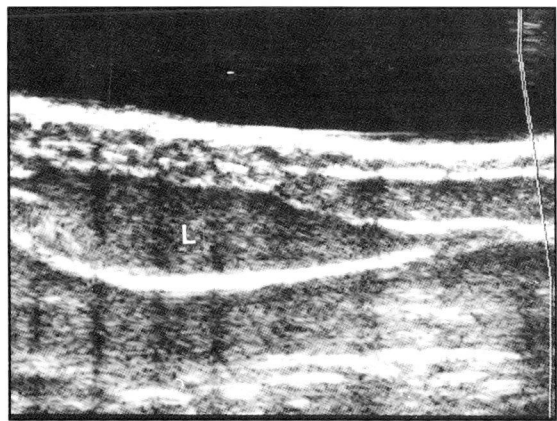

Figure 12
Sagittal image of the liver (L) of an indigo snake
(*Drymarchon corais*). Acoustic shadows are produced by
the ribs which reflect the ultrasound beam.

For small reptiles, 7.5 and 10MHz transducers will be required to obtain sufficient resolution. A waterpath standoff (literally a rubber bag of fluid which separates the transducer from the animal) may be required to improve the focussing of superficial structures. Some manufacturers have standoffs made to fit transducers (Kitecko Pad, 3M Health Care) but, if unavailable, a pad of thick gel can be used, though with more difficulty. 5.0 and 3.5MHz transducers can be used in larger animals where a greater depth of penetration is required.

Linear array transducers are more applicable in the majority of lizards and snakes due to the small size of these reptiles. In view of the shape of the beam produced by sector transducers superficial structures are not easily imaged in small animals. In chelonians the need for a small transducer which can be manoeuvred into the limited space available for scanning, and yet produce an image of a wide field, means that sector scanners are ideal.

APPLICATIONS

Diagnostic ultrasound

In common with all imaging techniques it is essential to have a knowledge of the anatomical relationships between different organs and surrounding structures. There are few anatomical descriptions which give sufficient detail to be comprehensive for ultrasonographic examinations. The following references may help:– Owen (1866); Grasse (1952); Harris (1963); Gans *et al* (1969); Gans and Parsons (1970); Gans and Dawson (1976); Gans and Parsons (1977); Davies (1981); Chiodini *et al* (1982); Penninck *et al* (1991); Sainsbury and Gili (1991).

In broad terms, the same organs that can be usefully scanned in mammals can be scanned in reptiles, ie. heart, liver, pancreas, spleen, gastrointestinal tract, kidneys etc. However, there are probably many specific differences. For example, studies indicate that it is not possible to scan the spleen and pancreas in some chelonians (Penninck *et al*, 1991). In monitor lizards the spleen is difficult to locate due to its small size; however, the pancreas is superficial and the liver large (see Figure 13) and these two organs are relatively rewarding to scan (Sainsbury and Gili, 1991).

Landmarks of use in domestic animal ultrasonography may not be applicable. For example, the ribs of snakes and lizards are numerous and closely aligned. There is no diaphragm in reptiles (a diaphragm-like membrane occurs in some lizards and in chelonians but is not easily visible on a scan). Well-fed lizards may have large fat bodies which will fill a large part of the coelom and hinder imaging of the kidneys.

As in mammals, organs can be distinguished by their different acoustic impedance, although the latter is not necessarily similar to that of mammalian organs. For example, the reptilian kidney is hyperechoic compared with the mammalian and does not have areas of varying echogenicity, such as the cortex, medulla and pelvis (see Figure 14).

There have been few reports of the diagnosis of disease in reptiles using ultrasound. However, it has the potential, as in other animals, to locate abscesses, cysts, neoplasms, effusions and, with experience, other conditions which change the shape, consistency or position of organs. Figure 15 shows the image of an intestinal abscess in a Burmese python (*Python molurus*). Since pus in reptiles is usually inspissated, a non-homogeneous hyperechoic image is produced.

Reproductive scanning

It is possible to distinguish the gonads of chelonians (Penninck *et al*, 1991) and ultrasound has been found to be useful for monitoring reproductive function of female chelonians (Robeck *et al*, 1990; Rostal *et al*, 1990). Robeck *et al* (1990) identified follicles and eggs at different stages of development and found ultrasound was a safe and effective method of aiding the management of the endangered Galapagos tortoise (*Chelonoidis elephantopus*). It has advantages over radiography and laparoscopy: in the case of the former since early development of follicles can be visualised and in the latter because ultrasound is non-invasive. Ultrasound can be used similarly in other reptiles and, in the authors' experience, is useful for determining whether or not females are gravid.

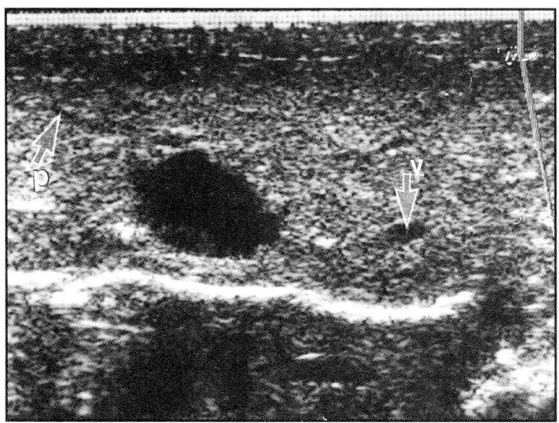

Figure 13
Transverse scan of the liver of a Bosc monitor (*Varanus exanthematicus*) including the gall bladder (large anechoic region), the portal vein (arrowed - p) and the cava (arrowed v).

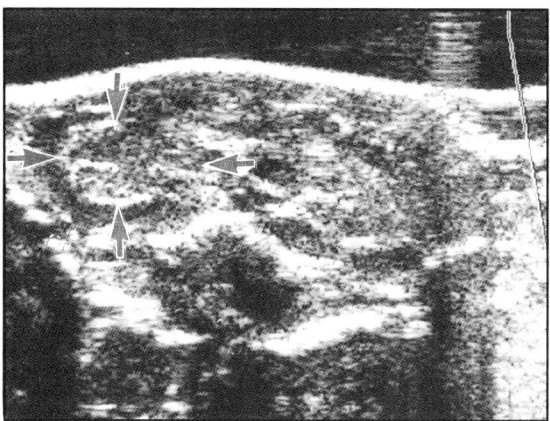

Figure 14
Transverse image of the kidneys of a Bosc monitor (*Varanus exanthematicus*). The boundary of the left kidney is marked by four arrows.

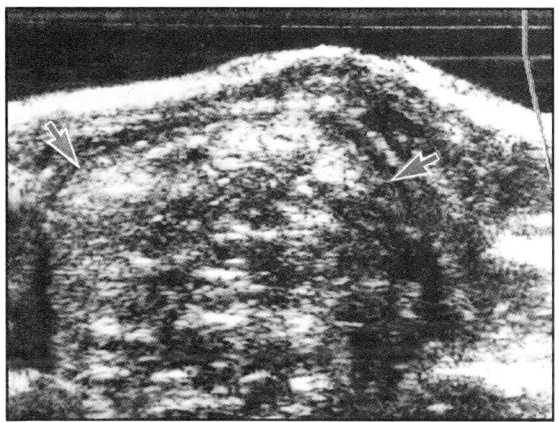

Figure 15
Sagittal scan of a Burmese python (*Python molurus*) with an intestinal abscess. The wall of the abscess is marked by arrows.

REFERENCES

AMBROSE, J. and HOUNSFIELD, G. (1973). Computerised transverse axial tomography. *British Journal of Radiology* **48**, 148.

AMERICAN INSTITUTE OF ULTRASOUND IN MEDICINE. (1990). *Recommended Ultrasound Terminology*. American Institute of Ultrasound in Medicine, Bethesda.

BARR, F.J. (1991). *Diagnostic Ultrasound in the Dog and Cat*. Blackwell Scientific Publications, Oxford.

BELLAIRS, A.d'A. (1969). *Life of Reptiles, Vol. 1*. Weidenfeld and Nicolson, London.

CHIODINI, R.J., SUNDBERG, J.P. and CZIKOWSKY, J.A. (1982). Gross anatomy of snakes. *Veterinary Medicine/Small Animal Clinician* **77**, 413.

DAVIES, P.M.C. (1981). Anatomy and physiology. In: *Diseases of the Reptilia, Vol.1*. (Eds. J E. Cooper and O.F. Jackson). Academic Press, London.

GANS, C., BELLAIRS, A.d'A. and PARSONS, T.S. (1969). Eds. *Biology of the Reptilia, Vol. 1*. Academic Press, London.

GANS, C. and DAWSON, W.R. (1976). Eds. *Biology of the Reptilia, Vol. 5.* Academic Press, London.

GANS, C. and PARSONS, T.S. (1970). Eds. *Biology of the Reptilia, Vol. 3.* Academic Press, London.

GANS, C. and PARSONS, T.S. (1977). Eds. *Biology of the Reptilia, Vol. 6.* Academic Press, London.

GRASSE, P.P. (1952). Ed. *Traite de Zoologie: Anatomie, Systematique Biologie, Vol.14.* Masson, Paris.

HARRIS, V.A. (1963). *The Anatomy of the Rainbow Lizard, Agama agama.* Hutchinson, London.

HERRING, D.S. and BJORNTON, G. (1985). Physics, facts and artefacts of diagnostic ultrasound. *Veterinary Clinics of North America: Small Animal Practice* **15**, 1107.

HOFFSTETTER, R. and GASC, J.P. (1969). Vertebrae and ribs of modern reptiles. In: *Biology of the Reptilia, Vol. 1.* (Eds. C. Gans, A.d'A. Bellairs and T.S. Parsons). Academic Press, London.

HOLT, P.E. (1978). X-ray equipment for veterinary practice. *Veterinary Record* **103**, 88.

JACKSON, O.F. (1981). Clinical aspects of diagnosis and treatment. In: *Diseases of the Reptilia, Vol. 2.* (Eds. J.E. Cooper and O.F. Jackson). Academic Press, London.

JACKSON, O.F. and FASAL, M.D. (1981). Radiology as a diagnostic aid in four chelonian conditions. *Journal of Small Animal Practice* **22**, 705.

LAMB, C.R., STOWATER, J.L. and PIPERS, F.S. (1988). The first twenty-one years of veterinary diagnostic ultrasound: a bibliography. *Veterinary Radiology* **29 (1)**, 37.

LEE, R. (1978). Contrast media and techniques 1. *Journal of Small Animal Practice* **19**, 589.

LEE, R. (1989). Ed. *Manual of Radiography and Radiology in Small Animal Practice.* BSAVA, Cheltenham.

MEIRE, H.B. and FARRANT, M. (1982). *Basic Clinical Ultrasound.* British Institute of Radiology Teaching Series No. 4. British Institute of Radiology, London.

OWEN, R. (1866). *Anatomy of Vertebrates, Vol. 1.* Longmans, Green and Co., London.

PENNINCK, D.G., STEWART, J.S., PAUL-MURPHY, J. and PION, P. (1991). Ultrasonography of the Californian desert tortoise (*Xerobates agassizi*): anatomy and application. *Veterinary Radiology* **23 (3)**, 112.

ROBECK, T.R., ROSTAL, D.C., BURCHFIELD, P.M., OWENS, D.W. and KRAEMER, D.C. (1990). Ultrasound imaging of reproductive organs and eggs in Galapagos tortoises (*Geochelone elephantopus*). *Zoo Biology* **9**, 349.

ROSTAL, D.C., ROBECK, T.R., OWENS, D.W. and KRAEMER, D.C. (1990). Ultrasound imaging of ovaries and eggs in Kemps Ridley sea turtles (*Lepidochelys kempi*). *Journal of Zoo and Wildlife Medicine* **21 (1)**, 27.

RÜBEL, G.A., ISENBÜGEL, E. and WOLVEKAMP, P. (1991). Eds. *Atlas of Diagnostic Radiology of Exotic Pets.* Wolfe, London.

SAINSBURY, A.W. and GILI, C. (1991). Ultrasonographic anatomy and scanning technique of the coelomic organs of the Bosc monitor (*Varanus exanthematicus*). *Journal of Zoo and Wildlife Medicine* **22 (4)**, 421.

CHAPTER SEVEN

INTEGUMENT

John E Cooper BVSc CertLAS DTVM FRCVS MRCPath FIBiol

INTRODUCTION

Skin diseases are prevalent in reptiles (Elkan and Cooper, 1980; Cooper and Jackson, 1981; Marcus, 1981; Ippen *et al*, 1985; Frye, 1991). They can be divided into two main groups:-

1. Infectious.

2. Non-infectious.

However, there is often overlap between the two groups and environmental factors frequently predispose to, or exacerbate, other conditions. Investigation of management is, therefore, always necessary: the veterinary surgeon will need to familiarise him/herself with the requirements of the species in captivity, especially temperature, relative humidity and substrate, and be prepared to visit the owner's premises in order to investigate these.

In this chapter the normal skin is discussed first, followed by a brief description of investigative techniques and information on some of the more prevalent or important conditions.

NORMAL FEATURES

Reptiles have a thick skin which is sometimes heavily keratinised, usually protected by scales (of ectodermal origin) but with scale pockets between. In addition, chelonians have a "shell" consisting of both osseous (dermal bone) and epithelial elements. The latter form shields or "scutes" (Zangerl, 1969).

Loss and regeneration of the tail (autotomy) occurs readily in certain lizards. The severed tail or limb of other reptiles will usually heal but not regenerate.

Skin shedding (sloughing or ecdysis) is normal. This is usually complete in many snakes and lizards but often only partial in other species: for example, chelonians may shed scales or shields (scutes). The sloughing pattern can be a useful guide to health in snakes and lizards.

Skin glands are rare in reptiles. There are, however, small cloacal glands and nuchal glands in a few species and the males of some lizards, eg. Iguanidae, have femoral pores.

Chromatophores are often abundant in the skin and make colour changes possible.

Osteoderms (bony structures in the skin) are well developed in some species, eg. crocodilians and certain lizards, and vary considerably in shape and size. They will be seen on radiography and can hamper surgery.

Dermal sensory receptors are often present in the skin of reptiles, including heat-sensitive pits in some snakes, eg. pit vipers (*Crotalus* spp.) and boids.

There are frequently well developed modifications to the skin, eg. horns, crests, spines and dewlaps. Often these are a feature of the species but sometimes there is sexual dimorphism: for example, the male of certain chameleons bears long horns while these are small or non-existent in the female. These and other "variations" in the skin may be associated with:-

1. Species and sub-species.

2. Gender.

3. Age and maturity.

4. Time of birth.

In recent years there has been increased interest in the "ecological" aspects of the skin of reptiles. Particular attention has focused on the chemicals, especially lipids, produced by the epidermis of snakes and their possible relationship with bacteria.

INVESTIGATION OF SKIN DISEASES

Evaluation of skin diseases necessitates a thorough investigation, paying particular attention to the whole animal and its environment and not just the presenting signs or lesions.

The following is the recommended approach:-

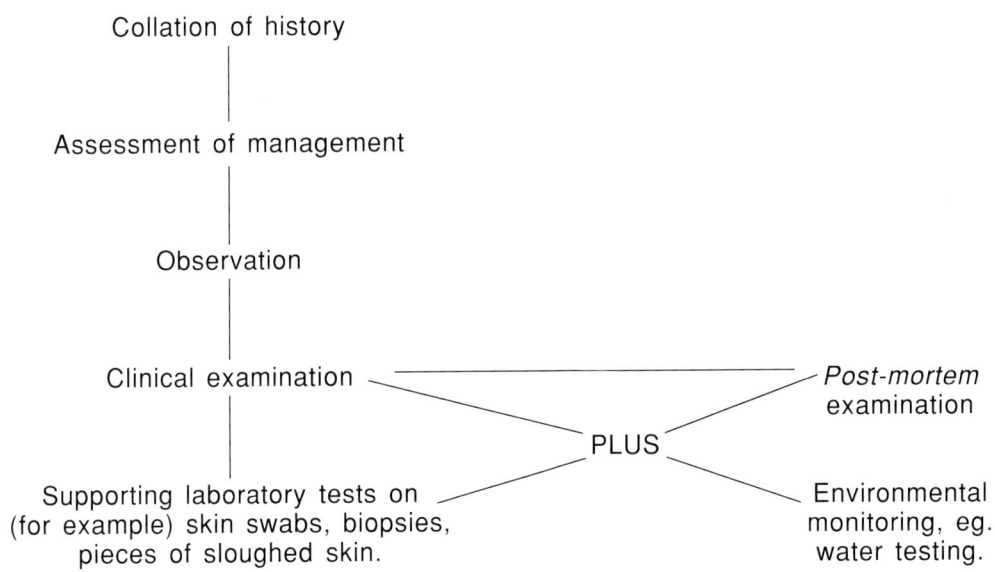

The collection of skin samples from live reptiles for laboratory investigation is discussed elsewhere (see "Laboratory Investigations") as are *post-mortem* techniques (see "*Post-Mortem* Examination"). The removal of shell biopsies from chelonians can prove particularly difficult.

Laboratory investigations are of great importance in diagnosis since clinical signs are not always pathognomonic (Cooper and Lawrence, 1982; Cooper, 1986). The practitioner dealing with skin diseases should assume from the outset that samples are likely to be needed for microbiological and histopathological examination and that other tests, eg. haematology, may be a useful aid to differential diagnosis and prognosis.

EXAMPLES OF SKIN DISEASES

Abscesses

These are usually raised, hard and well circumscribed. They are generally subcutaneous but can occur in the middle ear (see "Surgery") or under the spectacle (see "Ophthalmology"). The aetiology is usually a bacterial infection, often following trauma: a variety of organisms can be involved, including anaerobes.

Diagnosis may require aspiration and culture or histopathological examination of a biopsy. Differentiation from fungal, parasitic and neoplastic lesions is necessary. Abscesses are best removed *in toto,* including the capsule (see "Surgery"). Recurrence may be discouraged by the use of an appropriate antibacterial agent.

Beak and claw deformities

These are usually due to overgrowth/damage or a developmental abnormality. Nutritional and genetic factors are possibly involved. Clipping and shaping are required together with appropriate management changes.

"Blister disease"

Particularly prevalent in garter (*Thamnophis* spp.) and water (*Natrix* spp.) snakes, this condition is characterised by raised, subcutaneous fluid-filled lesions. The causes include poor ventilation or wet substrate, producing too high a relative humidity. The initial lesions are apparently sterile but if secondary infection occurs the fluid becomes cloudy and a bacteraemia may result (Cooper and Lawrence, 1982). Diagnosis is on clinical signs.

Treatment necessitates changes to the environment and attention to infected lesions.

Other causes of "blisters" are also recognised, including migrating nematodes, eg. *Kalicephalus* spp.

Burns

These usually occur in captivity because of a poorly protected heat source or when a large reptile is at liberty in the house and has access to lights or heaters. In the wild burns may be caused by fires. Surprisingly deep lesions may occur and (following ventral burns) the abdominal (coelomic) cavity may be perforated. *Pseudomonas* infections are often secondary.

Treatment consists of attention to the wounds and appropriate supportive therapy. Preventive measures must be instigated.

Damaged "shell"

The "shell" (carapace and plastron) of chelonians may be damaged by trauma or burning (Bourdeau, 1989). Changes can also be due to nutritional/metabolic/genetic factors or infection (see later).

Lesions can be repaired by cleaning, disinfection, application of plastic skin dressings (mild lesions) or epoxyresin (severe lesions). Healing may be prolonged (Jackson, 1978; Holt, 1981).

Before embarking on the treatment of "shell" damage the practitioner should acquaint him/herself with the normal anatomy, particularly the relationship between the keratinised, non-viable outer layers and the underlying live bone.

Dermatitis

Dermatitis, also known as necrotic dermatitis, ulcerative dermatitis or ventral dermal necrosis, is commonly referred to as "scale rot". It can take several forms and may be secondary to trauma or environmental factors. Foreign-body dermatitis is mentioned later.

The presenting clinical signs of dermatitis are discolouration or lesions of the skin. Severe cases will show ulceration which may extend deep into underlying musculature. Septicaemia can be a sequel.

Affected reptiles should be placed in a clean, dry vivarium and skin lesions treated using topical disinfectants, topical and/or systemic antibiotics (following sensitivity tests) and application of adhesive drapes (Cooper, 1981).

Dermatophilosis

This is due to *Dermatophilus congolensis* infection. It is characterised by raised lesions or subcutaneous abscesses over the skin of the body and limbs. Diagnosis is by culture or histopathological examination of a biopsy.

Treatment consists of surgical excision of lesions, topical application of povidone-iodine and administration of antibiotics.

Dysecdysis

Dysecdysis (difficulty in sloughing) may be due to a sub-optimum environment, eg. low relative humidity or no bathing facilities, but often there is a more deeply rooted problem - for example, systemic disease, old scars, endocrinological disorders etc.

Diagnosis demands systematic and detailed evaluation of the reptile and its environment. The owner's records can prove valuable.

Treatment is attempted initially by soaking the reptile in tepid water. Retained spectacles should be removed with great care (see "Ophthalmology"). If, despite treatment, dysecdysis persists or recurs, full investigation is justified. Sloughing can be stimulated by the administration of various drugs.

In aquatic chelonians failure to shed shell plates may be due to poor basking facilities or osteodystrophy, and inadequate lighting may contribute.

Ectoparasites

Mites and ticks are frequently seen on terrestrial reptiles and leeches on aquatic species.

While ticks are usually readily visible, mites can be difficult to locate, especially when present in small numbers. Careful searching may be necessary, particularly under the scales. Manual removal of ticks must be carried out with care: if mouthparts remain in the skin, infections may occur.

Dichlorvos (Vapona, Shell) can be used for treatment of ticks and mites, but with care. Alternatively, reptiles can be sprayed with a dilute solution of trichlorphon, or pyrethrum-based products may be applied. Ivermectin is effective by injection. Further information on therapy of parasitic conditions can be found in "Therapeutics".

The mite *Ophionyssus natricis* (which may transmit the bacterium *Aeromonas hydrophila*) is the most significant ectoparasite of reptiles. It is found mainly on snakes. It can cause poor shedding or even be a cause of death. *O. natricis* may, on occasion, be transmitted to and affect humans.

Other mites are also found on reptiles. Within the Cheyletoidea there are two Families, one living under the scales of snakes (Ophioptidae), the other in the cloaca of turtles (Cloacaridae). Trombiculid mites are found on reptiles but are not species-specific.

Ticks of many species, some host and region-specific, occur on reptiles, especially imported specimens. They vary in their pathogenicity.

Leeches (Hirudinea) are found on aquatic chelonians and crocodilians and sometimes (usually in the nasal and buccal cavity or attached to the cloaca) on snakes and lizards. They drop off the host once they have engorged with blood. Removal can be effected by touching the leech with a cotton bud which has been dipped in ivermectin.

Foreign-body dermatitis

This can occur as a result of exposure to fibreglass from the cage (Frye and Myers, 1985) or talc from gloves (Cooper, unpublished data). It usually manifests itself as one or more areas of chronic inflammation. Treatment is by excision of the lesions or, in superficial cases, repeated soaking and/or bathing of the affected reptile.

Fungal diseases (mycoses)

Fungal lesions of the skin have been reported in many species. Aquatic reptiles appear to be particularly susceptible. The fungi involved are often not isolated but *Aspergillus* has been reported. The relationship of fungal lesions to "scale rot" (see Dermatitis) is unclear. Fungi can also cause subcutaneous mycotic granulomas, which need careful differentiation from bacterial abscesses.

Diagnosis is usually based on the detection of fungi in histological sections of biopsies.

Treatment of fungal lesions can be attempted using topical agents, coupled with excision of badly affected tissues, but often proves unsuccessful.

Myiasis

Dipterous larvae may cause skin lesions, especially around the cloaca, following diarrhoea or cloacitis or after an attack by rodents. Diagnosis is by detection of the larvae. Treatment is by cleaning (flushing) - warm soapy water sprayed by syringe is often excellent - followed by disinfection and cautious application of insecticidal agents. Supportive therapy may be necessary and further attacks must be prevented by vigilance and regular examination.

Neoplasms

Neoplastic lesions involving the skin are sometimes seen in reptiles (Elkan, 1974; Cooper and Lawrence, 1982; Cooper *et al*, 1983). Fibrosarcomas appear to be particularly prevalent in snakes. A biopsy will help to distinguish a neoplasm from other lesions. Treatment is usually by surgical excision but cryotherapy may be of value (Baxter and Meek, 1988).

Osteodystrophy

This may be manifested by raised, usually bilaterally symmetrical, nodular lesions on the jaws. Osteodystrophy is not a primary disease of the integument and is, therefore, covered elsewhere in this manual (see "Nutritional Diseases").

Papillomatosis

This condition, characterised by the presence of raised papillomatous lesions, is particularly prevalent in European green lizards (*Lacerta viridis*). It is probably caused by a virus (Cooper and Lawrence, 1982). The lesions vary in distribution: in female lizards they predominate around the tail, while in males the base of the head appears to be the predilection site. Papillomata should be removed by excision or cryotherapy. A papillomatosis of Bolivian side-neck turtles (*Platemys platycephala*) has also been reported (Jacobson, 1991). It is possible that other species of reptile may also be affected.

Skin lacerations

These may be inflicted by other reptiles and predators, including live food (Rosskopf and Woerpel, 1981), or by physical damage. They should be treated as in other species. Attention to hygiene is important in order to reduce the risk of secondary infection. The application of an adhesive drape (Op-Site, Smith and Nephew) will assist (Cooper, 1981). Skin wounds may take several weeks to heal: the process is temperature-dependent (Smith *et al*, 1988).

Retention of slough

This can be a distinct entity from dysecdysis (see earlier). It is usually associated with a low relative humidity and/or lack of suitable items on which the reptile can rub itself. Lizards may lose the tips of digits or tail as a result of constriction of these extremities by bands of retained slough.

Treatment consists of soaking (see earlier) and attention to the environment.

Excessive sloughing may be due to endocrine disorders or hypervitaminosis A (see "Nutritional Diseases").

Rostral abrasions (erosions)

These commonly occur because reptiles do not recognise glass as a barrier and may repeatedly rub or bang against it. Sexual frustration or competition (trying to escape from another animal) may be an exacerbating factor.

Prevention is relatively simple - an opaque barrier should be provided and the reptile offered cover or hiding holes.

Treatment may be less easy. Topical treatment is sometimes adequate but in severe cases there may be permanent damage to the rostrum.

Ulcerative shell disease of chelonians

This is usually due to poor hygiene, intra-species aggression, low temperature or poor shedding. It may be confined to the shell or become septicaemic. A number of different organisms may be involved: culture and sensitivity should be performed.

Topical treatment consists of cleaning, debridement and application of antimicrobial agents as necessary.

A specific condition caused by *Citrobacter freundii* and termed "septicaemic cutaneous ulcerative disease" (SCUD) has been reported in terrapins (freshwater turtles, *Trionyx* spp.) in the UK and in the USA. The lesions consist of necrotic ulcerations of the shell and skin. A number of different bacteria are possibly involved and synergism may occur. Improvement of water quality and topical attention to the lesions are the recommended treatment.

Viral diseases

A number of viral diseases may affect the skin of reptiles (Jacobson, 1991). Papillomatosis was discussed earlier.

Other conditions include "gray patch disease" of green sea turtles (*Chelonia mydas*) and poxvirus skin disease of crocodilians and other reptiles (Stauber and Gogolewski, 1990).

It is likely that other diseases of reptiles caused by viruses will be described in due course. The clinician should be aware of the possibility of viral involvement and, especially in cases where conventional therapy proves ineffective, consider submitting biopsy material for transmission electron microscopy and/or virus isolation.

REFERENCES

BAXTER, J.S. and MEEK, R. (1988). Cryosurgery in the treatment of skin disorders in reptiles. *British Herpetological Journal* **1**, 227.

BOURDEAU, P. (1989). Pathologie des tortues. 2e partie: affections cutanées et digestives. *Point Veterinaire* **20**, 871.

COOPER, J.E. (1981). Use of a surgical adhesive drape in reptiles. *Veterinary Record* **108**, 56.

COOPER, J.E. (1986). The role of pathology in the investigation of diseases of reptiles. *Acta Zoologica at Pathologica Antiverpiensia* **2**, 15.

COOPER, J.E. and JACKSON, O.F. (1981). Eds. *Diseases of the Reptilia*. Academic Press, London.

COOPER, J.E., JACKSON, O.F. and HARSHBARGER, J.C. (1983). A neurilemmal sarcoma in a tortoise (*Testudo hermanni*). *Journal of Comparative Pathology* **93**, 541.

COOPER, J.E. and LAWRENCE, L. (1982). Pathological studies on skin lesions in reptiles. In: *Proceedings of the 1st International Colloquium on Pathology of Reptiles and Amphibians*. (Eds. C. Vago and G.Matz). Angers.

ELKAN, E. (1974). Malignant melanoma in a snake. *Journal of Comparative Pathology and Therapeutics* **84**, 51.

ELKAN, E. and COOPER, J.E. (1980). Skin biology of reptiles and amphibians. *Proceedings of the Royal Society of Edinburgh* **79B**, 115.

FRYE, F.L. (1991). *Biomedical and Surgical Aspects of Captive Reptile Husbandry*. 2nd Edn. Krieger, Malabar.

FRYE, F.L. and MYERS, M.W. (1985). Foreign-body dermatitis in a snake. *Modern Veterinary Practice* **66 (3)**, 204.

HOLT, P.E. (1981). Healing of a surgically induced shell wound in a tortoise. *Veterinary Record* **108**, 102.

IPPEN, R., SCHRÖDER, H-D. and ELZE, K. (1985). Eds. *Handbuch der Zootierkrankheiten, Band 1. Reptilien*. Akademie-Verlag, Berlin.

JACKSON, O.F. (1978). Tortoise shell repair over two years. *Veterinary Record* **102**, 184.

JACOBSON, E.R. (1991). Diseases of the integumentary system of reptiles. In: *Dermatology for the Small Animal Practitioner*. (Eds. L. Ackerman and G. Nesbitt). Veterinary Learning Systems Company, Trenton.

MARCUS, L.C. (1981). *Veterinary Biology and Medicine of Captive Amphibians and Reptiles*. Lea and Febiger, Philadelphia

ROSSKOPF, W.J. and WOERPEL, R.W. (1981). Rat bite injury in a pet snake. *Modern Veterinary Practice* **62**, 871.

SMITH, D.A., BARKER, I.K. and ALLEN, O.B. (1988). The effect of ambient temperature and type of wound on healing of cutaneous wounds in the common garter snake (*Thamnophis sirtalis*). *Canadian Journal of Veterinary Research* **52**, 120.

STAUBER, E. and GOGOLEWSKI, R. (1990). Poxvirus dermatitis in a tegu lizard (*Tupinambis teguixin*). *Journal of Zoo and Wildlife Medicine* **21 (2)**, 228.

ZANGERL, R. (1969). The turtle shell. In: *Biology of the Reptilia, Vol.1. Morphology: A*. (Eds. C. Gans, A. d'A. Bellairs and T.S. Parsons). Academic Press, London.

CHAPTER EIGHT

CARDIOVASCULAR SYSTEM

David L Williams MA VetMB CertVOphthal MRCVS

ANATOMY AND PHYSIOLOGY

Heart and great vessels

The hearts of lizards, snakes and chelonians are sufficiently similar to be considered together. Reptilian hearts differ from those of mammals in several ways but the main difference is that the reptilian heart has only one ventricle into which blood flows from the right and left atria. The ventricle has three sub-chambers, the cava pulmonale, venosum and arteriosum. The cavum pulmonale leads to the pulmonary artery and the cava venosum and arteriosum receive blood from the right and left atria respectively. The cavum venosum supplies the left and right aortic arches. It is thought that right atrial blood is almost exclusively directed to the pulmonary artery with the pulmonary venous return being directed both to the systemic circulation and back to the lungs. It is suggested that both right and left atrial blood may pass to the aortic arches or the pulmonary circulation (White, 1976).

In terrapins the percentage of total cardiac output reaching the systemic circulation changes when the animal is breathing or apnoeic during diving. When breathing, the red-eared terrapin (*Trachemys scripta elegans*) directs 60% of its cardiac output into the pulmonary artery and 40% into the systemic circulation, while in apnoea the systemic circulation is favoured at the expense of the pulmonary blood supply. Pressure measurements have shown that functionally the ventricle acts as if it were two chambers, with sequential ejection of first the pulmonary and then the systemic arterial supply; changes in the relative contributions to cardiac output occur by varying the ejection fractions. These changes have potentially major implications for anaesthetised reptiles, especially in animals that may be apnoeic for a considerable time while being anaesthetised without intubation or intermittent positive pressure ventilation (IPPV). The apnoeic reptile will fail to transfer a gaseous anaesthetic to its systemic circulation unless it breathes or is forced to by IPPV.

Crocodilians have a more advanced circulation with clear evidence of a dual circulation in which the pulmonary and systemic blood flows are functionally separate (Davies, 1981). There are two almost completely separate chambers with the left aortic arch originating from the right ventricle. The blood flow into the left aortic arch depends on the relative pressure in the right ventricle and the aortic arches. The left and right ventricles are connected only by the foramen of Panizza, and normally the pressure in the left ventricle is higher than that in the right, so that no right ventricular systemic outflow occurs. It would appear that only when diving - when the pulmonary vascular bed resistance increases dramatically - does a right to left shunt occur. In these circumstances the pulmonary outflow is not particularly well oxygenated compared with the blood in the inflow. The significance of this change under anaesthetic, or in conditions where pulmonary pathology would increase the resistance of the pulmonary vascular bed, has not been investigated.

Renal portal system

One element of the peripheral circulation with suggested clinical implications is the renal portal system. This consists of a vein arising at the confluence of the epigastric and external iliac veins which extends cranially and dorsally to reach the kidney before bifurcating into the vertebral vein draining the dorsal body wall and the hypogastric vein receiving tributaries from the bladder, cloaca and sex organs. It is possible that renal clearance of drugs injected into hindlimb muscle masses may occur before escape into the systemic circulation if these drugs are eliminated by renal excretion. This would have clinical implications but there is debate between Europe and the USA as to whether or not a renal clearance effect actually occurs.

CARDIOVASCULAR INVESTIGATIVE PROCEDURES

Cardiovascular disease can be the result of dietary or metabolic conditions as well as systemic infections or parasitic diseases, and it may manifest itself with a number of systemic and organ-system signs. Therefore, a full clinical examination is mandatory for any reptile with suspected cardiovascular disease and this should include auscultation, assessment where possible of peripheral blood flow and observation for peripheral oedema, gross ascites, ecchymoses and other skin or mucosal lesions. This should be followed by radiography (plain and contrast angiography), ultrasonography and electrocardiography if available.

Auscultation and palpation

In snakes the heart is generally identified at a point about one third of the length of the animal from the rostrum to the cloaca. Any increase in the size of the heart can be seen by noting heart movements during beating. Auscultation is not difficult but the use of a damp towel may be helpful in reducing the noise of the stethoscope against the snake's scales (Frye and Himsel, 1988). Location of an artery for taking a pulse is more difficult but in anaesthetised snakes the artery at the base of the glottis can be used if a Doppler flow detector is not available.

In lizards the heart is in the midline thorax; auscultation is possible but finding the peripheral pulse is not easy.

In chelonians the carapace and plastron clearly limit access to the heart but auscultation is possible if a damp towel is placed around the shell to avoid exogenous sounds interfering with audible heart sounds (see "Respiratory System"). Normal heart rates even within one individual can vary markedly according to external temperature, as can intervals in the electrocardiogram (ECG). For example, the patched-nosed snake (*Salvadora hexalepis*) has a heart rate of 6.15 beats per minute (bpm) with a P-R interval of 1.63 seconds at 6°C, but a heart rate of 131bpm and a P-R interval of 0.12 seconds at 40°C. For this reason, when recording heart rates the temperature should also be noted. With slower beats it may be necessary to count rates over more than one minute to obtain an accurate measurement. Heart rate is also related to the size of the animal, with a range from 100bpm in a one gram lizard (*Scincella* spp.) to 20bpm in a 4.4kg monitor lizard (*Varanus* spp.), both at 30°C (Templeton, 1970). As has been discussed earlier, gross changes in cardiac physiology occur during apnoea and changes in heart rate during such periods are also marked. These points must be remembered when heart monitors are used to record changes during anaesthesia, as well as during normal measurements taken with ECG's in a clinical investigation. The use of conventional heart monitors is difficult to assess; although the monitors pick up electrical activity, it is sometimes difficult to decide whether it is cardiac in origin, or due to some other muscle source.

Reptile ECGs

The reptile ECG is similar to that of the mammal, ie. there are defined P, QRS and T waves (McDonald, 1976). A sinus venosus (SV) wave has been described originating in the upper portion of the caudal vena cava and sinus venosus. Surface electrodes can be used on moderate sized reptiles and small silver chloride electrodes can be attached directly to the scales of snakes and lizards with adhesive tape and conducting gel. In chelonians internal electrodes made from 25G hypodermic needles can be fixed subcutaneously on the limbs using the same placing of the leads as for quadruped mammals. It has been suggested that tiny holes can be drilled in the carapace, just deep enough to expose bone, through which ECG electrodes can be placed, but such extreme measures are not advised. Cheap but acceptable long-term surface electrodes for anaesthetic monitoring can be produced by attaching multistrand wire with the end splayed to the reptile's skin with conducting silver paint, which is quick drying and easily removed. Signal voltages from most reptiles, particularly chelonians, are fairly small – often under 1.0mV – so a battery powered pre-amplifier is useful. However, interference from muscular activity can be deleterious in these small animals, giving high background 'noise' as can be seen in Figure 1.

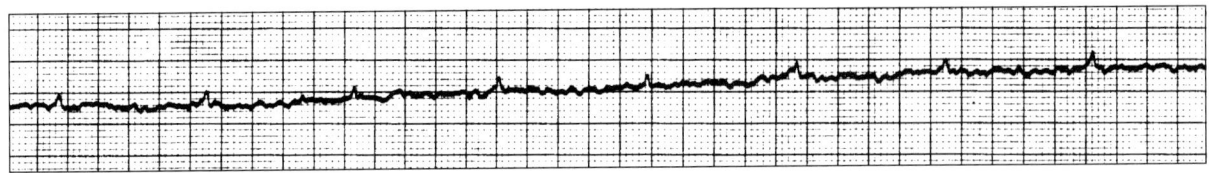

Figure 1

In small reptiles, eg. *Iguana iguana*, the background 'noise' of muscle activity makes it difficult to define changes in waveforms associated with diseases, in this case cardiomegaly. 20mm/mV, 25mm/sec.

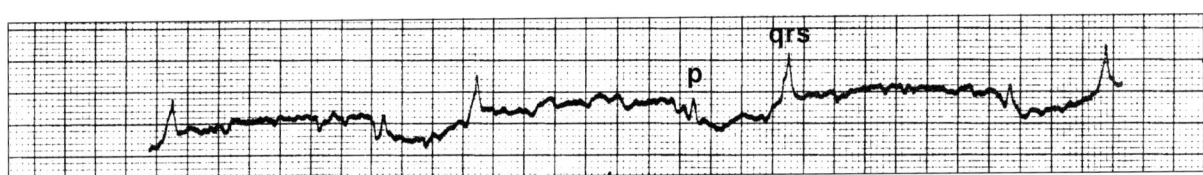

Figure 2

An electrocardiogram from a Childrens python (*Liasis childreni*) with chronic aortic valvular stenosis and ventricular hypertrophy. The tracing shows a sinus bradycardia and first degree heart block with an abnormally long P-R interval. 10mm/mV, 25mm/sec. Courtesy of Dr. F.L. Frye.

In snakes electrodes need only be placed two heart lengths cranial and caudal to the heart. In some lizards, where the heart is positioned cranially just below the pectoral girdle, a neck electrode may give better results than a 'forelimb' placement. As conventional limb positions are not available, a mean electrical axis is difficult to estimate in snakes.

Because of the wide spectrum of reptile sizes and changes with ambient temperature, it is difficult to give values for normal wave amplitudes and durations. Comparison with a normal, similarly sized member of the species will be important. While considerable work has been undertaken on normal reptile ECGs at different temperatures and in different species, ECGs of animals with cardiovascular disease are rarely reported in the literature.

Imaging techniques

Standard plain radiography in the ventrodorsal and lateral planes will give some information on heart size in lizards, snakes and crocodilians, although normal data for such measurements are not readily available. Ultrasonography can be very useful in these species. However, the bony carapace and plastron make such investigations in chelonians difficult; a small window only is available through the axilla or inguinal fossa. This is not the case with the specialised techniques of computerised (axial) tomography (CT) and nuclear magnetic resonance/magnetic resonance imaging (NMR/MRI), where cardiovascular size and contraction real time studies can be investigated. However, these techniques are not widely available at present. More information can be found elsewhere in this manual (see "Radiological and Related Investigations").

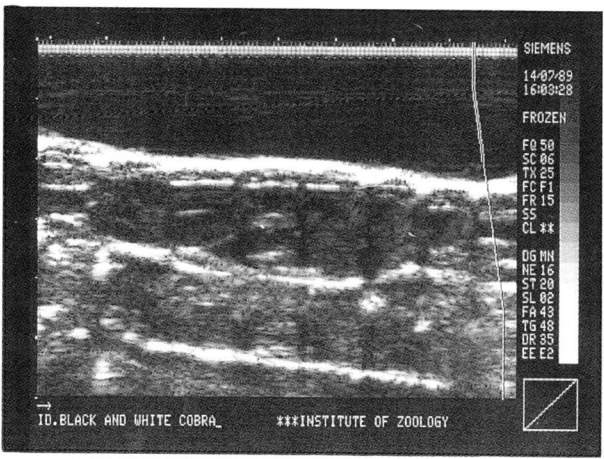

Figure 3
Ultrasound scan of a cobra (*Naja* sp.) heart. The cardiac valves can be seen and, in real time, would be obvious by their movement. The excursion of the ventricular walls would also be evident and a reduction in cardiac contraction would be seen in cardiomyopathy. Mural thrombi, atrioventricular valve endocardiosis, stenosis and similar structural defects would also be diagnosed by such scans. Courtesy of Mr A.W. Sainsbury, Zoological Society of London.

Post-mortem examination

Reptiles which die from suspected cardiovascular disease should be subjected to *post-mortem* evaluation to investigate and gain further understanding of these diseases. In routine necropsies cardiac or vessel lesions may also be noted as incidental findings. Immediately after opening the body cavity, the heart and great vessels should be examined *in situ* for evidence of pathological changes such as vessel calcification. Detection of vessel calcification and mineralisation of (for example) parasitic lesions, can be facilitated if the material is radiographed, preferably using a low kV (Cooper and Jackson, 1981).

CARDIOVASCULAR THERAPEUTICS

Little is known of the basic pharmacology and pharmacokinetics of cardiac glycosides, sympatholytics, vasodilators and diuretics in most reptiles. Some pharmacological studies have shown the presence of beta-adrenergic receptors mediating the positive ionotropic and chronotropic effects of catecholamines on the saurian heart, and vagal stimulation giving negative chronotropic effects which can be blocked with atropine. However, clinical data on the use of these drugs are lacking. It should be possible to use the same combination of cardiac glycosides and diuretics as in mammals to treat congestive heart failure in dilated cardiomyopathy if this is diagnosed. Although there appears to be no specific dose rate for the use of cardiac glycocides in reptiles, frusemide (5mg/kg) can be used (Holt, 1981; Frye, 1991). Allometric scaling of dosages is advised (see "Therapeutics").

It should be noted that major differences exist between mammals and reptiles: the coronary arteries of some chelonians have been shown to constrict with adrenaline - the exact opposite to mammals - and the pulmonary circulation of aquatic chelonians is well innervated allowing almost total vascular bed shutdown in periods of apnoea. Therefore, making extrapolations from mammalian studies is unlikely to be helpful.

SPECIFIC DISEASES

Viral diseases

Few viral infections have been reported to cause cardiovascular lesions in reptiles, but herpesvirus infection in Iguanidae has been recorded as producing a histiocytic lymphoid infiltrate in the myocardium amongst other organs (Hoff and Hoff, 1984).

Bacterial diseases

Bacterial endocarditis has been reported in a number of reptiles. A vegetative thrombus from which *Vibrio damsela* was isolated was reported in a leatherback turtle (*Dermochelys coriacea*) (Obendorf et al, 1987) and *Salmonella arizona* and *Corynebacterium* spp. were cultured from endocardial thrombi in the sinus venosus and right atrium in a Burmese python (*Python molurus bivittatus*) (Jacobson et al, 1991). The leatherback turtle was found dead, so no clinical history was available. In the python clinical signs included distension of the thoracic body wall and pitting oedema of the integument cranial to the heart, as well as severe oedema of the mucosa of the oral cavity and mild cyanosis. Ultrasound investigation proved useful, showing right atrial dilatation and a large heteroechoic mass in the area of the right atrioventricular valve. Contrast radiography suggested incomplete ventricular filling and precaval venous stasis.

Chlamydial diseases

Chlamydial infection has been reported as causing myocarditis in snakes (Jacobson et al, 1989). Four puff adders (*Bitis arietans*) exhibited widespread pathology on *post-mortem* examination, including myocarditis, purulent pericarditis, hepatic granulomas and inflammatory lesions in a number of other tissues. Electron microscopy revealed numerous basophilic inclusion stages of *Chlamydia* spp., although multiple attempts to culture the organism were unsuccessful. The cardiac lesions ranged from myocardial necrosis to myocardial granulomas while epicardial surfaces were covered with fibrocellular necrotic debris. The only clinical sign reported, apart from occasional regurgitation, involved one snake which exhibited mild *ante-mortem* respiratory distress with mouth gaping and oral exudate.

Fungal diseases

Fungi can also cause infectious myocardial lesions, eg. chronic fibrotic endocarditis noted in a chelonian with widespread mycotic infection (Hamerton, 1935).

Non-specific infections

In a number of reptiles with signs of inflammatory epicardial and myocardial disease, no significant organisms can be isolated. In one series, thirty saurians exhibited inflammatory lesions of epimyo- and myocarditis, myocardial abscessation and mural thrombi, while 3% of snakes examined had myocardial abscesses and myocarditis (Cooper and Elkan, unpublished data). Epicardial plaques and micro-abscesses were also noted in chelonians. This may reflect the high incidence of septicaemia spreading to the heart in dying reptiles rather than primary cardiovascular disease.

Peripheral vascular pathology associated with *Pseudomonas* spp. – related vasculitis has been reported in inflammatory skin lesions, and necrotic vasculitis was noted in a mycobacterial salpingitis in a coach-whip snake (*Masticophis flagellum*) in which heterophilic infiltrate and vessel wall oedema were the main vascular lesions (Reichenbach-Klinke and Elkan, 1965).

Parasitic diseases

A number of documented *post-mortem* cases reveal evidence of parasitic and presumed post-parasitic lesions in reptiles (Ardlie and Schwartz, 1965; Finlayson, 1965; Finlayson and Woods, 1977). Focal, segmented scarring, often limited to the aortic arch or pulmonary trunk, was the most common lesion in snakes, and is most probably a result of post-parasitic scar tissue.

A number of nematodes are found in the cardiovascular system. Members of the Dracunculoidea and Filaroidea have larval microfilarial forms which can block small vessels and lymphatics. Subsequent pathology includes thrombosis and necrosis of tissues supplied by these arterioles. This was particularly noted in a group of snakes with dermal lesions subsequent to capillary microfilarial infestation and concurrent parasitic mesenteric arteritis, caused by adult and microfilarial *Macdonaldius oschie*. A similar parasitic burden of *Thamugadia physignathi* was reported in an Australian water dragon (*Physignathus lesueuri*), but very little tissue damage related to intraluminal parasites was noted (Reichenbach-Klinke and Elkan, 1965).

Digenetic trematodes of the Spirorchidae occur in the mesenteric veins of marine turtles (*Chelonia* spp.) but, as with many other parasitic infestations, the life-cycle involves an invertebrate secondary host and, therefore, the trematodes should only be seen in wild-caught animals.

Congenital abnormalities

While numerous congenital skeletal and soft tissue abnormalities have been recorded in reptiles, cardiac and large vessel abnormalities do not appear to have been reported, although this by no means indicates that they do not occur. A neonatal Indian python (*Python molurus*) with an exceptionally dilated cardiomyopathy (about four times the normal size) characterised by myocardial hypoplasia has been noted recently (Frye, personal communication). This animal had a grossly swollen pericardial region. An ECG revealed a first degree heart block and an anomalous pre-P wave. *Post-mortem* examination revealed extremely thin atrial and ventricular walls and retained yolk masses in the coelomic cavity. Histopathology showed myocardial hypoplasia and a plasmacytic pericarditis.

Metabolic and dietary diseases

The most common cause of cardiovascular disease in larger saurians is medial calcification of the major arteries (Finlayson and Woods, 1977). While this could be a degenerative change, it is often related to an imbalance in the dietary calcium and phosphorus levels. In several studies large numbers of iguanas and other reptiles, either at *post-mortem* examination or on radiography, showed calcification of the aorta and other great vessels. These lesions are confined to the media and, in most cases, by the time the animal has died, gross destruction of medial elastic tissue and obvious calcification has occurred. In some animals which have yet to be grossly affected there are areas of cystic change in the vessel wall which may be a precalcification change. These medial changes with aberrant calcium levels have been associated either with food sources low in calcium and high in phosphorus or from overzealous supplementation with vitamin D_3-rich additives (Wallach and Hoessle, 1966; Jackson and Cooper, 1981).

Clinical signs in these animals can be minimal unless myocardial calcification occurs, in which case ventricular wall compliance may be reduced markedly with a consequent fall in cardiac output. Sudden death may occur through rupture of an aneurysmic defect in a large vessel wall; in smaller vessels the resulting haemorrhage may not be fatal and clinical signs may be evident before death, eg. rupture of an ophthalmic artery in an iguana producing a large periorbital haematoma in the living animal (Reichenbach-Klinke and Elkan, 1965).

While medial calcification is common in reptiles, atherosclerosis or atheroma is very rarely seen. This is surprising since a large number of aged snakes, in particular, are obese. A study of lipid profiles in obese and non-obese rattlesnakes showed that obese snakes had higher total cholesterol and triglyceride levels and a higher beta-lipoprotein distribution than non-obese animals (Bauer and Jacobson, 1989). Animals in the collection had died of aneurysms of the dorsal aorta, ventricular aneurysms and/or medial calcification of the great vessels.

Deposition of uric acid tophi in the myocardium has been observed in reptiles being fed a high protein diet; this condition may also be a result of gentamicin toxicity (Jacobson, 1976; Montali *et al*, 1979) (see "Urogenital System").

Cardiomyopathy may occur in water snakes (*Thamnophis* and *Natrix* spp.) with thiamine deficiency.

Degenerative and ageing changes

Amyloidosis of cerebral blood vessels has been reported in chelonians (Trautwein and Pruksaraj, 1967) and these changes may occur in the vessels of a number of ageing animals. Its relationship to mammalian amyloidosis is unclear.

A number of myocardial degenerations have been seen in aged snakes (Jacobson *et al*, 1989; Wagner, 1989) and in iguanas (Frye, personal communication) but this predisposition may be a consequence of the greater ages which these reptiles attain in captivity and the ease with which they can be examined *post mortem* compared with the larger chelonians.

Neoplastic diseases

Primary cardiovascular tumours are rare in reptiles. Of the few reported most are haemangiomas, haemangioepitheliomas and angiosarcomas (Wadsworth, 1956; Effron et al, 1977; Harshbarger, 1980). Rhabdomyosarcoma of the ventricles has been reported (Reichenbach-Klinke and Elkan, 1965). The most common cardiovascular malignancies are metastatic lymphosarcomas reported by several investigators (Cowan, 1968; Harshbarger, 1974).

Acknowledgement

The author is grateful to Dr. F.L. Frye for helping provide material for this chapter.

REFERENCES

ARDLIE, N.G. and SCHWARTZ, C.J. (1965). Arterial pathology in the Australian reptile: a comparative study. *Journal of Pathology and Bacteriology* **90,** 487.

BAUER, J.E. and JACOBSON, E.R. (1989). Hyperlipidemia and cardiovascular disease in obese rattlesnakes. In: *Third International Colloquium on the Pathology of Reptiles and Amphibians.* (Ed. E.R. Jacobson). Orlando.

COOPER, J.E. and JACKSON, O.F. (1981). Eds. *Diseases of the Reptilia.* Academic Press, London.

COWAN, D.F. (1968). Diseases in captive reptiles. *Journal of the American Veterinary Medical Association* **153 (8),** 848.

DAVIES, P.M.C. (1981). Anatomy and physiology. In: *Diseases of the Reptilia, Vol.1.* (Eds. J.E. Cooper and O.F. Jackson). Academic Press, London.

EFFRON, M., GRINER, L. and BERIRSCHKE, K. (1977). Nature and rate of neoplasia in captive wild mammals, birds and reptiles at necropsy. *Journal of the National Cancer Institute* **59,** 185.

FINLAYSON, R. (1965). Spontaneous arterial disease in exotic animals. *Journal of Zoology* **147,** 239.

FINLAYSON, R. and WOODS, E. (1977). Arterial disease in reptiles. *Journal of Zoology* **183,** 397.

FRYE, F.L. (1991). *Biomedical and Surgical Aspects of Captive Reptile Husbandry.* 2nd Edn. Krieger, Malabar.

FRYE, F.L. and HIMSEL, C.A. (1988). The proper method for stethoscopy in reptiles. *Veterinary Medicine* **December,** 1250.

HAMERTON, A.E. (1935). Report on deaths occurring in the Society's Gardens for the Year 1934. In: *Proceedings of the Zoological Society of London.* Zoological Society of London, London.

HARSHBARGER, J.C. (1974). *Activities Report Registry of Tumors in Lower Animals, 1965–1973.* Smithsonian Institute, Washington DC.

HARSHBARGER, J.C. (1980). *Activities Report Registry of Tumors in Lower Animals, 1980 Supplement.* Smithsonian Institute, Washington DC.

HOFF, G.L. and HOFF, D.M. (1984). Salmonella and Arizona. In: *Diseases of Amphibians and Reptiles.* (Eds. G.L. Hoff, F.L. Frye and E.R. Jacobson). Plenum Press, New York.

HOLT, P.E. (1981). Drugs and dosages. In: *Diseases of the Reptilia, Vol. 2.* (Eds. J.E. Cooper and O.F. Jackson). Academic Press, London.

JACKSON, O.F. and COOPER, J.E. (1981). Nutritional diseases. In: *Diseases of the Reptilia, Vol. 2.* (Eds. J.E. Cooper and O.F. Jackson). Academic Press, London.

JACOBSON, E.R. (1976). Gentamicin-related visceral gout in two boid snakes. *Veterinary Medicine/Small Animal Clinician* **71,** 361.

JACOBSON, E.R., GASKIN, J.M. and MANSELL, J. (1989). Chlamydial infection in puff adders (*Bitis arietans*). *Journal of Zoo and Wildlife Medicine* **20 (3),** 364.

JACOBSON, E.R., HOMER, B. and ADAMS, W. (1991). Endocarditis and congestive heart failure in a Burmese python (*Python molurus bivittatus*). *Journal of Zoo and Wildlife Medicine* **22,** 245

McDONALD, H.S. (1976). Electrocardiography. In: *Biology of the Reptilia, Vol. 5: Methods for the Physiological Study of Reptiles.* (Ed. C. Gans). Academic Press, London.

MONTALI, R.J., BUSH, M. and SMELLER, J.M. (1979). The pathology of nephrotoxicity of gentamicin in snakes. *Veterinary Pathology* **16,** 108.

OBENDORF, D.L., CARSON, J. and McMANUS, T.J. (1987). *Vibrio damsela* infection in a stranded leatherback turtle. *Journal of Wildlife Diseases* **23**, 666.

REICHENBACH-KLINKE, H. and ELKAN, E. (1965). *The Principal Disease of Lower Vertebrates, Book 3: Diseases of Reptiles.* Academic Press, London.

TEMPLETON, J.R. (1970). Reptiles. In: *Comparative Physiology of Thermoregulation.* (Ed. G.C. Whitlow). Academic Press, New York.

TRAUTWEIN, C. and PRUKSARAJ, D. (1967). Uber amyloidose bei Schildkröten. *Deutscher Tierärztliche Wochenschrieber* **74**, 184.

WADSWORTH, J.R. (1956). Serpentine tumors. *Veterinary Medicine* **51,** 326.

WAGNER, R.A. (1989). Clinical challenge, case 1. *Journal of Zoo and Wildlife Medicine* **20**, 238.

WALLACH, J.D. and HOESSLE, C. (1966). Hypervitaminosis D in green iguanas. *Journal of the American Veterinary Medical Association* **149**, 912.

WHITE, F.N. (1976). Circulation. In: *Biology of the Reptilia, Vol. 5: Methods for the Physiological Study of Reptiles.* (Ed. C. Gans). Academic Press, New York.

CHAPTER NINE

RESPIRATORY SYSTEM

Lynne C Stoakes BVetMed MRCVS

INTRODUCTION

Respiratory disease is a major cause of morbidity and mortality in reptiles (Jacobson, 1988), often seen in newly acquired specimens when the stress of handling and transport lowers their resistance to disease. Respiratory disease may also be seen in established collections when the environment in which they are kept is inadequate or when there is a change in the environmental conditions, such as the temperature or relative humidity, or if the husbandry is changed.

All reptiles breathe primarily by means of lungs, although some aquatic turtles, eg. soft-shelled turtles (Trionychidae), and some aquatic species of snakes can obtain oxygen across the pharynx and skin (Jacobson, 1988). Reptiles lack a functional diaphragm which prevents active, expulsive coughing. The exudate produced in respiratory infections of reptiles is thick and tenacious and this, coupled with the anatomy and lack of a cough reflex, means that reptiles are prone to severe and often fatal respiratory diseases.

Some reptiles, especially aquatic chelonians and crocodilians, show some degree of anaerobic respiration. They have higher levels of sodium bicarbonate, haemoglobin and other serum proteins to counter the effects of acidosis caused by the buildup of carbon dioxide (Frye, 1981a).

ANATOMY

Chelonians

The external nares lead into a nasal sinus and then open into the mouth (internal nares) so there is communication between the mouth and nasal chambers. This is important when the cause of a nasal discharge is being investigated: excess saliva can exit through the nostril.

The glottis is situated at the base of the muscular tongue and the trachea continues only a short way down the neck before bifurcating. The bronchi enter the lungs dorsally. The lungs are located in a dorsal position beneath the carapace and aid buoyancy in aquatic species.

The rigid shell of most chelonians hampers expansion of the lungs during respiration. Air movement in and out of the lungs is accomplished by movements of the head and limbs which alter the pressure within the body cavity. During hibernation even small changes in pressure caused by the heartbeat may be sufficient for gaseous exchange to take place (Jackson, 1991). The snapping turtle (*Chelydra serpentina*) uses four pairs of muscles for respiration. The diaphragmaticus and transversus abdominis muscles compress the visceral cavity and cause exhalation. The testocoracoideus and obliquus abdominis muscles expand the visceral cavity and cause inhalation. The snapping turtle has a reduced plastron and can also use external forces to aid respiration. The hydrostatic pressure when in water compresses the lungs and gravity acts to expand the lungs (Gaunt and Gans, 1969). The lungs of chelonians are multicompartmented and saccular; there are no bronchioles (Fowler, 1980a). The bronchi enter the compartments directly on the dorsal surface. The surface on which gaseous exchange takes place is reticular in nature. Due to the compartmented nature of the lungs, exudates formed in infections cannot drain but are allowed to pool (Fowler 1980a).

Snakes

The glottis is situated far forward in the mouth and may be visible protruding from the mouth when the snake is ingesting its prey (Jacobson, 1978). The colubrid snakes (rat snakes) have one functional lung; the left lung is usually vestigial or absent. In boids (pythons and boas) two lungs are usually present, the left being smaller (Lawrence, 1985). The right lung courses dorsally to the liver and may extend as far caudally as the cloaca. Only the first third is functional; the remainder is avascular air sac and may act as a reserve of air during periods of apnoea. The functional part of the lung has a reticular surface. The air sac is lined with simple squamous epithelium.

In some snakes there is a tracheal lung which is situated on the dorsal aspect of the trachea so that the snake can continue to breathe when the main lung is compressed by swallowed prey (Davies, 1981).

Lizards

The internal nares are usually far forward in the mouth except in those lizards with a longer snout, eg. the monitors (*Varanus* spp.) (Davies, 1981). The glottis is variable in position. Unlike snakes, lizards have two lungs which are of equal size. The lungs of lizards are the closest in structure to mammalian lungs.

Crocodilians

The internal nares open caudally in the mouth so that crocodilians have a relatively well formed hard palate. There are flaps of tissue which can seal off the buccal cavity when the crocodilian is under water, so that it may still breathe when grappling with its prey (Davies, 1981). The coelomic cavity is bisected by a muscular septum, and the heart, lungs and liver lie cranial to this septum. This arrangement is the nearest to the mammalian diaphragm seen in reptiles (Jacobson *et al*, 1983).

CLINICAL SIGNS OF RESPIRATORY DISEASE

Nasal discharge. This is the most common sign of rhinitis. It is often seen in cases of pneumonia but, because of their structure, it is unlikely to originate from the lungs. The discharge is more likely to be due to a concurrent rhinitis.

Gasping. In severe respiratory disease the reptile will exhibit open mouth breathing in an attempt to obtain as much oxygen as possible (see Figure 1). This is often accompanied by noisy respiratory sounds.

Posture. Snakes will often hold their head up with the mouth open. Tortoises extend the neck and terrapins may swim in a lopsided manner in unilateral cases of pneumonia, as the affected lung is consolidated and less buoyant. Aquatic species will often spend more time out of the water.

Figure 1
Royal python (*Python regius*) exhibiting open mouth breathing.

Cyanosis. Reptiles with severe respiratory embarrassment will have cyanotic mucous membranes. This is especially visible in the mouth and always carries a very guarded prognosis.

Depression. Lethargy is the result of impaired respiration.

Anorexia. Weight loss may be seen in chronic respiratory disease.

Hypopyon. This was reported as a secondary complication in a tortoise exhibiting signs of respiratory disease (Tompson *et al*, 1976). The hypopyon cleared when the respiratory disease was successfully treated with antibiotics. Hypopyon has also been seen concurrently with respiratory disease in Tokay gecko lizards (*Gekko gecko*) (Bonney *et al*, 1978). Three geckos in a collection died and pneumonia was diagnosed *post mortem* on histological examination of the lungs. It is likely that the hypopyon occurred as a result of infection of the lacrimal system.

Sub-spectacular abscesses. These are found in snakes associated with infection of the nasolacrimal system (see "Ophthalmology").

BACTERIAL DISEASES

Rhinitis

Rhinitis occurs mainly in terrestrial tortoises (*Testudo* spp.) and in the American box-tortoise (*Terrapene carolina*). Many causes have been suggested; it is likely that rhinitis is multifactorial with an environmental component. It is difficult to interpret the results of cultures from nasal discharge as the bacteria found may be cultured from apparently healthy tortoises (Fowler,1977).

Rhinitis caused by bacteria is highly contagious and can spread rapidly through entire collections. Jackson (1991) described three types:-

1. Sinusitis - the discharge is often unilateral and of thick consistency.

2. Excess salivation - because of the open internal nares the saliva flows out through the nostrils.

3. True rhinitis - this is nearly always bilateral and, in the early stages, the discharge is watery, although later it becomes thick and cloudy. It is continuous day and night and there may be accompanying ocular discharge due to blockage of the nasolacrimal system (see Figure 2).

Diagnosis - clinical signs are typical, although a concurrent pneumonia should be ruled out by radiography before treatment is initiated.

Treatment - the discharges must be physically removed by simultaneously pressing on the dorsal external surface of the face and the hard palate internally, and wiping away any material that is pushed out of the nostrils. Antibiotic drops, with or without steroids, are applied to the nostrils (steroids must NOT be used if there is evidence of pneumonia).

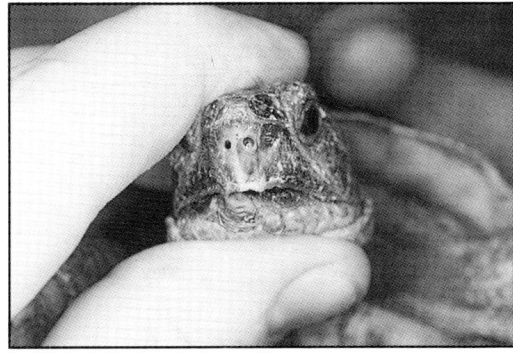

Figure 2
Tortoise with rhinitis.

Prevention - improving the environment to simulate closely the free-living state and ensuring that the diet is correct are important for therapy and prevention of future problems (Fowler, 1980b). Rhinitis is often a recurring problem and difficult to clear completely. Affected individuals should be isolated.

Pneumonia

The bacteria implicated in cases of pneumonia may often be cultured from the respiratory tract of asymptomatic reptiles (Frye, 1981a). *Aeromonas* spp. are considered the classic opportunistic bacteria (Shotts, 1984). Other bacteria found include *Klebsiella* spp., *Pasturella* spp., *Proteus* spp. and *Pseudomonas* spp. It is important to look for the factors which may have made the reptile susceptible to the disease in the first place, as successful recovery will depend on correction of these factors.

Pneumonia may be a primary condition or may be secondary to some other condition. It is often seen as a sequel to stomatitis due to aspiration of infected material or, possibly, to septicaemic spread of the causative organisms (Jacobson, 1978).

Infection in the lung leads to the accumulation of exudate within the airways and eventual consolidation of one or more segments of the lung (see Figures 3 and 4).

Diagnosis - may be confirmed by radiography in advanced cases where there is significant consolidation (see Figures 5 and 6). Swabs for culture and sensitivity can be taken by passing a sterile cannula into the trachea (usually no anaesthetic is necessary), instilling a small amount of sterile saline and immediately aspirating.

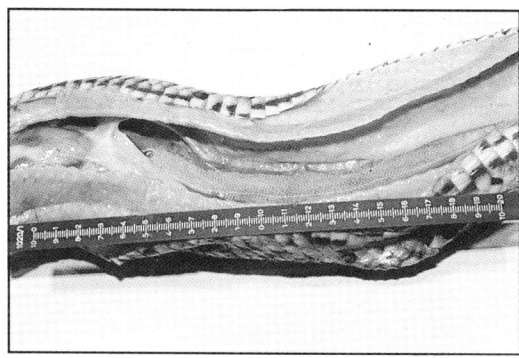

Figure 3
Post mortem of a snake
showing normal lung.

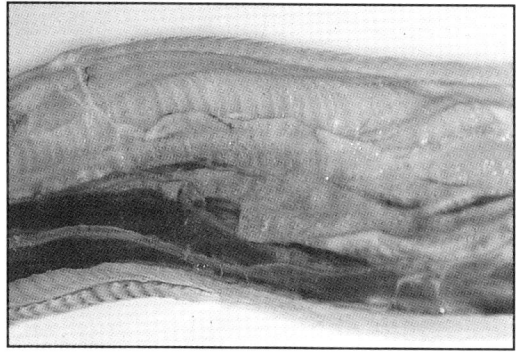

Figure 4
Post mortem of a snake
showing pneumonia.

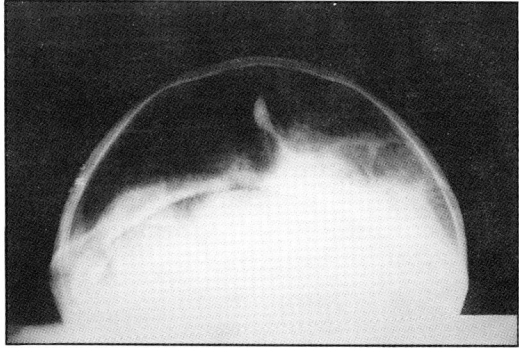

Figure 5
Craniocaudal radiograph of a tortoise
showing density in the left lung field.

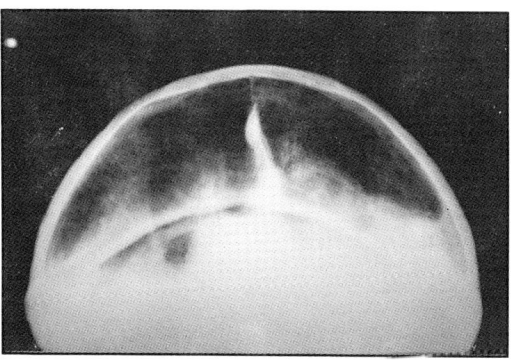

Figure 6
Same tortoise as in Figure 5 after
treatment with antibiotic.

In chelonians a swab may be taken directly from the lesion (located by radiography) by drilling a hole (under general anaesthesia) in the appropriate area of the carapace and inserting a sterile swab into the site (see Figures 7 and 8). The hole is covered with a sterile waterproof dressing.

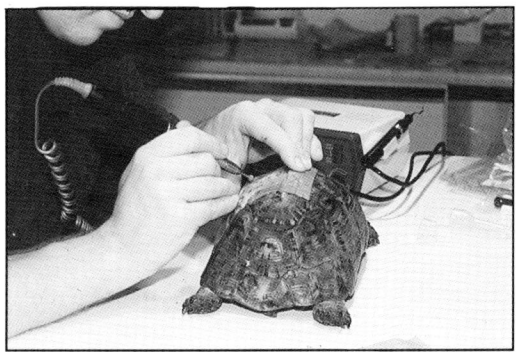

Figure 7
Drilling a hole to gain access to a pneumonic lung in an anaesthetised tortoise.

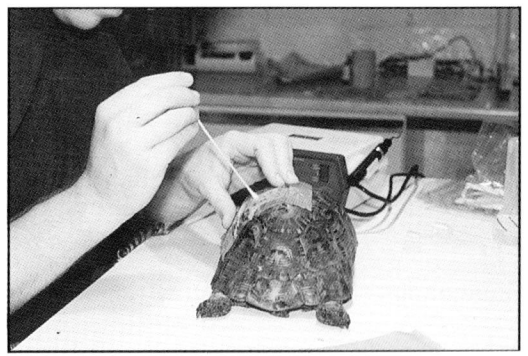

Figure 8
Taking a swab from the lesion for culture.

Treatment - the appropriate antimicrobial agent should be given by injection and the reptile kept at the appropriate temperature for the agent being used (see "Therapeutics"). In chelonians, if a hole has been drilled for taking a swab, this may also be used for instilling antibiotics directly into the site of the infection.

In some instances it may be necessary to start antimicrobial therapy before the results of the sensitivity tests are known. In such cases a broad-spectrum agent should be chosen to initiate the treatment and this may then be modified as necessary.

VIRAL DISEASES

Paramyxo-like virus

Paramyxo-like virus was initially reported in viperid snakes (Clark *et al*, 1979). It has since been reported as causing deaths in elapids, boids and colubrids (Jacobson *et al*, 1980; Ahne *et al*, 1987). Typical respiratory signs are seen and a brown mucus may be expelled from the trachea. CNS signs, including loss of righting reflex and convulsions, may occur (see "Neurological Diseases").

The virus is contagious and fatal. *Post mortem* the respiratory tract is filled with mucoid exudate, and histologically there is an interstitial pneumonia. On transmission electron microscopy, virus particles typical of paramyxovirus may be seen budding from the cell surface (Jacobson *et al*, 1981). Other snakes examined were found to have antibodies to the virus but remained unaffected; thus, there is a possibility of a reservoir host state (Jacobson *et al*, 1981).

Diagnosis - the history and *post-mortem* findings are suggestive of this condition and it may be confirmed by electron microscopy. Serology using a haemagglutination inhibition test is possible (Gaskin *et al*, 1989).

Treatment - vaccines have been tried experimentally (Gaskin *et al*, 1989). Many snakes die from secondary bacterial infections. Affected and incontact animals should be isolated.

Sendai virus

This virus has been implicated in rhinitis of chelonians (Jackson and Needham, 1983). However, it has not been proved to cause rhinitis and a positive titre may be only an incidental finding (Lawrence and Needham, 1985).

PARASITIC DISEASES

Protozoa

Entamoeba invadens - the tissue cysts of this parasite can result in an inflammatory reaction in almost any organ, although the intestinal tract and liver are most often affected.

Flukes

"Renifers" (*Dasymetra* spp.)

These inhabit the oral cavity and respiratory tract of snakes. Completion of their life-cycle requires an amphibian intermediate host. Infections are usually asymptomatic, although if there is tissue damage by the migrating larvae, debility may result.

Diagnosis - the eggs can be found in the faeces on microscopy or they may be an incidental *post-mortem* finding.

Treatment - as the infections are mostly asymptomatic and the drugs which are effective against flukes are fairly toxic, treatment is not recommended (Marcus, 1977).

Nematodes

Ascarids

There are many species of ascarid worm found in reptiles. The life-cycle may be direct or indirect. Frogs, lizards or mammals are the intermediate hosts for the ascaroid species found in predatory reptiles. Ingested larvae migrate through the lungs and adults are found in the alimentary tract. The larvae can cause respiratory problems during their migration through the lungs (Sprent, 1984).

Diagnosis - the characteristic thick-shelled eggs may be found in faeces.

Treatment - oxfendazole (Systamex, ICI; Bandit, Pitman-Moore) (60mg/kg orally as a single dose) is effective (Lawton, 1991).

Kalicephalus spp.

These are hookworm-like nematodes found in snakes. The life-cycle is direct; infective larvae are ingested. Percutaneous migration is possible (Cooper, 1971). *Kalicephalus* sp. is, therefore, a parasite which can increase in numbers in captive snakes sufficient to cause problems. The larvae also undergo a visceral larva migrans and can cause respiratory problems.

Diagnosis - the embryonated eggs or free larvae may be found in faecal smears or on microscopy of tracheal washings.

Treatment - oxfendazole (60mg/kg) or ivermectin (Ivomec, MSD Agvet) (200mg/kg s/c). Ivermectin should not be used in chelonians as it is toxic even at very low doses (Teare and Bush, 1983; Jacobson, 1988).

Rhabdias spp.

This lungworm is found in the respiratory tract of snakes. The adult parthenogenetic females live in the lung. The eggs migrate up the trachea and are swallowed. They develop into infective larvae in the faeces and enter the new host by being ingested, or they may migrate through the skin and find their way to the lungs by tissue migration. Often they cause minimal damage; however, they may result in an inflammatory reaction and secondary bacterial infection.

Diagnosis - The larvae may be found in preparations of faeces, or larvated eggs may be found in a tracheal wash.

Treatment - levamisole (Levacide, Norbrook) (l0mg/kg intracoelomically repeated after two weeks) (Brannian, 1984) or ivermectin (200mg/kg s/c).

Pentastomes

Pentastomes are known as tongue worms or linguatulids. These are annulate metazoan parasites of the reptile respiratory tract, especially that of snakes. Many species occur in the wild, so they may be seen in newly captured specimens. The large tongue-shaped adult lives in the lung, trachea or nasal passages of the snake and feeds on tissue fluids. The embryonated eggs are swallowed, pass out in the faeces and become larvae. These are ingested by an intermediate host, which may be a rodent, carnivore, non-human primate or even human. As such it is a potential zoonosis (Hendrix and Blagburn, 1988). In the intermediate host the larvae undergo tissue migration and encyst, during which time they metamorphose and become infective nymphs. The life-cycle is completed when the intermediate host is ingested by the final host. Pathological effects are minimal and are due to tissue migration, but they may predispose to secondary bacterial infection, blockage or foreign body reaction.

Diagnosis - the ova may be seen in a tracheal wash. Radiography may reveal the calcified cysts or thickened body exoskeleton. Bronchoscopy may assist in detection.

Treatment - generally ineffective. Levamisole (5mg/kg intracoelomically) or thiabendazole (110mg/kg by stomach tube) has been suggested (Hendrix and Blagburn,1988) and Jacobson (1988) proposed the use of ivermectin (200mg/kg). Surgical removal of the adult worms is possible (Jackson, personal communication) and is recommended if this condition is diagnosed.

FUNGAL DISEASES

Fungal infections occur but are not commonly diagnosed *ante mortem*. Of the fungi isolated from reptiles, *Aspergillus* spp. is the commonest. *Aspergillus* spp. are found in the environment and the respiratory tract is a common site of infection (Migaki et al, 1984). In chelonians the terrestrial species seem more susceptible to this problem (Jacobson, 1978). The response of the reptile to fungal infection is to form granulomatous nodular lesions. Clinical signs are not usually noticed until the disease is advanced.

Diagnosis - if it is practicable to biopsy the lesion, microscopy will reveal the fungal hyphae. However, fungal infections of internal organs are usually only diagnosed *post mortem*.

Treatment - fungal infections are often the result of poor environmental conditions. Improvement of these may be the only course of action required. Jacobson (1980) reported the use of 5mg amphoteracin B nebulized in 150ml saline (bid for one hour each time for 7 days) for the treatment of pneumonitis in snakes.

NON-INFECTIOUS CAUSES

Foreign bodies

A laryngeal foreign body has been reported as a cause of respiratory distress in lizards (Anderson, 1976). The signs included forced inspiration and expiration. The pharynx was red and inflamed. The foreign body consisted of fibres which had caused an exaggerated inflammatory response resulting in oedema. Surgical removal of the fibres was required for successful treatment.

Shedding problems in snakes and lizards can cause obstruction of the nostrils if old skin is retained. Treatment involves removal of the dead skin by bathing in warm water.

Space-occupying lesions

In chelonians, due to the rigid shell, any large space-occupying lesion, including obesity, ascites and abscesses, will compress the lungs and cause respiratory embarrassment (see Figures 9 and 10).

Diagnosis may be made by radiography or endoscopy and treatment of the underlying condition is required.

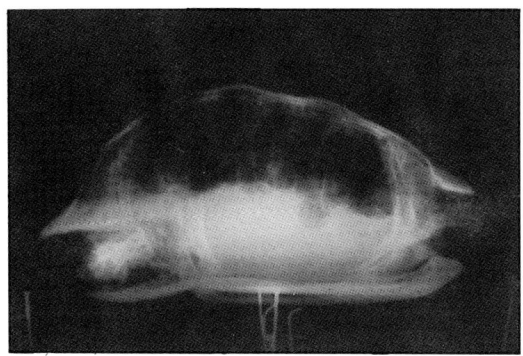

Figure 9
Lateral radiograph of a normal tortoise (*Testudo* sp.).

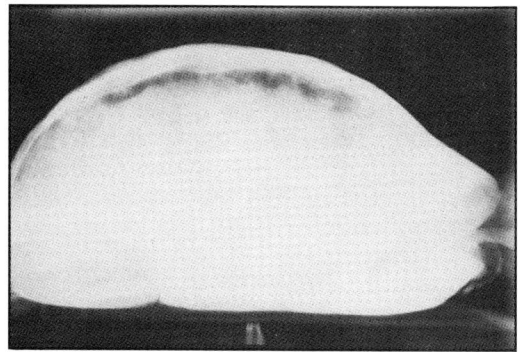

Figure 10
Radiograph of an obese tortoise (*Testudo* sp.) presented with dyspnoea.

Hypovitaminosis A

Deficiency of vitamin A causes squamous metaplasia of epithelial surfaces (Elkan and Zwart, 1967). The respiratory epithelium becomes cornified and ciliary function is lost. The lacrimal and Harderian glands are also affected and this leads to ocular signs. Focal white areas may become visible in the mouth and pharynx. These resemble tiny abscesses but are in fact areas of squamous metaplasia. The signs of hypovitaminosis A cannot be distinguished grossly from respiratory infection. The two conditions often co-exist as the affected epithelium is prone to secondary infection (Fowler, 1980b).

Terrapins and tortoises, particularly the American-box tortoise (*Terrapene* spp.), are most commonly presented with this condition. It is caused by feeding foods, such as Iceberg lettuce, which are deficient in vitamin A (Fowler, 1980b).

Diagnosis - this is difficult for the reasons mentioned earlier. It is not easy to measure vitamin A levels in reptiles due to the large volume of blood or section of liver required. Therefore, diagnosis is often made retrospectively based on response to treatment.

Treatment - weekly injections of vitamin A (1,500 IU i/m for four weeks) (Fowler, 1980b). In addition, it is necessary to correct the diet by including dark green vegetables and alfalfa hay, which are rich in vitamin A, or by supplementation with a vitamin supplement, eg. ACE-High (Vetark).

Antimicrobial agents may be required to control secondary infection. However, it is important to remember that the kidneys may be affected by the squamous metaplasia and nephrotoxic drugs should be avoided.

In chronic cases a guarded prognosis should be given as the changes become irreversible and response to treatment can be poor.

Trauma

Damage to the dorsal part of the carapace of chelonians often involves the lungs. Damage may be caused by lawnmowers which cause depressed fractures of the shell or fire damage which causes necrosis of the shell and underlying bone.

Robinson (1973) reported the surgical repair of a herniated lung which was caused by a fight between a common iguana (*Iguana iguana*) and a rhinoceros iguana (*Cyclura cornuta*).

Drowning

Tortoises and semi-aquatic terrapins are most likely to suffer this fate (Frye, 1981b). Tortoises often fall into ponds and semi-aquatic terrapins that are not provided with a ledge on which to rest will eventually become exhausted and drown. It is important to know whether the reptile was immersed in fresh or salt water: this is reflected in the treatment. Most captive reptiles drown in fresh water. The aspirated water is rapidly absorbed into the circulation resulting in haemodilution and hypervolaemia. Haemolysis occurs upsetting the balance of electrolytes still further. Drowning in salt water results in the opposite situation. Fluid from the circulation enters the lungs causing pulmonary oedema (Frye, 1981a).

Treatment - fresh water - replacement of plasma electrolytes (especially sodium) is necessary and diuretics such as frusemide (Lasix, Hoechst) (5mg/kg i/m) should be given (Holt, 1981).

salt water - hypotonic or isotonic fluids should be used.

In both cases positive pressure ventilation, antibiotic cover and general nursing care are essential (Frye, 1981a). The prognosis is poor due to the difficulty in administering intravenous fluids to chelonian species.

AIDS TO DIAGNOSIS

Observation

A clinically healthy reptile at rest breathes approximately every 30 seconds; however, smaller species will have faster respiratory rates. It should be borne in mind that stress, handling and environmental temperature will alter the respiratory rate.

Auscultation

This is difficult in reptiles due to the nature of the skin. Many extraneous sounds are heard caused by the scales or shell. It is possible, however, to reduce these extraneous noises by wrapping the reptile in a wet towel thereby allowing the lung sounds to be heard (Frye and Himsel, 1988) (see Figure 11).

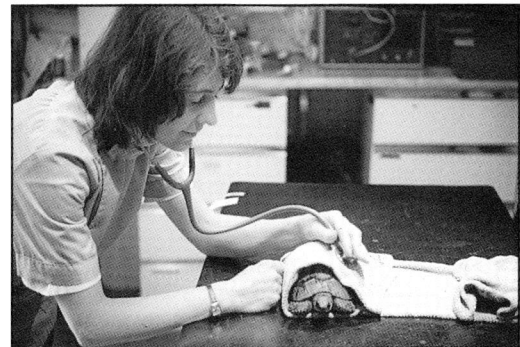

Figure 11
Auscultating the lung of a tortoise (*Testudo* sp.).

Radiography

Chelonians - rarely need sedation. Three views are useful:-

1. Dorsoventral, vertical beam.

2. Lateral, horizontal beam.

3. Craniocaudal, horizontal beam.

These three views will often allow localisation of the lesion, although early lesions may not be visible.

Snakes and lizards may be persuaded to lie still by placing a folded towel over their heads while dorsoventral views are taken but sedation may be needed for lateral views. Some species, eg. monitors, can be placed on their back and will assume "tonic immobility".

Tracheal washes

It is possible to pass a sterile cannula through the glottis of a conscious reptile. 1 – 2ml of sterile saline are instilled and immediately aspirated. A sample is sent for culture and sensitivity testing and the rest of the fluid is used for microscopic examination. On an unstained preparation the embryonated eggs of *Rhabdidas* spp. or larvae of ascarids or *Kalicephalus* spp. may be seen. Bacteria may be found on an appropriately stained slide.

GENERAL TREATMENT

Antibiotics

Reptiles are ectothermic (poikilothermic); therefore, the rate at which they metabolise a drug is related to the ambient temperature. It is important to keep the reptile at a constant temperature while it is receiving antibiotic treatment so that the drug is metabolised at a constant rate (see "Therapeutics"). When dose rates of antibiotics for reptiles are quoted in the literature the optimum ambient temperature is also stated.

Bacterial infections in reptiles are often resistant to the common antibiotics used in mammals and the importance of culture and sensitivity testing cannot be overstressed. The following is a list of some antibiotics which have proved useful in respiratory diseases of reptiles:-

Tylosin - 25mg/kg i/m daily at 30°C. This is a bacteriostatic antibiotic (Murphy, 1973).

Gentamicin - 2.5mg/kg i/m every third day at 24°C. Gentamicin is nephrotoxic (Montali *et al*, 1979). Care must be taken when using this antibiotic in debilitated reptiles. It is important to ensure that the reptile is adequately hydrated.

Oxytetracycline - 50mg/kg i/m at 26°C. A long-acting preparation should be given every third day.

Cephaloridine - 7mg/kg i/m daily at 24°C.

Tobramycin - 2mg/kg i/m daily at 26°C. In severe infections doses as high as l0mg/kg have proved useful (Lawton and Stoakes, unpublished data). Tobramycin may also be used in combination with cephaloridine: the combination seems to be synergistic.

Temperature and relative humidity

The reptile should be kept at its preferred body temperature (or at the temperature required for its antibiotic treatment). If the temperature is too high respiratory embarrassment may ensue and the reptile can become dehydrated.

Fluid therapy

A reptile with respiratory disease is usually anorexic and so it is important to supply nutrients in a readily assimilated form. Duphalyte (Solvay-Duphar Veterinary) can be given daily by stomach tube at a dose rate of 3–4% of bodyweight in divided doses. In severely debilitated cases Hartmann's solution may be given intracoelomically at the same dose rate. Many of the drugs used are nephrotoxic: to prevent kidney damage the reptile must be kept well hydrated.

Vitamins

Hypovitaminosis A has been implicated in respiratory disease as this vitamin is important for healthy mucous membranes (Fowler, 1980b). It may be administered by injection (1,500 IU/kg weekly i/m), by oral supplementation using products such as ACE-High (Vetark), or by correction of the diet, which should include dark green plant matter, in an attempt to aid recovery.

Steroids

Chronic, inflammatory, non-infectious cases diagnosed on lung biopsy may benefit from the use of steroids (Lawton and Cooper, unpublished data). A dose of dexamethasone (0.625–0.125mg/kg i/m) is suggested by Frye (1981a).

Diuretics

Frusemide (5mg/kg i/m or s/c sid or bid) has been recommended by Frye (1981a).

Mucolytics

Bromhexine (Bisolvon, Boehringer Ingelheim) injection may prove useful in severe cases of respiratory disease in an attempt to loosen the exudate.

PREVENTION

Isolation

All newly acquired reptiles should be isolated for a period of at least one month before being mixed with an existing collection. This is particularly important if boarding reptiles for friends while they are on holiday, as this is an opportunity for infections to spread. During this time of isolation it is wise to check faecal samples for parasites and to treat if necessary.

Temperature and relative humidity

It is important to ensure that the environment closely resembles that in nature. If it does not, the reptile will be subject to stressors and be prone to disease.

Husbandry

When in captivity reptiles usually have limited space in which to move, compared with their natural environment. It is important, therefore, to remove excreta from the vivarium at regular intervals to prevent a buildup of bacteria, which are potential pathogens and which increase the likelihood of infection.

Diet

Diet is equally important in preventing respiratory disease. Captive reptiles are entirely dependent on the diet provided by the keeper and if incorrect or insufficient food is provided deficiencies soon occur, eg. vitamin A deficiency. Obesity can also lead to difficulty in breathing and is often seen in terrestrial tortoises due to incorrect feeding, eg. being fed bananas, cake or jam.

REFERENCES

AHNE, W., NEUBERT, W.J. and THOMPSEN, I. (1987). Reptilian viruses: isolation of myxovirus-like particles from the snake, *Elaphae oxycephala*. *Journal of Veterinary Medicine* **34**, 607.

ANDERSON, M.P. (1976). Laryngeal foreign body as a cause of acute respiratory distress in lizards. *Veterinary Medicine/Small Animal Clinician* **71**, 940.

BONNEY, C.H., HARTFIEL, D.A. and SCHMIDT, R.E. (1978). *Klebsiella pneumoniae* infection with secondary hypopyon in Tokay gecko lizards. *Journal of the American Veterinary Medical Association* **79 (9)**, 1115.

BRANNIAN, R.E. (1984). Lungworms. In: *Diseases of Amphibians and Reptiles*. (Eds. G.L. Hoff, F.L. Frye and E.R.Jacobson). Plenum Press, New York.

COOPER, J.E. (1971). Disease in East African snakes associated with *Kalicephalus (Nematoda diatheanocetphalidae)*. *Veterinary Record* **89**, 385.

CLARK, H.F., LIEF, F.S. and LUNGER, P.D. (1979). Fer-de-Lance virus: a probable paramyxovirus isolated from a reptile. *Journal of General Virology* **44**, 405.

DAVIES, P.M.C. (1981). Anatomy and physiology. In: *Diseases of the Reptilia*, Vol. 1. (Eds. J.E. Cooper and O.F. Jackson). Academic Press, London.

ELCAN, E. and ZWART, P. (1967). The ocular diseases of young terrapins caused by vitamin A deficiency. *Pathologica Veterinarian* **4 (3)**, 201.

FOWLER, M.E. (1977). Respiratory disease in desert tortoises. *American Association of Zoo Veterinarians Annual Proceedings.* Honolulu.

FOWLER, M.E. (1980a). Respiratory disease in reptiles. In: *Current Veterinary Therapy VII. Small Animal Practice.* (Ed. R.W. Kirk). W.B. Saunders, Philadelphia.

FOWLER, M.E. (1980b). Comparison of respiratory infection and hypovitaminosis A in desert tortoises. In: *Comparative Pathology of Zoo Animals.* (Eds. R.J. Montali and G. Migaki). Smithsonian Institute. Washington DC.

FRYE, F.L. (1981a). Respiratory disease. In: *Biochemical and Surgical Aspects of Captive Reptile Husbandry.* 1st Edn. Veterinary Medical Publishing Company, Edwardsville.

FRYE, F.L. (1981b). Traumatic and physical diseases. In: *Diseases of the Reptilia, Vol.2.* (Eds. J.E. Cooper and O.F. Jackson). Academic Press, London.

FRYE, F.L. and HIMSEL, C.A. (1988). The proper method for stethoscopy in reptiles. *Veterinary Medicine* **December**, 1250.

GASKIN, J.M., HASKELL, M., KELLER, N. and JACOBSON, E.R. (1989). Serodiagnosis of ophidian paramyxovirus infections. *Third International Colloquium on the Pathology of Reptiles and Amphibians.* (Ed. E.R. Jacobson). Orlando.

GAUNT, A.S. and GANS, C. (1969). Mechanics of respiration in the snapping turtle *(Chelydra serpentina).* *Journal of Morphology* **128**, 195.

HENDRIX, C.M. and BLAGBURN, B.L. (1988). Reptilian pentostomiasis: a possible emerging zoonosis. *Compendium on Continuing Education for the Practicing Veterinarian* (*North American Edn.*) **10 (1)**, 46.

HOLT, P.E. (1981). Drugs and dosages. In: *Diseases of the Reptilia, Vol. 2.* (Eds. J. E. Cooper and O.F. Jackson). Academic Press, London.

JACKSON, O.F. (1991). Chelonians. In: *Manual of Exotic Pets.* New Edn. (Eds. P.H. Beynon and J.E. Cooper). BSAVA, Cheltenham.

JACKSON, O.F. and NEEDHAM, J.R. (1983). Rhinitis and virus antibody titres in chelonians. *Journal of Small Animal Practice* **24**, 31.

JACOBSON, E.R. (1978). Diseases of the respiratory system in reptiles. In: *Veterinary Medicine/ Small Animal Clinician* **73 (9),** 1169.

JACOBSON, E.R. (1980). Infectious diseases of reptiles. In: *Current Veterinary Therapy VII. Small Animal Practice.* (Ed. R.W. Kirk). W.B. Saunders, Philadelphia.

JACOBSON, F.R. (1988). Chemotherapeutics used for bacterial pathogens. In: *Exotic Animals.* (Eds. E.R. Jacobson and G.V. Kollias). Churchill Livingstone, New York.

JACOBSON, E.R., GASKIN, J.N., SIMPSON, C.F. and TERRELLT, G. (1980). Paramyxo-like virus infection in a rock rattlesnake. *Journal of the American Veterinary Medical Association* **177 (9)**, 796.

JACOBSON, E.R., CRISPIN, P., SPENCER, C.P. and POULOS, P.W. (1983). Radiographic evaluation of reptiles. *American Association of Zoo Veterinarians Annual Proceedings.*

JACOBSON, E.R., GASKIN, J.M., PAGE, D., IVERSON, W.D. and JOHNSON, J.W. (1981). Illness associated with paramyxo-like virus infection in a zoologic collection of snakes. *Journal of the American Veterinary Medical Association* **179 (11)**,1227.

LAWRENCE, K. (1985). Snakes. In: *Manual of Exotic Pets.* Revised Edn. (Eds. J.E. Cooper, M.F. Hutchison, O.F. Jackson and J.R. Maurice). BSAVA, Cheltenham.

LAWRENCE, K. and NEEDHAM, J.R. (1985). Rhinitis in long-term captive Mediterranean tortoises. *Veterinary Record* **25/26**, 662.

LAWTON, M.P.C. (1991). Lizards and snakes. In: *Manual of Exotic Pets.* New Edn. (Eds. P.H. Beynon and J.E. Cooper). BSAVA, Cheltenham.

MARCUS, L.C. (1977). Parasitic diseases of captive reptiles. In: *Current Veterinary Therapy VI. Small Animal Practice.* (Ed. R.W. Kirk). W.B. Saunders, Philadelphia.

MIGAKI, G., JACOBSON, E.R. and CASEY, H.W. (1984). Fungal disease in reptiles. In: *Diseases of Amphibians and Reptiles.* (Eds. G.L.Hoff, F.L. Frye and E.R. Jacobson). Plenum Press, New York.

MONTALI, R.J., BUSH, M. and SMELLER, J.M. (1979). The pathology of nephrotoxicity of gentamicin in snakes. *Veterinary Pathology* **16**, 108.

MURPHY, J.B. (1973). The use of the macrolide antibiotic tylosin in the treatment of reptilian respiratory infections. *British Journal of Herpetology* **4**, 317.

ROBINSON, P.T. (1973). Surgical repair of a herniated lung in a common iguana. *Journal of the American Veterinary Medical Association* **163**, 655.

SHOTTS, E.B.J. (1984). Aeromonas. In: *Diseases of Amphibians and Reptiles.* (Eds. G.L. Hoff, F.L. Frye and E.R. Jacobson). Plenum Press, New York.

SPRENT, J.F.A. (1984). Ascaridoid nematodes. In: *Diseases of Amphibians and Reptiles.* (Eds. G.L. Hoff, F.L. Frye and E.R. Jacobson). Plenum Press, New York.

TEARE, J.A. and BUSH, M. (1983). Toxicity and efficacy of ivermectin in chelonians. *Journal of the American Veterinary Medical Association* **183**, 1195.

TOMPSON, F.N., MCDONALD, S.E. and WOLF, E.D. (1976). Hypopyon in a tortoise. *Journal of the American Veterinary Medical Association* **169**, 942.

CHAPTER TEN

GASTROINTESTINAL SYSTEM

Robin D Bone BVMS MRCVS

ANATOMY AND PHYSIOLOGY

This is covered more fully in Pritchard (1979), Davies (1981) and Frye (1991).

Oral cavity

There is little mastication of food by most reptiles and it is directly swallowed. The saliva is lubricatory in function with little evidence of proteolytic activity (Davies, 1981). Snake and lizard venom is produced by modified salivary glands and is often rich in proteolytic enzymes. The venom is delivered by specialised teeth, either grooved or hollow, depending on the species.

The tongue is variable in reptile Families. Chelonians have a largely immobile, fleshy tongue. Most snakes have a thin, highly mobile, protrusible tongue which has lost most of its mechanical capabilities in the interests of chemosensory specialisation. Lizards often have mobile protrusible tongues which are important mechanical aids in feeding and drinking.

In chelonians the jaws have a limited range of movements. However, in some snakes and lizards, depending on the species, there is considerable modification of the bones and joints of the skull. This allows movement of both the upper and lower jaws in relation to the cranium. This is called "kinesis" and results in a much increased gape to cope with ingestion of large prey. There is the ability, in some lizards, to move the lower jaw forward and backwards to assist further with the ingestion of prey. Chelonians have no teeth and rely on a horny (keratinous) beak, while other reptiles shed and replace teeth throughout their lives.

In egg-eating snakes, eg. *Dasypeltis* spp., there are ventral extensions of the cervical vertebrae adapted to break the eggshell which is often regurgitated, the contents being ingested.

Stomach and intestinal tract

The stomach is similar to mammals and produces a variety of enzymes, hydrochloric acid and pepsin (Davies, 1981). The intestines vary in length depending on the diet. Snakes have relatively short intestines whilst herbivorous chelonians have a longer intestinal tract. There are often extensive longitudinal folds in the reptilian alimentary tract, especially in snakes; this increases surface area for absorption and allows distension to accommodate bulky food. A range of enzymes and bile salts, similar to mammals, is produced by the pancreas and liver (Davies, 1981). There is differentiation into small intestine and colon.

Cloaca

This is a structure which is not present in mammals but is seen in reptiles and birds. It consists of three consecutive compartments: the copradaeum into which faeces are discharged, a urodaeum which receives the urogenital tract ducts and a proctodaeum which acts as a common collecting area prior to evacuation.

In snakes the location of organs can be expressed as a percentage of the rostrum to cloaca length (Lawrence, 1985a) - for example:-

    liver   38 – 56%
    stomach   46 – 67%
    intestines 68 – 81%
    colon   81 – 100%.

DISEASES OF THE ORAL CAVITY

STOMATITIS

Stomatitis, usually known as "mouth rot" or "canker", is a common condition which can affect all reptiles, especially chelonians and snakes. It is probably one of the most frequently seen reptile conditions (Holt *et al*, 1979; Phillips, 1986). Stomatitis is an infection of the oral cavity involving bacterial, viral or fungal agents or, more usually, a combination of these. Stomatitis is not generally a primary condition but is often secondary to stress, trauma, weakness, malnourishment or poor husbandry, eg. sub-optimal temperature or inadequate feeding. Trauma is more significant in snakes and stomatitis may occur if the snake persistently rubs its nose or strikes at the glass of the vivarium (Cooper, 1981). Traumatic lesions can occur following repeated force-feeding. It has been reported that a deficiency of vitamin C predisposes to this condition in snakes (Wallach, 1969), although this has yet to be proven.

In tortoises stomatitis is commonly seen following hibernation (Holt *et al*, 1979), especially if body condition or hibernating techniques are poor; it is an important cause of the post-hibernation anorexia syndrome (Lawrence, 1987a). White blood cell counts are lowest in the spring after hibernation, making the tortoise more susceptible to disease at this time of year (Lawrence and Hawkey, 1986). Most cases of bacterial stomatitis are associated with *Pseudomonas* spp. and *Aeromonas* spp. and these are always regarded as significant isolates. Many other, mainly Gram-negative, organisms can be isolated and it may prove difficult to implicate them as primary pathogens. The most frequently reported organisms are listed in Table 1. Draper *et al* (1981) reported that 60.3% of isolates from the oral cavities of healthy snakes were Gram-positive organisms, while 80.3% of isolates from stomatitis cases were Gram-negative oganisms; this represents a significant difference. There have also been reports of herpesvirus isolations from tortoises with stomatitis (Jacobson *et al*, 1985a; Cooper *et al*, 1988). Fungi and yeast elements may be found on direct smears from oral lesions and on culture but these are usually secondary invaders.

Chelonians

The commonest presenting sign of stomatitis in chelonians is anorexia. The typical bacterial case starts with one or more whitish, raised plaques on the dorsum of the tongue or lower gums. In the early stages these can often be wiped away with a cotton bud leaving no discrete lesions, although the mucosa may be inflamed. This condition can, however, become much more extensive, spreading to involve the hard palate, pharynx and glottis with white or yellow diphtheritic plaques. This necrotic material is more difficult to remove and, when removed, may reveal ulcerated or bleeding lesions. Eventually, the oesophagus or even the trachea and lungs can become involved. In severe cases there is a foul-smelling, purulent discharge from the mouth. Osteomyelitis of the jaw is always a possible sequel. The prognosis is guarded and the condition will rapidly become fatal if not treated promptly.

The herpesvirus-like type of stomatitis that has been reported affecting Mediterranean tortoises (*Testudo graeca*) (Cooper *et al*, 1988) involved mainly the lower gums, which became very swollen and inflamed, with white debris appearing later over the affected areas. In some cases, however, the spread of infection caused massive oedematous swelling of the neck region and rapid death, without any necrotic material being visible. Viral inclusion bodies were found in sections of trachea and mucous membranes from *post-mortem* material. Diagnosis did not prove possible from *ante-*

Table 1
Bacterial isolates from stomatitis cases.

Reference numbers for isolates from:	General reptiles	Tortoises	Snakes	Lizards
Pseudomonas spp.	6,9	5	1,2,7	1
Pseudomonas aeruginosa	10	4	7,8	8
Pseudomonas maltophila			7	
Pseudomonas fluorescens	10		1	1
Aeromonas spp.	6,9	3,5	1,2	1
Aeromonas hydrophila	10		7,8	8
Aeromonas aerophila	10			
Aeromonas aerogenes	10			
Proteus spp.		4	1,2	1
Proteus vulgaris			7	
Pasteurella spp.			1	1
Pasteurella haemolytica			7	
Providencia rettgeri	9		7	
Citrobacter freundii			7	
Flavobacterium spp.			7	
Morganella morganii			7	
Enterobacter spp.		5		
Escherichia coli	9	5	7	
Neisseria spp.		5		
Klebsiella spp.		5	7	
Hafnia spp.		5		
Streptococcus spp.	10		7	
Staphylococcus spp.	10			
Staphylococcus epidermidis		5		
Corynebacterium spp.	9	5		

References: 1 - Wallach, 1969; 2 - Cooper, 1973; 3 - Holt and Cooper, 1976; 4 - Keymer, 1978a; 5 - Holt *et al*, 1979; 6 - Cooper, 1981; 7 - Draper *et al*, 1981; 8 - Marcus, 1981; 9 - Phillips, 1986; 10 - Fry, 1991.

mortem oral swabs. A herpesvirus was isolated from an outbreak of necrotic stomatitis in a large colony of Argentine tortoises (*Chelonoidis chilensis*) (Jacobson et al, 1985a). The distribution of lesions was different from the preceding account: necrotic cellular debris was found around the glottis, hard palate and internal nares.

Snakes and lizards

Snakes also show anorexia as the main presenting sign of stomatitis, together with weight loss and lethargy. Excess salivation may be seen at this stage. Stomatitis is not as common in lizards but presents a similar picture. Examination of the snake often shows only petechiation in the early stages of the condition progressing to swollen gums and the appearance of diphtheritic material on the hard palate and tongue. The infection may spread into the tooth sockets and, eventually, the jawbones leading to loss of teeth and osteomyelitis.

Complications of stomatitis can prove life threatening: aspiration of infective debris may cause pneumonia and invasion of the bloodstream will lead to septicaemia. The necrotic exudate may block the nasolacrimal duct and cause obstruction of flow from the Harderian gland. This causes swelling of the sub-spectacular space (see "Ophthalmology"). There may be direct spread of infection from the mouth causing a sub-spectacular abscess (Marcus, 1981).

Treatment

In all but the earliest cases a swab should be taken for bacterial culture and sensitivity. The organisms isolated are often resistant to many antibiotics (Needham, 1981; Lawrence et al, 1984). Framycetin (Framomycin, C-Vet) (10mg/kg i/m every 48 hours at 26°C) (Jackson, personal communication) can be used whilst awaiting results from the culture. All necrotic debris should be removed from the oral cavity by cleaning with a dilute (1 in 4) aqueous solution of povidone-iodine - this contains no detergent or alcohol. Dilute (3%) hydrogen peroxide can be used on intractable lesions. If very extensive lesions are present sedation, analgesics or general anaesthesia may be required for a thorough and painless debridement. In the earliest cases it is probably sufficient to clean the mouth twice daily as described; this should be carried out for a few days after no more lesions have been seen to ensure that all infection has been eliminated. The more severe cases require appropriate systemic and topical antibiotics based on sensitivity tests. If framycetin is indicated it can be applied locally as a paste (Framomycin Anti-Scour Paste, C-Vet); this is effective in most cases. Maintenance of the reptile at its preferred body temperature (PBT) or the temperature recommended for a particular antibiotic is essential (see "Therapeutics"). Reptiles will not eat while their mouths are sore; debilated animals will need stomach tubing with nutrients and fluids. This must be done with care to avoid passing infective material down the oesophagus. Tubing is best carried out after cleaning and treatment have been performed. If the mouth is full of antibiotic paste this will both coat and lubricate the tube. Supplementation with vitamin A (10,000 IU/kg) and vitamin C (10–20mg/kg as sodium ascorbate) can be useful if healing appears to be slow (Frye, 1991); this is because both vitamins are involved in tissue regeneration. The treatment of sub-spectacular swellings or abcesses in snakes is covered in "Ophthalmology".

Stomatitis can become a recurring problem in large pythons and boas and in these cases an autogenous vaccine may be of value (Lawrence, 1987b). The snake should be anaesthetised and material removed from the base of the lesion for culture from which to prepare the autogenous vaccine. An effective regime is 0.5ml intramuscularly every three days for 18 treatments, weekly for 12 treatments and then a single dose one month later. Localised reactions, such as vesicles which often rupture yielding a clear fluid, can occur at the injection site. A similar protocol was used, apparently successfully, to treat a reticulated python (*Python reticulatus*) by Addison and Jacobson (1974).

Cryotherapy can be used in chronic cases in snakes, especially if granulation tissue is present. An Indian rock python (*Python molurus molurus*) was successfully treated by Green et al (1977). Complete resolution took four months.

Herpesvirus-type "mouth rot" in tortoises responded well to acyclovir 5% ointment (Zovirax, Wellcome) applied to the mouth twice a day (Cooper et al, 1988).

ABNORMAL DEVELOPMENT OF MOUTHPARTS

This is a problem which particularly affects chelonians. It is usually the maxilla which is overgrown rostrally to a point (Holt *et al*, 1979). This can be easily trimmed with nail clippers. More extensive overgrowth laterally may require several attempts at clipping and filing over a period of months to correct the malalignment.

Overgrowth of the mandible causes greater distortion of the mouthparts. It is more difficult to correct, if not impossible, due to increased subluxation of the tempero-mandibular joint and alteration in the conformation of the mandibular rami (Frye, 1991). These conditions are usually caused by osteodystrophy, with associated weakness of the jaw bones (see "Nutritional Disease"), or by malalignment of the maxilla and mandible. They may also be initiated by soft foods, eg. lettuce and tomatoes, that are commonly offered to captive tortoises with little foraging effort being required. Feeding tougher foods, eg. dandelions and cabbage, can help prevent this condition.

PARASITIC CONDITIONS OF THE ORAL CAVITY OF SNAKES

Trematodes

Trematodes are occasionally seen in the oral cavity of snakes; they are digenetic trematodes of the Renifer group, which includes the genera *Renifer*, *Ochetosoma*, *Mesocoelium* and others (Marcus, 1981). They have an amphibian intermediate host which is ingested and followed by migration of adult trematodes, via the oral cavity, to the lungs where they can cause focal lesions. The yellow-orange egg with a polar cap can be found in faeces and in tracheal washes. Although not a serious pathogen, treatment should be considered, eg. manual removal of those visible in the mouth or use of praziquantel (Droncit, Bayer) (100mg/kg orally or 3.5 - 7.5mg/kg s/c) (Lawrence, 1985b).

Rhabdias spp.

Thin-shelled eggs and free first stage (L_1) larvae can be found during microscopic examination of oral secretions. Infective larvae (usually L_3) penetrate the skin or oral mucosa and migrate via the blood stream to the lungs where they mature. Parthenogenetic females deposit eggs in the lungs; these develop into first stage larvae which are expelled in oral mucus or faeces to produce a free-living sexual generation. This is primarily a respiratory parasite but it has been known to cause skin lesions and inflammation of the oral mucosa with production of a mucous exudate (Frank, 1981). In garter snakes (*Thamnophis* spp.) this has been associated with dental pocket abscesses (Frank, 1981). Treatment is with levamisole (Levacide, Norbrook) (10mg/kg s/c) or ivermectin (Ivomec, MSD Agvet) (200mcg/kg s/c) (Lawrence, 1985b).

Kalicephalus spp.

Adults and eggs can occasionally be found in oral mucus following expulsion from the oesophagus. This parasite is discussed more fully later (see Diseases of the Intestinal Tract).

Pentastomids

These large worm-like arachnids are principally respiratory parasites (see "Respiratory System") but they are associated with the production of a viscous oral mucus (Frank, 1981). The characteristic embryonated and, in some species, double-shelled egg can be found in oral mucus and faeces. The larvae in the egg usually have four leg-like appendages. Surgical removal may be attempted but this can be difficult due to embedded mouthparts (Cooper, personal communication). This is a potential zoonosis, the larval stage being associated with serious human illness affecting the liver and lungs (Frank, 1981).

PARASITIC CONDITIONS OF THE ORAL CAVITY OF CHELONIANS

Trematodes

Trematodes of the genus *Polystomoidella* have been found in the oropharynx and oesophagus of various freshwater chelonians (Frank, 1981). They are monogenetic trematodes with a direct life-cycle and no intermediate host. They are rarely more than 3mm in length. No clinical signs have been reported.

Leeches

Leeches (Class Hirudinea) can be found in and around the oral cavity of aquatic chelonians. They are common, especially in North America (Cooper, 1990). Heavy infestations may lead to anaemia; however, they may also transmit bacteria and protozoan parasites, eg. Haemogregarines and Trypanosomes (Frank, 1981).

DISEASES OF THE STOMACH

REGURGITATION

Snakes

Because snakes have a poorly developed cardiac sphincter, regurgitation in snakes is a common problem and is usually caused by poor husbandry or parasitism. Endoscopy or even coeliotomy may be required for diagnosis. Biochemistry, bacteriology and cytology may also be useful.

A snake may regurgitate if it is maintained at a sub-optimal temperature. The PBT is required for optimum enzyme function and normal digestive processes; these will be retarded if the snake is too cold and the food may be regurgitated 2 – 3 days later as a foul smelling, discoloured and mucus-covered mass.

Force-feeding of a snake which has not eaten for some time can cause regurgitation due to reduced enzyme function, a poorly functioning intestinal tract, dehydration and toxaemia caused by long-term cachexia. Stomach tubing with liquid foods can be attempted, eg. Dulphalyte (Solvay-Duphar Veterinary); this will also correct any dehydration.

Handling a snake within 3 – 4 days of feeding can cause regurgitation. This is a reflex action, advantageous in the wild, which is triggered by stress (Frye, 1991). Foreign bodies can result in regurgitation - sand or substrate ingested with prey may become impacted and cause a blockage. Bones (especially if small fish-eating snakes are fed pieces of a large fish) may be too large to pass through the oesophagus and stomach. Lubrication with liquid paraffin can be attempted - otherwise, surgery may be required. Removal of the foreign body may be possible via an endoscope with retriever forceps.

Gastritis and gastric ulceration can cause regurgitation. The former can be treated with a kaolin/antibiotic mixture; the latter is often associated with parasitism or long-term cachexia. Spasm of the pylorus is a possibility and this may be stress related (Frye, 1991).

Mycotic granulomata can cause regurgitation due to external pressure on the stomach. These are common in reptiles but can only be diagnosed by coeliotomy or at necropsy. Tumors can cause similar problems (Frye, 1991).

Parasitism in various forms can cause regurgitation. If the husbandry factors mentioned earlier have been ruled out then vomitus and faeces should be checked for evidence of parasites.

Nematodes (*Kalicephalus* spp.) can cause ulceration of the oesophageal and gastric mucosa; heavy burdens with secondary bacterial infections are associated with a high mortality rate (Cooper, 1971). This parasite is dealt with more fully later (see Diseases of the Intestinal Tract). Ascarids in large numbers have been associated with a necrotic and ulcerative gastritis which can lead to abscess formation (Frank, 1981).

Tapeworms (*Bothridium* spp.), especially in pythons, can be a problem if large numbers are present. The regurgitated food is often covered with proglottides. Diagnosis and treatment are described later (see Diseases of the Intestinal Tract). Although dysentery is a more usual sign, entamoebiasis should also be considered here.

Chronic hypertrophic gastritis associated with *Cryptosporidium* spp. infection can cause regurgitation prior to death over a period varying from four days to two years (Brownstein *et al*, 1977). The other notable clinical finding is a palpably enlarged stomach, felt as a firm mid-body swelling; this is caused by a massive thickening of the stomach wall, almost occluding the lumen. Diagnosis is by finding oocysts on a faecal smear stained with periodic acid-Schiff (PAS). Biopsy of the stomach can be performed. Staining with toluidine blue shows the organisms attached to the microvillar borders of the surface, pit and glandular epithelium, although this is not always easy to demonstrate. The condition is generally regarded as untreatable but one case, diagnosed on biopsy, has responded to metronidizole (260mg/kg two doses two weeks apart) (Lawton and Cooper, personal communication). The zoonotic implications should be considered.

Lizards

Regurgitation is not common in lizards because, unlike snakes, they have a well developed cardiac sphincter. Vomiting is a serious sign and is usually associated with septicaemia and imminent death.

Chelonians

Regurgitation and vomiting in chelonians are uncommon signs and both carry a grave prognosis, usually being associated with terminal septicaemia.

ATROPHIC GASTRITIS

This condition has been reported in one Hermann's tortoise (*Testudo hermanni*) and two red-eared terrapins (*Trachemys scripta elegans*). The three reptiles were either euthanased or died following renal disease, with the flagellate *Hexamita parva* being found in two cases (Zwart and Gaag, 1981). Chronic atrophic gastritis was present with marked atrophy of the fundic mucosa and alterations in the secretory glands. This appeared to be related to areas of calcification of the stomach wall which were probably secondary to the renal disease. No clinical signs specific for gastric disease were recorded and a diagnosis of renal disease with gastric pathology as secondary findings was made.

FOREIGN BODIES

Gravel and stones can often be seen as an incidental finding on radiographs of chelonians and lizards. This is dealt with more fully later (see Diseases of the Intestinal Tract).

DISEASES OF THE INTESTINAL TRACT

CHELONIANS

Parasitic diseases

Flagellates

Flagellate infections are a common problem, particularly in colonies ("herds") of tortoises where diseases can spread very quickly. They are generally associated with sudden onset anorexia, faeces containing undigested food or, occasionally, diarrhoea. Polydipsia is seen, caused by renal damage, oedema and fluid retention in the inflamed intestinal mucosa. The organisms involved are *Trichomonas* spp. and *Hexamita* spp. (Keymer, 1981). These can be seen in fresh faecal smears under microscopic examination as fast-moving ovoid/round organisms with obvious flagella. The bodies of the organisms from both genera measure less than 10μm (Reichenbach-Klinke and Elkan, 1965; Zwart and Truyens, 1975). Desquamated tall columnar epithelial cells may be found in increased numbers; they will also often be seen in normal faeces. The flagellates are found in faeces from apparently healthy chelonians but there are usually very few present. On examining faeces from a clinical case, numerous organisms will be seen in most fields under microscopic examination. The flagellates will rapidly dry out and die in small samples and this can make diagnosis difficult. However, they can be found after several days in large, moist samples, although often in reduced numbers. Urine samples should be examined as *Hexamita parva* is an important renal pathogen (Zwart and Truyens, 1975).

Treatment needs to be effective and rapid to prevent irreversible damage to the intestinal mucosa. The drug of choice is metronidazole (160mg/kg daily for four days by stomach tube) (Jackson, 1991). This regime can be associated occasionally with CNS disturbances and a single dose of 260mg/kg is less likely to cause problems (Lawton, personal communication). This single dose treatment can be useful for patients that are difficult to stomach tube. Flagyl S Suspension (May and Baker) or crushed tablets can be used. Dimetridazole (Emtryl, Rhône Mérieux) can be used (40mg/kg for four days) (Lawrence, 1985a).

A faecal sample should be examined at the end of treatment to check for clearance. The treatment can often disrupt normal intestinal flora, and stomach tubing with liquidised green foods with some live yoghurt may help restore the balance. Probiotics, eg. Transvite (Gosforth Veterinary Products) or Prozyme (Univet), can also be added to the food to help re-establish normal bowel flora. The prognosis is poor if there has been extensive mucosal damage.

Ciliates

Ciliate infections involving the *Paramecium*-like organism *Balantidum coli* occur fairly commonly in tortoises, with similar clinical signs to those seen in flagellate infections. They are found in the colon and, if in large numbers, can cause a severe colitis; an occasional organism will be seen in faeces from healthy tortoises. The direct life-cycle results in resistant cysts being passed in the faeces. A heavy burden can kill hatchling tortoises. Diagnosis is by finding the organism, which measures 50-150μm (Soulsby, 1982). Treatment is difficult as the drug of choice, paromomycine (Humatin, Parke-Davis) is not available in the UK. However, metronidazole can be used at a dosage as indicated for flagellates. Mixed flagellate/ciliate infection can occur. This can lead to a more serious disease with severe gastrointestinal tract disturbance.

Helminths

Various nematode eggs and larvae can be found in chelonian faeces. They are mainly from the ascarid and oxyurid groups. The former, especially *Angusticaecum* spp., can be up to 20cm in length, the latter much smaller (a few mm). They are not generally a serious problem but any tortoise with worms should be treated to prevent contamination of the ground and infection of other tortoises. If a tortoise is already debilitated then a heavy worm burden may become significant. Large infestations of ascarids can cause anaemia and blockage of the bowel. In colonies of tortoises worm numbers can buildup rapidly due to contamination of the ground with eggs; this factor should be discussed with the owners. Faeces should be checked every year and the tortoises treated as required, ideally after hibernation. Treatment is with oxfendazole 2.265% (Synanthic, Syntax; Bandit, Pitman-Moore) (60-66mg orally) (Jackson, 1991). Fenbendazole (50-100mg/kg orally) has been used (Holt and Lawrence, 1982) but is not as effective. Ivermectin should not be used in tortoises as it has been associated with paralysis and death (Teare and Bush, 1983).

Colic

Colic is caused by fermentation of food within the intestinal tract leading to abnormal gas production. This is commonly seen following stomach tubing and feeding sick chelonians with inappropriate food. Colic can be caused by the administration of too large a volume to a sick patient, especially if at sub-optimal temperatures. Colic will also occur following the use of foods containing too much protein or sugar, especially if a milk base has been used (dairy products should not be fed to terrestrial chelonians). Colic can also be secondary to foreign bodies, such as stones and bones. Treatment is with small frequent doses of kaolin orally and metoclopramide by injection (Emequell, SmithKline Beecham) (2 – 6mg i/m as a single dose) (Jackson, 1991). Antibiotics may also be required. Surgery may be required for foreign bodies. Colic associated with enteritis in terrapins can cause an abnormal swimming position. Treatment is with antibiotics.

Salmonellosis

Both tortoises and terrapins may excrete *Salmonella* spp. and *Arizona* spp. in their faeces (Keymer, 1978a,b; Savage and Baker, 1980). The reptiles are usually clinically normal but the zoonotic implications should be considered, especially with terrapins and the potentially contaminated water in the tank (Jephcott *et al*, 1969).

Identification of carriers is difficult because the excretion rate is variable and one culture may fail to show organisms either in faeces or water (Marcus, 1981). In a USA report (Fox, 1974) 39 groups of terrapins were certified *Salmonella* – free but 15 (38%) were later found to be excreting again. A latent infection can be made patent by stress, eg. dehydration (DuPonte *et al*, 1978). Attempts at eliminating carriers are rarely successful (Cooper, 1981). Treatment may lead to resistant strains and, therefore, apparently increased virulence.

Foreign bodies

The commonest foreign bodies seen on radiographs are stones, although other objects, including bones, can be found. They are not normally a problem and usually pass through uneventfully. Liquid paraffin can be used to lubricate the passage of the stones. It should be remembered that gut passage time in tortoises can be two to four weeks (Lawrence and Jackson, 1982). A bowel blockage may need surgery, especially if associated with colic.

SNAKES

Enteritis with diarrhoea is a common finding in snakes and is usually caused by parasitic infections. The snakes may appear bloated due to fluid-filled intestines which can be palpated. Renal disease can also cause diarrhoea and will need to be excluded as a diagnosis.

Parasitic diseases

A faecal sample will always be required for diagnosis of parasitic diseases. If this is not available, a saline enema can be administered to obtain a reasonable sample.

Entamoebiasis. *Entamoeba invadens* is an important protozoan parasite affecting most species of snakes in captivity. It can produce a severe disease with high mortality and can spread rapidly through a group of susceptible snakes. An epizootic causing the death of 16 exhibition snakes was reported by Donaldson *et al* (1975).

Although amoebic cysts can be found in faeces of herbivorous reptiles, eg. terrestrial chelonians, the parasite is regarded as a commensal organism causing no clinical disease. The high proportion of plant material in the diet provides starch for the amoebae to accumulate glycogen in the cytoplasm, to encyst and so complete their life-cycle. In snakes the amoebae have to feed on the mucosa and mucous coating, which leads to tissue invasion and clinical disease (Meerovitch, 1958).

The infective cyst is ingested and passes unaffected from the stomach to the small intestine. Here it excysts and releases one quadrinucleate trophozoite which divides to produce eight uninucleate amoebae. These mature in the colon, eventually encysting and passing out in the faeces. The infective cysts are able to survive for 7 – 14 days (Keymer, 1981). The migrating amoebae (possibly carrying bacteria from the intestinal tract) cause a necrotising enterohepatitis, spreading to the liver via the portal circulation. Clinical signs may not be noticed until the terminal phase, but non-specific signs, such as lethargy and anorexia, will be seen earlier. Mucoid, bloody diarrhoea and vomiting may be observed, as well as polydipsia and CNS signs. The cloacal region may feel swollen and firm on palpation. The enlarged colon may be palpated. Progression can be variable from a slow deterioration over several weeks to death within 24 hours of clinical signs being shown.

Diagnosis is by finding the amoebae or cysts in a fresh faecal sample using a high power magnification (x400). They measure, respectively, approximately 16μm and 11 – 20μm when chemically fixed (Barnard, 1986). The cysts can be seen on direct smears but they will be detected more easily if stained with either 1% aqueous eosin or double strength Lugol's iodine (Keymer, 1981).

Treatment is with metronidizole orally (160mg/kg orally for three days); however, Donaldson *et al* (1975) used an effective single dose of 275mg/kg. This latter regime is more suitable for snakes which are fractious or difficult to stomach tube. All in-contact snakes should be treated. Environmental temperature can be a method of controlling the infection (Donaldson *et al*, 1975): at higher temperatures (35° – 37°C) infestation is prevented and at low temperatures (<15°C) pathogenicity is absent. The recommended temperature for metronidizole treatment is 38°C for 48 hours (Lawrence, 1985a), although the snake will need to be observed for heat stress. In mixed reptile collections, hygiene must be very strict in order to avoid the spread of infective cysts from asymptomatic herbivorous reptile carriers to the highly susceptible snakes and carnivorous lizards.

Hookworms. *Kalicephalus* spp. cause a wide range of signs including lethargy, regurgitation, diarrhoea (with or without bloat), anorexia and debility (Cooper, 1971). Infection is by ingestion of contaminated food and water or by a percutaneous route. In water snakes, eg. garter snakes (*Thamnophis* spp.), cutaneous migration can cause blister formation. The life-cycle is direct, with a prepatent period of 2 – 4 months. The adults (1 – 1.5cm) become embedded in the oesophageal, gastric and intestinal mucosa causing ulceration, usually with a secondary bacterial infection. The buildup of necrotic debris can cause occlusion of the oesophagus (Cooper, 1971). The larval migration phase is often associated with respiratory disease. Diagnosis is by finding a thin-shelled larvated egg in a fresh faecal sample. However, often only the free-living larvae will be seen. Worms, eggs and larvae can be found in oral and oesophageal mucus. Treatment is often unsuccessful but oxfendazole 2.265% (3ml/kg by stomach tube) can be given and repeated every two weeks until no eggs/larvae are seen. Recovery can be very protracted. There is a low host specificity and many species of snake can be infected (Frank, 1981). This is especially important where several species are kept together. All in-contact snakes should be treated. Good husbandry is very important in controlling and preventing this condition.

Ascarids. Ascarids can cause diarrhoea and other non-specific signs. Diagnosis is by finding thick-shelled embryonated eggs in the faeces. Treatment is with oxfendazole, as described previously.

Strongyles. *Rhabdias* spp. and *Strongyloides* spp. can also be associated with enteritis. *Rhabdias* spp. are primarily respiratory parasites; *Strongyloides* spp. are found in the intestinal tract. The larvated eggs of these species, found in faeces, are indistinguishable (and are similar to those of *Kalicephalus* spp.). If the larvae are cultured they may then be identified. A tracheal wash can be more useful. Eggs and larvae may be found with *Rhabdias* spp. infection but only larvae are seen with *Strongyloides* spp. infections (eggs are found only in the faeces). Both nematodes can be treated with levamisole (10mg/kg s/c), oxfendazole 2.265% (3ml/kg orally) or ivermectin (200mcg/kg s/c).

Flagellates. Flagellates can be important in snakes. Several genera have been implicated including *Monocercomonas* spp. and *Trichomonas* spp. (Keymer, 1981; Zwart *et al*, 1984). Details of diagnosis are as outlined under chelonians. As with chelonians, a few flagellates may be found in the faeces of healthy individuals. However, stressors can initiate a massive increase in numbers, causing clinical disease. Clinical signs include anorexia, lethargy, polydipsia and watery diarrhoea. The snake can become bloated with fluid accumulation in the bowel. Treatment is with a single dose of metronidazole (100 – 275mg/kg orally) maintaining the temperature at the snake's PBT (Lawton, 1991). The lower dosage is for larger snakes and the total dose should not exceed 400mg (Lawton, personal communication). Response to treatment can be rapid.

Tapeworms. Tapeworm infections are common in small fish-eating snakes, eg. garter snakes (*Thamnophis* spp.) and grass snakes (*Natrix* spp.). The genus *Ophiotaenia* has been found to affect several species in this group (Jackson and Muller, 1976). They are not usually pathogenic but heavy burdens can cause debility in smaller species. Diagnosis is by finding an egg (with an onchosphere) in the faeces. The genus *Bothridium* is found in boids (boas and pythons) and diagnosis is by finding operculated eggs in the faeces or proglottides on regurgitated food. Clinical signs are rare. However, even with low burdens the site of attachment can act as a portal for secondary bacterial infection, potentially leading to septicaemia. Treatment is with praziquantel (Droncit, Bayer) (3.5 – 7.5mg/kg s/c or 20 – 30mg/kg orally) (Lawrence, 1983, 1985a), giving the lower doses for large snakes.

Coccidia. Coccidiosis caused by *Eimeria* spp. and *Isopora* spp. has been reported occasionally in snakes (Keymer, 1981). Clinical signs are diarrhoea, loss of condition and anorexia. Diagnosis is by finding oocysts in faecal samples. Treatment is covered under lizards.

Viral diseases

Various viruses have been associated with intestinal disease (Heldstab and Bestetti, 1984), causing a variety of clinical signs including sudden death, diarrhoea, vomiting and neurological features. Parvoviruses, herpesviruses, adenoviruses and picornaviruses have been isolated, mainly from the duodenum, stomach and spleen. Little is known about the precise nature of these diseases and treatment, at present, can only be supportive.

Bacterial diseases

Salmonella spp. and *Arizona* spp. are frequently isolated from snakes - 55% in one report (Cambre *et al*, 1980) but no overt clinical disease was described. However, the implications as a potential zoonosis must be considered (Cooper, 1981). These organisms may act as opportunists, eg. following parasitic infection, stomatitis or pneumonia, with serious consequences causing septicaemia or death (Boever and Williams, 1975). Treatment to eliminate carriers is unlikely to be successful.

Other simple bacterial diarrhoeas can be treated with kaolin or a kaolin/neomycin mixture. Fluid replacement with electrolyte mixtures, such as Hartmann's solution or Lectade (SmithKline Beecham), is very effective.

Constipation

This may occur when large snakes are fed heavily furred rodents, eg. guinea pigs or long-haired hamsters, or when a snake is dehydrated. Treatment involves making sure that the snake is properly hydrated and then giving liquid paraffin by stomach tube. An enema, using warm saline, can be used. The faeces may be milked out under general anaesthesia; however, if this is not successful surgery may be required. Urate calculi in the cloaca should also be considered and eliminated as a possible cause. Cloacal prolapse can occur following prolonged straining or physical intervention.

LIZARDS

Enteritis

Coccidiosis is an important problem in lizards, particularly chameleons (Marcus, 1981), and various species of *Eimeria*, *Isopora* and other genera have been implicated (Keymer, 1981). It is usually seen if husbandry is poor or there is overcrowding. This condition should always be suspected in any sick chameleon. Clinical signs are non-specific and include anorexia, lethargy and diarrhoea or dysentery. Diagnosis is based primarily on finding oocysts in a faecal sample. Treatment is difficult but sulphadimidine (Sulphamethazine, ICI) (50mg/kg for 3 days by stomach tube) or amprolium (Amprol-Plus, MSD Agvet) (dilute 14.4ml in 100ml water, then give 1ml/kg orally) can be used.

Entamoeba invadens can be a problem in carnivorous and omnivorous lizards with similar clinical signs to those described for snakes, with the addition that colic is often seen (Keymer, 1981).

Helminth infestations have been diagnosed but no specific condition or clinical signs have been reported (Frank, 1981). The severity of signs will depend on the burden. Treatment, on finding eggs or larvae in a faecal sample, is as for snakes.

Foreign bodies

Lizards, such as iguanas and monitors, often ingest gravel and other substrates from the vivarium floor. Geckos can ingest sand, which may cause constipation and blockages. This can be demonstrated on radiographs. Unless severe, when liquid parafin can be used, this condition rarely needs treatment. These reptiles may eat gravel in a misguided attempt to obtain calcium; therefore, supplementation with calcium should also be considered in this situation. Reptiles may also swallow plastic tubing and small syringes if care is not taken when dosing orally.

Intussusception

A fatality related to intussusception in an iguana has been reported (Saik *et al*, 1987). This may have resulted from a parasitic or bacterial infection. Clinical signs were abnormal bright yellow faeces, acute depression, twitching of the entire body and a change in skin colour from green to patchy brown. Surgery could be attempted if the condition was diagnosed *ante mortem*; contrast media radiographs would be helpful.

DISEASES OF THE LIVER

There are various specific diseases affecting the liver in reptiles, although the organ is also often affected by systemic diseases, especially parasitism, where the main clinical signs may be associated with different organs, eg. entamoebiasis. The liver may also be affected by non-specific factors, eg. bad husbandry.

"Fatty liver" in chelonians

Although a certain amount of intrahepatic fat can be normal, excessive amounts can be associated with over feeding or the feeding of unsuitable foods. "Fatty liver" is an important cause of liver disease in tortoises (Jackson, 1991). The abnormal diets that can cause this condition may include dog or cat food, milk, jam, bread, potato chips or cheese. It cannot be emphasised too strongly that these foods should not be fed to tortoises. The tortoise will probably be obese - as determined from Jackson's weight for length ratio curve (Jackson, 1980; see "Examination and Diagnostic Techniques"). There are often palpable fat pads between the neck and forelimbs or cranial to the hindlimbs. The liver is damaged by fat infiltration and, if severe, this will cause anorexia, although the tortoise may still be quite active. Mild jaundice may be seen. Questioning the owner about the diet is important in the diagnosis of this condition. Blood tests are very useful for diagnosis and monitoring progression of the condition. There will be elevation of the liver enzymes and also of triglyceride and cholesterol levels (see "Appendix Three - Haematological and Biochemical Data" for normal levels). Retesting every 2 – 3 months may reveal a gradual improvement. Lawrence (1985b) reported the use of laparoscopy in the diagnosis of this condition and of radiography to show craniodorsal displacement of the stomach - a lateral view after a barium meal. The displacement is caused by the enlarged 'fatty' liver.

It takes a long time to restore normal hepatic function and therapy may require thyroid tablets (Tetroxin, Pitman-Moore) (20mcg every 2 days orally) and anabolic steroids, (eg. Nandoral, Intervet) (0.5mg every 7 days orally) for several months (Jackson, 1991). Stomach tubing will be required if anorexia lasts more than a week or so, using only liquidised green vegetable material with added vitamins and calcium, eg. Vionate, Squibb. The tortoise may have to be kept awake and metabolically active during the winter months to allow treatment to continue, or else have a much reduced period of hibernation.

Post-hibernation jaundice

A transient jaundice can be seen in some tortoises after hibernation. This often disappears rapidly once the tortoise is eating and is active again. Any persistence may indicate liver damage and should be investigated as to cause and degree of damage.

Hepatitis

Various viruses and bacteria have been implicated in hepatic damage. They include:–

1. Hepatitis associated with *Salmonella* spp. and *Aeromonas hydrophila* septicaemia in Mediterranean tortoises (*Testudo* spp.) was reported by Keymer (1978a) as a necropsy finding. General signs of septicaemia included anorexia, lethargy, enteritis, petechiation of the skin and mucosa and ecchymotic bleeding beneath the scutes (Marcus, 1971). Abscesses are a common finding in the liver at necropsy, often secondary to septicaemia or parasitic spread (Frye, 1991).

2. A herpesvirus was isolated from swollen and pale livers in two Map turtles (*Graptemys* spp.). They died a week after initial clinical signs of lethargy, anorexia and subcutaneous oedema (Jacobson *et al*, 1982). More recently, viral particles have been seen in Mediterranean tortoises (*Testudo* spp.) with liver damage in the UK (Cooper, personal communication).

3. A herpesvirus-like agent was the cause of death in two Pacific pond turtles (*Clemmys marmorata*) within 24 hours of clinical signs of anorexia, lethargy and muscle weakness. Petechial and ecchymotic haemorrhages were noticed beneath the skin of the head and neck and beneath the plastron scutes (Frye *et al*, 1977).

4. An adenovirus caused hepatic necrosis in an adult boa constrictor (*Boa constrictor constrictor*) (Jacobson *et al*, 1985b). Two days before death the snake developed a head tilt. The liver was grossly enlarged with swollen borders and pale areas scattered throughout.

5. A herpesvirus-like agent caused death in two young boa constrictors associated with anaemia, acute hepatitis and pancreatitis (Hauser *et al*, 1983).

6. An iridovirus was isolated from the liver of a Hermann's tortoise (*Testudo hermanni*). The tortoise died after two days of anorexia. There were multiple grey spots disseminated throughout the hepatic tissue (Heldstab and Bestetti, 1982).

Parasitic diseases

Several parasitic diseases, although not primarily located in the liver, can cause hepatic damage. These include:–

1. Hexamitiasis, caused by the flagellate *Hexamita parva*, although mainly a renal or enteric disease, has been associated with liver damage in chelonians (Zwart and Truyens, 1975). Hexamitae were found in the bile at necropsy and there was damage to the epithelium of the bile ducts, with degeneration and necrosis in some areas. The main clinical signs were vague with lethargy, loss of weight and failure to thrive for several weeks or months prior to death. The organism can be found in the urine and faeces of affected chelonians (see Diseases of the Intestinal Tract).

2. Entamoebiasis in snakes is associated with hepatic damage and abscess formation following migration of the amoebae to the liver from the intestinal lumen (Marcus, 1981).

3. Coccidiosis – *Eimeria bitis* and *E. cascabeli* have been recovered from the bile ducts and gall bladders of various snakes (Keymer, 1981). They do not appear to invade liver tissue, although they may cause a cholecystitis. No clinical signs are recorded.

4. Trematodes can be found frequently in liver tissue of snakes at necropsy but there is often no associated pathological change. Blockage of the main bile duct with trematodes caused a fatal jaundice in a water snake (*Erpeton tentaculatum*) (Frank, 1981).

5. Nematodes of *Capillaria* spp. have been found in hepatic tissue of lizards and snakes (Telford, 1971). Pathogenicity is assumed when the liver is heavily parasitised.

6. Cestode larvae of the genus *Mesocestoides* have been found in massive numbers in the livers of lizards and snakes and may cause a reduction in the functional capacity of the liver (Telford, 1971).

Mycoses

Fungal infections of the liver are usually chronic and granulomatous, often associated with pulmonary lesions. The subject is reviewed by Austwick and Keymer (1981) but few specific signs are recorded and the condition is usually a necropsy diagnosis (Frye, 1991).

Nutritional diseases

Calcification of liver tissue has been seen in lizards associated with imbalances of calcium metabolism (Frye, 1991). Excess vitamin D in the diet can lead to calcification of soft tissues, including the liver (see "Nutritional Diseases").

Metabolic and nutritional diseases may result in fatty change and/or glycogen depletion of the liver, other causes of fatty change are discussed earlier.

REFERENCES

ADDISON, J.B. and JACOBSON, E.R. (1974). Use of an autogenous bacterin to treat a chronic mouth infection in a reticulated python. *Journal of Zoo Animal Medicine* **5**, 10.

AUSTWICK, P.K.C. and KEYMER, I.F. (1981). Fungi and actinomycetes. In: *Diseases of the Reptilia, Vol. 1.* (Eds. J.E. Cooper and O.F. Jackson). Academic Press, London.

BARNARD, S.M. (1986). Color atlas of reptilian parasites. Part 1, protozoans. *Compendium on Continuing Education for the Practicing Veterinarian (North American Edn.)* **8**, 145.

BOEVER, W.J. and WILLIAMS, J. (1975). *Arizona* septicemia in three boa constrictors. *Veterinary Medicine/Small Animal Clinician* **70**, 1357.

BROWNSTEIN, D.G., STRANDBERG, J.D., MONTALI, R.J., BUSH, M. and FORTNER, J. (1977). Cryptosporidium in snakes with hypertrophic gastritis. *Veterinary Pathology* **14**, 606.

CAMBRE, R.C., GREEN, D.E., SMITH, E.E., MONTALI, R.J. and BUSH, M. (1980). Salmonellosis and arizonosis in the reptile collection at the National Zoological Park. *Journal of the American Veterinary Medical Association* **177**, 800.

COOPER, J.E. (1971). Disease in East African snakes associated with *Kalicephalus* worms. *Veterinary Record* **89**, 385.

COOPER, J.E. (1973). Treatment of necrotic stomatitis at the Nairobi Snake Park. *International Zoo Yearbook* **13**, 268.

COOPER, J.E. (1981). Bacteria. In: *Diseases of the Reptilia, Vol. 1.* (Eds. J.E. Cooper and O.F. Jackson). Academic Press, London.

COOPER, J.E. (1990). A veterinary approach to leeches. *Veterinary Record* **127**, 226.

COOPER, J.E., GSCHMEISSNER, S. and BONE, R.D. (1988). Herpes-like virus particles in necrotic stomatitis of tortoises. *Veterinary Record* **123**, 554.

DAVIES, P.M.C. (1981). Anatomy and physiology. In: *Diseases of the Reptilia, Vol. 1.* (Eds. J.E. Cooper and O.F. Jackson). Academic Press, London.

DONALDSON, M., HEYNEMAN, D., DEMPSTER, R. and GARCIA, L. (1975). Epizootic of fatal amebiasis among exhibited snakes. *American Journal of Veterinary Research* **36**, 807.

DRAPER, C.S., WALKER, R.D. and LAWLER, H.E. (1981). Patterns of oral bacterial infection in captive snakes. *Journal of the American Veterinary Medical Association* **179**, 1223.

DuPONTE, M.W., NAKAMURA, R.M. and CHANG, E.M.L. (1978). Activation of latent *Salmonella* and *Arizona* organisms by dehydration in red-eared terrapins. *American Journal of Veterinary Research* **39**, 529.

FOX, M.D. (1974). Recent trends in salmonellosis epidemiology. *Journal of the American Veterinary Medical Association* **165**, 990.

FRANK, W. (1981). Endoparasites and ectoparasites. In: *Diseases of the Reptilia, Vol. 1.* (Eds. J.E. Cooper and O.F. Jackson). Academic Press, London.

FRYE, F.L. (1991). *Biomedical and Surgical Aspects of Captive Reptile Husbandry.* 2nd. Edn. Krieger, Malabar.

FRYE, F.L., OSHIRO, L.S., DUTRA, F.R. and CARNEY, J.D. (1977). Herpesvirus-like infection in two Pacific pond turtles. *Journal of the American Medical Association* **171**, 882.

GREEN, C.J., COOPER, J.E. and JONES, D.M. (1977). Cryotherapy in the reptile. *Veterinary Record* **101**, 525.

HAUSER, B., METTLER, F. and RUBEL, A. (1983). Herpesvirus-like infection in two young boas. *Journal of Comparative Pathology* **93**, 515.

HELDSTAB, A. and BESTETTI, G. (1982). Spontaneous viral hepatitis in a spur-tailed Mediterranean land tortoise. *Journal of Zoo Animal Medicine* **13**, 113.

HELDSTAB, A. and BESTETTI, G. (1984). Virus associated gastrointestinal diseases in snakes. *Journal of Zoo Animal Medicine* **15**, 118.

HOLT, P.E. and COOPER, J.E. (1976). Stomatitis in the Greek tortoise. *Veterinary Record* **98**, 156.

HOLT, P.E. and LAWRENCE, K. (1982). Efficacy of fenbendazole against the nematodes of reptiles. *Veterinary Record* **110**, 302.

HOLT, P.E., COOPER, J.E. and NEEDHAM, J.R. (1979). Diseases of tortoises: a review of seventy cases. *Journal of Small Animal Practice* **20**, 269.

JACKSON, O.F. (1980). Weight and measurement data in tortoises and their relationship to health. *Journal of Small Animal Practice* **21**, 409.

JACKSON, O.F. (1991). Chelonians. In: *Manual of Exotic Pets*. New Edn. (Eds. P.H. Beynon and J.E. Cooper). BSAVA, Cheltenham.

JACKSON, O.F. and COOPER, J.E. (1981). Nutritional diseases. In: *Diseases of the Reptilia, Vol. 2*. (Eds. J.E. Cooper and O.F. Jackson). Academic Press, London.

JACKSON, O.F. and MULLER, T.A. (1976). Pathogenicity and diagnostic signs of tapeworm infestation in small snakes. *Veterinary Record* **99**, 375.

JACOBSON, E.R., GASKIN, I.E. and WAHQUIST, H. (1982). Herpesvirus-like infection in map turtles. *Journal of the American Veterinary Medical Association* **181**, 1322.

JACOBSON, E.R., GASKIN, J.M. and GARDINER, C.H. (1985a). Herpesvirus-like infection in Argentine tortoises. *Journal of the American Veterinary Medical Association* **187**, 1227.

JACOBSON, E.R., GASKIN, J.M. and GARDINER, C.H. (1985b). Adenovirus-like infection in a boa constrictor. *Journal of the American Veterinary Medical Association* **187**, 1226.

JEPHCOTT, A.E., MARTIN, D.R. and STALKER, R. (1969). Salmonella excretion by pet terrapins. *Journal of Hygiene, Cambridge* **67**, 505.

KEYMER, I.F. (1978a). Diseases of chelonians: (1) necropsy survey of tortoises. *Veterinary Record* **103**, 548.

KEYMER, I.F. (1978b). Diseases of chelonians: (2) necropsy survey of terrapins and turtles. *Veterinary Record* **103**, 577.

KEYMER, I.F. (1981). Protozoa. In: *Diseases of the Reptilia, Vol. 1*. (Eds. J.E. Cooper and O.F. Jackson). Academic Press, London.

LAWRENCE, K. (1983). Praziquantel as a taeniacide in snakes. *Veterinary Record* **113**, 200.

LAWRENCE, K. (1985a). Snakes and Addendum (Lizards and Snakes). In: *Manual of Exotic Pets*. Revised Edn. (Eds. J.E. Cooper, M.F. Hutchison, O.F. Jackson, and R.J. Maurice). BSAVA, Cheltenham.

LAWRENCE, K. (1985b). An introduction to haematology and blood chemistry of the reptilia. In: *Reptiles: Breeding, Behaviour and Veterinary Aspect*. (Eds. S. Townson and K. Lawrence). British Herpetological Society, London.

LAWRENCE, K. (1987a). Post-hibernation anorexia in captive Mediterranean tortoises. *Veterinary Record* **120**, 87.

LAWRENCE, K. (1987b). Common conditions of lizards and snakes. *In Practice* **9**, 64.

LAWRENCE, K. and HAWKEY, C. (1986). Seasonal variations in haematological data from Mediterranean tortoises in captivity. *Research in Veterinary Science* **40**, 225.

LAWRENCE, K. and JACKSON, O.F. (1982). Passage of ingesta in tortoises. *Veterinary Record* **111**, 492.

LAWRENCE, K., MUGGLETON, P.W. and NEEDHAM, J.R. (1984). Preliminary study on the use of ceftazidime in snakes. *Research in Veterinary Science* **36**, 16.

LAWTON, M.P.C. (1991). Lizards and snakes. In: *Manual of Exotic Pets*. New Edn. (Eds. P.H. Beynon and J.E. Cooper). BSAVA, Cheltenham.

MARCUS, L.C. (1971). Infectious diseases of reptiles. *Journal of the American Veterinary Medical Association* **159**, 1626.

MARCUS, L.C. (1981). *Veterinary Biology and Medicine of Captive Amphibians and Reptiles.* Lea and Febiger, Philadelphia.

MEEROVITCH, E. (1958). Some biological requirements and host-parasite relations of *Entamoeba invadens. Canadian Journal of Zoology* **36**, 513.

NEEDHAM, J.R. (1981). Microbiology and laboratory techniques. In: *Diseases of the Reptilia, Vol. 1.* (Eds. J.E Cooper and O.F. Jackson). Academic Press, London.

PHILLIPS, I.R. (1986). Reptiles encountered in practice: a survey of two hundred and forty cases. *Journal of Small Animal Practice* **27**, 807.

PRITCHARD, P.C.H. (1979). *Encylopedia of Turtles.* TFH Publications, New Jersey.

REICHENBACH-KLINKE, H. and ELKAN, E. (1965). *The Principal Diseases of Lower Vertebrates, Book 3: Diseases of Reptiles.* Academic Press, London.

SAIK, J.E., BAROW, B. and DITERS, R.W. (1987). Intestinal intussusception in an iguana. *Companion Animal Practice* **1**, 43.

SAVAGE, M. and BAKER, J.R. (1980). Incidence of *Salmonella* in recently imported tortoises. *Veterinary Record* **106**, 558.

SOULSBY, E.J.C. (1982). *Helminths, Arthropods and Protozoa.* Bailliére Tindall, London.

TEARE, J.A. and BUSH, M. (1983). Toxicity and efficacy of ivermectin in chelonians. *Journal of the American Veterinary Medical Association* **183**, 1195.

TELFORD, S.R. (1971). Parasitic diseases of reptiles. *Journal of the American Veterinary Medical Association* **159**, 1644.

WALLACH, J.D. (1969). Medical care of reptiles. *Journal of the American Veterinary Medical Association* **155**, 1017.

ZWART, P. and TRUYENS, E.H.A. (1975). Hexamitiasis in tortoises. *Veterinary Parasitology* **1**, 175.

ZWART, P. and van der GAAG, I. (1981). Atrophic gastritis in a Hermann's tortoise and two red-eared terrapins. *American Journal of Veterinary Research* **42**, 2191.

ZWART, P., TEUNIS, S.F.M. and CORNELISSEN, T.M.M. (1984). Monocercomoniasis in reptiles. *Journal of Zoo Animal Medicine* **15**, 129.

CHAPTER ELEVEN

UROGENITAL SYSTEM

Peernel Zwart DVM PhD

INTRODUCTION

The kidneys in reptiles are paired organs. The ureters open into the proctodaeum of the cloaca. In chelonians and most lizards there is a urinary bladder. Urine enters the bladder via the proctodaeum. In the urinary bladder, the urine can undergo alterations; therefore, the urine is not a straightforward indicator of the functioning of the kidneys, as it may be in mammals.

GOUT

As in all individuals of the Animal Kingdom, the urinary organs contribute to homeostasis. The renal function includes removal of the end products of nitrogen metabolism. In most reptiles the main end product is uric acid: these are referred to as uricotelic reptiles. In most aquatic reptiles the relative amount of uric acid is less (turtles and terrapins even allow themselves to produce ammonia (NH_3)) (Moyle, 1949). The blood level of uric acid varies depending on whether an animal has eaten recently. This is especially pronounced in snakes, where the level can be 5 – 10 times higher than the basic concentration of 119 – 238µmol/l uric acid over a period of 1 – 4 days after feeding.

For the main part, uric acid is excreted in the form of urates by active secretion via the proximal part of the renal tubules. Therefore, damage to the epithelium of these proximal parts leads to a diminution in the excretion of urates and a rise in the blood levels of uric acid and urates. If the blood level exceeds a certain amount (about 1,487µmol/l), recrystallisation takes place, mainly in less metabolically active parts or cells of the body, producing localised accumulations of uric acid, known as gout-tophi. These tophi are macroscopically visible as whitish foci, the size of pinheads, and consist of a centre of radiating urate crystals surrounded by macrophages and other inflammatory cells. Urates are mainly deposited in the damaged renal epithelia and the interstitial tissues of the kidneys, in the pericardial space and in the joints.

In chelonians gout may be found in 3.9% of deaths. In snakes, which are mainly uricotelic, gout accounts for 12.3% of deaths, while in lizards gout occurs in 11.7% of the deaths (Ippen and Schröder, 1977).

Gout is mainly caused by nephritis; dehydration is a minor causative factor. Too much protein in the food has never been found to be correlated with gout despite often being incriminated. In some terrapins vitamin A deficiency may contribute to gout as this affects the epithelia of the nephrons and the collecting ducts. It is well known that gentamicin is highly nephrotoxic and daily doses of 5mg/kg bodyweight over a period of 4 weeks may lead to gout (Montali et al, 1979). In the author's experience administration of 9mg/kg to members of the Pythoninae twice only, may result in severe renal damage.

Clinical signs

As in birds, gout in reptiles can be differentiated into visceral and articular forms. Clinical cases of gout may be difficult to diagnose. In chelonians there are no specific signs of the disease. In snakes clinical signs are also generally unspecific. The animals are less active or even lethargic, and patients may occasionally bite themselves at the site of the most severe urate deposits in the internal organs. In rare, though convincing cases, urate tophi may be visible under the oral mucosa as bright whitish nodules. In extraordinary cases, urate crystals may be excreted by the epithelia of the lungs and the intestinal tract or even deposited in the scales. Urates deposited in the scales must be differentiated from mites. Lizards with gout may also be lethargic. In thin-skinned species, eg. phelsumas (*Phelsuma* spp.) and geckos (*Gekko* spp.), urates can be recognised as subcutaneous whitish deposits around the joints.

Diagnosis

Diagnosis may be verified by determining the uric acid levels in the blood. Blood concentrations between 720 and 4,840µmol/l were determined in animals which died from gout (Frank, 1978). Blood samples from snakes should preferably be taken 12 days after the last meal. Values in excess of 1,500µmol/l are indicative of gout. Laparoscopic examination of the kidneys may provide direct proof of renal pathology. In this case the sex of the snake should also be determined. In male snakes, seasonally induced sexual activity coincides with hyperplasia of the sexual segments in the renal tubules. The epithelial cells of the active sexual segments are filled with a whitish secretion. These are seen as whitish coiled tubules, mainly on the surface of the kidneys, whereas urate deposits are usually in the form of irregularly distributed whitish tophi. In tortoises and lizards the kidneys cannot be found easily endoscopically.

Differential diagnosis

Pseudogout, renal hyperparathyroidism, arthritis and tuberculosis of muscles of the extremities should be considered. In pseudogout swellings around joints are caused by calcium deposits. A diagnosis can be made radiographically; the calcium is seen as radiodense deposits. In renal hyperparathyroidism due to a severe nephritis, the calcium metabolism may be severely disturbed. Calcium deposits can be observed in muscles and can be differentiated from urate deposits by their elongated appearance, coinciding with bundles of muscles. Calcium deposits in the subcutis may give rise to a firm collar around the neck, or to islands of calcium which can be identified by palpation or scratching with a scalpel. Occasionally, the skin degenerates and whitish calcium deposits are exposed and are visible to the naked eye. Arthritis is seen as a localised swelling, occasionally with radiodense exostoses or deformities. Tuberculosis of the muscles leads to circumscribed proliferative inflammatory processes which resemble profane abscesses.

Treatment

In cases of gout there is no therapy and the outcome is always fatal. Some veterinarians consider that allopurinol is useful.

RENAL DISEASES

Renal diseases in the form of glomerular, tubular and/or interstitial changes can be recognised on histology (Zwart, 1985a) but, although severe and often fatal, they cannot be diagnosed by clinical examination.

Tubulonephrosis of green (common) iguanas (*Iguana iguana*)

In green iguanas and related species, such as the black iguana (*Ctenosaura acanthura*), aged 5 – 6 years, tubulonephrosis of unknown origin is fairly frequent. At the end stage this may lead to disturbance of calcium metabolism, with metastatic calcification of the large vessels near the heart (in which case radiographic examination is a useful diagnostic tool). Calcium deposits can also occur in the mucosa of the stomach fundus.

The condition is clinically recognised because the animal refuses to eat or even vomits, while food that is forcefed is excreted in a poorly digested state. Occasionally, calcium may be deposited in the musculature of the heart, leading to congestion of peripheral blood vessels which can be seen on the sclera of the eye. Poor blood supply to the tail may lead to reduced resistance to bacterial infections, or even sterile necrosis of the tail.

As the tubulonephrosis cannot be treated, the prognosis is grave. A low calcium diet would lead to a rapid aggravation of the situation, as the patient already mobilises the calcium from the skeleton.

PARASITIC DISEASES OF THE KIDNEYS

Renal trematodiasis

This disease occurs mainly in pythons and boas. In pythons two species of trematodes are found, *Styphlodora renalis* and *S. elegans*, while in boas there is only *S. horridum* (Frank, 1985). In pythons the proximal parts of the ureters and the collecting ducts are affected. In boas, in addition to the infection of the ureters, the parasites enter deeper into the kidneys.

Clinical signs are non-specific. A possible sign is anorexia. At necropsy, the parasites can be found in the ureters where they appear as greyish interruptions in the whitish strand of urates filling the ureters. In unilateral infections atrophy of the affected kidney with compensatory hypertrophy of the other kidney may occur.

Parasites and their eggs can occasionally be found in the masses of urates produced at urination (Frank, 1985).

For therapy, praziquantel (Droncit, Bayer) (5mg/kg bodyweight s/c or orally) may be tried. Parasites which do not seem to be associated with distinct pathology or disease include larval stages of acanthocephalans and cestodes found around the kidneys of snakes.

Protozoan infections of boid snakes

The only true renal protozoon in reptiles is a coccidian parasite, *Klossiella boae*, which is found in the boa constrictor (*Boa constrictor constrictor*) (Zwart, 1964). This parasite has never been reported as causing severe infections and seems to be of no clinical importance (Keymer, 1981).

Other protozoa found in the urinary tract are secondary invaders. The amoeba, *Entamoeba invadens*, can penetrate the intestinal wall (see "Gastrointestinal System") and proceed to the adjacent kidneys, or reach these organs via the haematogenous route in the event of sepsis. In sporadic cases amoebae can ascend the ureters.

Monocercomonas sp. invades via the ureters and can cause an urethritis; it occasionally reaches the kidneys.

Hexamitiasis in chelonians

This is seen mostly in terrapins and sometimes also in tortoises. Amongst genera affected are *Clemmys*, *Cuora*, *Emys*, *Geoemyda*, *Trachemys*, *Terrapene* and *Testudo* (Keymer, 1981).

Hexamitiasis is caused by the small flagellate *Hexamita parva*. It is primarily an intestinal parasite but disease is due to an infection of the kidneys which ascends the ureters and invades the collecting ducts (Zwart and Truyens, 1975). The infection predominantly occurs in colonies of terrapins in which new animals are frequently being introduced. There is great diversity in the development of the disease. Rapidly spreading epizootics with a number of acute cases may occur. In most instances the spread and development of the disease is slow. Animals that have been kept alone for long periods may contract the disease if, for instance, they are taken to another colony during a holiday. In isolated Hermann's tortoises (*Testudo hermanni*) it may take as long as 8–10 years before a fatal nephritis develops.

Clinical signs. Clinically, the animals may show anorexia and, in very severe and acute cases, polydipsia and polyuria may be observed. The nephritis which occurs leads to a disturbance in calcium metabolism. In more prolonged cases hypertrophy of the parathyroid glands occurs, inducing metastatic calcifications of soft tissues. In such cases calcium deposits in the mucosal lining of the stomach lead to anorexia and even vomiting. Calcium deposits in the muscles may be diagnosed on radiographical examination. In groups of animals one or more cases of softening of the carapace (and plastron) may indicate the presence of underlying hexamitiasis. The urinary bladder is occasionally affected, this being direct from the proctodaeum. In such cases the turbidity and mucosity of the urine are distinct signs.

Diagnosis. Diagnosis is made by direct microscopic examination of wet preparations of urine or faeces. The parasites are small and oval shaped and can be detected at 100 times magnification: 400 times magnification is usually required for an exact diagnosis. The main characteristic of the parasites is that they are able to swim in a straight line due to their symmetric anatomy, although quick loops and curves can also be made.

Prognosis. Prognosis depends on the situation. In more acute outbreaks the prognosis is good for the majority of the animals, although individuals may die. In clinically normal infected animals the prognosis is also good. In markedly affected animals, especially when a long-standing disturbance of calcium metabolism and softening of the carapace are present, the prognosis is very poor.

Treatment. Therapy is with imidazole compounds, preferably metronidazole (Flagyl, May and Baker) (75mg/kg bodyweight daily), or with ronidazole (Duodegran, MSD Agvet) (10mg/kg bodyweight). For each of these two medicaments a complete treatment takes 10–12 days. With dimetridazole (Emtryl, Rhône Mérieux) at a daily dose of 30mg/kg bodyweight over a period of 8–10 days, there is a risk of intoxication leading to central nervous signs. The owner should be instructed to pay careful attention and to stop therapy immediately if signs of intoxication become apparent.

The medicaments are best administered by stomach tube (terrapins may occasionally regurgitate after tubing). In the case of aggressive animals, eg. larger specimens of the snapping turtle (*Chelydra serpentina*), cloacal dosing may be used.

In groups of animals the mode of therapy also depends on the environment. Terrapins, which are largely aquatic, can be bathed in medicated water. It is wise to use a smaller, separate tank; 600mg ronidazole (6 grams of the 10% formula, Duodegran, MSD Agvet), are dissolved in 10 litres of water. The water is renewed every day and medication continued for 10 days. Tortoises can be bathed daily in medicated water of double concentration over a period of 14 days.

DISEASES OF THE GENITAL ORGANS

The copulatory organs of male snakes and lizards are paired and situated ventrolaterally at the base of the tail just caudal to the vent. At rest they are retracted and a pocket-like sac exists, into which a probe can be inserted to ascertain the sex of the animal (positive in males, negative in females, except for pythons where the females have a pocket of about half the depth of that of the males). In chelonians and crocodilians the hemipenes are situated in the ventral wall of the cloaca. The hemipenes can be extruded and introduced into the female's cloaca to facilitate the transport of sperm.

PLUGGING OF THE RETRACTED AND INVERTED HEMIPENES

The sacs which are present in the base of the tail of snakes and lizards can become filled with accumulations of desquamated cornified epithelial cells. Their exact origin is not known, but such plugs disturb the normal copulatory activity. The male approaches the female, but withdraws at the moment of copulation.

The tip of these plugs can be seen by lifting the cloacal scale. In general the tip of the plug is dried and hardened. The plug can be grasped with a pair of fine forceps and pulled in a cranial direction. In some cases manual assistance, by pushing forward the caudal end of the plug, is necessary. The sac of the retracted hemipenis should then be filled with an antibiotic ointment. Complications, eg. abscessation or perforation, have not been observed to date.

PARAPHYMOSIS

Paraphymosis is seen most frequently in terrapins and tortoises but also occurs in snakes and lizards. In terrapins it is caused by trauma, eg. bites occurring in an overcrowded terrarium, or it may be the result of an inflammatory process. In snakes and lizards lesions of one of the copulatory organs are mainly due to traction during attempted copulation, when a female drags the male behind her.

In terrapins with bite wounds, the penis may be lacerated or can be completely torn apart. Secondary infections can lead to either superficial or deeper lesions of the swollen copulatory organ and, occasionally, may spread into the cloaca. In the case of more generalised superficial trauma, such as may occur in snakes and tortoises, the hemipenis can be covered with inflammatory exudate.

Therapy in uncomplicated cases is by reposition of the penis. Reduction of oedema and hyperaemia is achieved by constant pressure with cold compresses during manipulation, if necessary under anaesthesia.

In tortoises and terrapins the cloaca can be closed with a purse-string suture (see "Surgery"). In cases of minor, very fresh lesions the wounds can be sutured using 4/0 or 6/0 absorbable material, such as Dexon or Vicryl (Ethicon). Superficial infection can be treated topically with antibiotics.

In tortoises the hemipenis can be positioned on a layer of gauze, taped to the plastron, embedded in antibiotic ointment and covered with a second layer of gauze. The cloacal opening should be free so that urination and defaecation are possible.

In snakes the copulatory organ can occasionally be repositioned in the base of the tail by gentle, equally distributed pressure of long duration. The opening of the socket in the cloaca can be partially closed with a purse-string suture.

If an animal is presented with a severely traumatised or lacerated copulatory organ, it is important to remember that, in chelonians, this organ is a hemipenis and has no closed urethra. This situation facilitates amputation (see "Surgery). The organ is pulled forward until healthy tissue is visible, a ligature is placed around the organ and the inflamed part is cut away. Finally, an antibiotic ointment can be introduced into the cloaca. The prognosis is fair.

EGGBINDING (DYSTOCIA)

Eggbinding is seen relatively frequently and can occur under a number of circumstances:-

1. Lack of a suitable nesting place. A nesting place is of importance for all reptiles. Chelonians and lizards generally dig holes to deposit their eggs. For terrapins in a terrarium a suitable fairly dry sandy, land area should be provided. If this is lacking, many individuals drop their eggs in the water. Others, however, swim restlessly around with a complete clutch of eggs in their oviducts and these may become eggbound.

2. Stress caused by disturbance of the environment - for instance, when the owner digs in the sand after a female has inspected her future nesting area.

3. Competition for nesting places may lead to psychogenic (emotional) eggbinding (Zwart, 1985b). Amongst lizards living in a terrarium with restricted places suitable for the deposition of eggs, one female may prevent others from depositing eggs. Several days delay may result in eggbinding.

4. Stress caused by transport of pregnant females. However, under comparable situations, eg. a long journey, eggs have occasionally been produced.

5. Disturbances in mineral metabolism, especially of calcium. This may occur in terrapins or lizards fed on meat or insects. Under these circumstances the calcium deficiency leads to atony of the oviduct. A somewhat different process occurs in renal or nutritional hyperparathyroidism; this may lead to the deposition of calcium in the musculature of the oviduct and thus to malfunctioning.

6. Eggbinding can also coincide with, or be complicated by, infections of the oviducts. Infections generally are considered to be secondary to eggbinding and are principally bacterial. In one boa constrictor (*Boa constrictor constrictor*) the author identified *Monocercomonas* sp. in the oviduct. In this case fetal deaths occurred in the affected part but fetuses were still alive in the (cranial) healthy part.

Some abnormalities in the formation of the eggshell have been noticed, eg. focal absence of the calcified outer layer of the shell in *Epicrates* spp.; doubling of the calcified layer in only one egg of a clutch produced by Hermann's tortoises (*Testudo hermanni*); or the production of masses of calcium which simulated in size and form an egg of the Aldabran giant tortoise (*Megalochelys gigantea*).

Clinical signs

Clinical signs can be very varied. Eggbound terrapins are restless and swim around sporadically stretching their hindlimbs as they would during deposition of eggs in the nesting hole. On occasion it has been noticed that when swimming they incline to one side. After some days these clinical signs disappear; the only indication of an abnormality is that the animal is less active and is not thriving; some patients lose weight.

Lizards may be either restless or subdued. The quiet animals hide; their thermoregulation is disturbed and they fade away.

In snakes the animal may remain in a reasonable state for weeks or months after the end of the gestation period before finally becoming ill. In some cases persistent anorexia is the only sign of disease.

Diagnosis

Diagnosis of eggbinding in terrapins and tortoises can often be made by ballotment. The animal should be held in a vertical position (head up), with the index fingers on either side in the shallow excavation cranial to the hindlimbs. The eggs can be felt as smooth, firm, rounded structures resting on the fingertips.

In chelonians and Squamata, diagnosis can be confirmed by radiographical examination, computer tomography or by ultrasonography (see "Radiological and Related Investigations").

Therapy and medication

Initially, one should try to encourage the eggs to be passed naturally. As the calcium requirement of pregnant reptiles is high, a relative calcium deficiency may exist. In all reptiles treatment with injections of calcium (0.5ml/kg i/m daily for up to 5 days) has a favourable effect on the propulsive muscular activity of the oviducts. This in itself may induce labour. If, however, no success is achieved, the animal can be given oxytocin (3iu i/m). Further calcium and oxytocin injections can be given if required (see "Neurological Diseases" for further details).

Surgical treatment

If medical therapy is ineffective, surgical salpingotomy may be performed (see "Surgery"). In tortoises and terrapins a square can be cut out of the plastron, using an orthopaedic drill. The size of the square ("trap door") is based on the size of the eggs as assessed by radiographical examination. The pelvic muscles must be avoided. Orientation is also based on radiographs. Care must be taken to cut only the bony tissues because, in terrapins and tortoises, two caudocranially orientated large veins are present immediately under the plastron. Avoidance is facilitated by the use of an oscillating saw. The peritoneal membrane is incised, salpingotomy performed and the eggs taken out. In some cases the oviduct can be retracted and exposed on the plastron which makes the procedure easier. In animals which are not exclusively intended for breeding, ovariohysterectomy may be performed. The "trap door" is replaced and held in position temporarily with needles. The incision is filled with bone-wax and then covered with an epoxyresin. In terrapins, which should be put back into water, it may be wise to fix the corners of the square permanently with metal sutures, though this carries a risk of local infection. In terrapins living in an aquatic environment the plastron is carefully degreased and the area is sealed water-tight with silicone-glue (the type used to make aquaria).

In snakes, occasionally, an egg can be brought very near to or into the cloaca by gentle massage. In some cases the (soft-shelled) egg can be punctured using a needle introduced via the cloaca and emptied. The egg can than be pulled out of the cloaca but, occasionally, the procedure may be complicated by a subsequent prolapse.

If these measures fail, salpingotomy may be performed in snakes under general anaesthesia. The incision in the skin and abdominal musculature is made over the egg(s), avoiding the midline where the abdominal vena cava is situated. If the distance between eggs is extremely large, two or more incisions can be made. The oviduct is incised and the eggs are gently taken out. The oviduct is sutured using non-traumatising material. Occasionally, it helps to inject oxytocin shortly before suturing in order to let the oviduct contract and, therefore, produce a thicker layer of musculature[1]. Skin sutures are best placed between the scales. In uncomplicated cases fertility will be unimpaired. If infection is present, antibiotics can be given either systemically or as local intracloacal therapy, depending on the situation.

INFLAMMATION AND DYSFUNCTION OF THE SALPINX

On occasion proteinaceous material in the form of caseous yellowish brown plugs may be passed via the cloaca. This is indicative of an irritation or inflammation of the mucosa of the salpinx, with erroneous production of egg albumen. The protein may be manipulated by the salpinx to form egg-shaped masses. The material becomes inspissated. The plugs can be recognised on ultrasonography.

The plugs cannot be removed by injections of oxytocin (Zwart, unpublished data) and surgical removal is indicated. Long delays may result in peritonitis.

Salpingitis can be the result of an incidental bacterial infection due to *Edwardsiella tarda*, as has been recognised in alligators (Wallach *et al*, 1966), or to other bacteria, or be due to eggbinding, which may occur after transport or other factors which prevent the deposition of eggs.

---

[1] The original idea was formulated by Mr. David Jones BVetMed BSc MRCVS FIBiol, of the Zoological Society of London, to whom I am greatly indebted. It has further been used on several occasions. In the presence of eggs, the wall of the oviduct is extremely thin. Contraction of the muscular wall is both circular and longitudinal; therefore, a thicker mass is produced, which facilitates suturing of the oviduct wall.

DISTURBANCES OF THE OVARIES

A large number of small oocytes are present in the normal ovaries of Squamata. They are irregularly distributed and grouped in nests. Months or weeks before mating some of the oocytes develop into mature follicles in each ovary. When the point of maturation is reached, these egg follicles are released from the ovary into the salpinx where they are fertilised. The small oocytes in the ovary increase in size by the accumulation of yolk. The yolk is composed of specific proteins, called vitelline proteins or vitellinogens, which are produced by the liver.

Two types of vitelline can be recognised. In the early stage the follicles enlarge slowly by taking up (by pinocytosis) a colourless and homogenous type of vitelline. At a later stage the follicles enlarge rapidly and are filled with a yellowish tinged vitelline, which in Giemsa stained smears and on histology appears granulated. Follicles at the first stage are known as primary vitellinogenic follicles and those at the later stage as secondary vitellinogenic follicles.

The production of vitelline is triggered by the production of gonadotrophin-like follicle stimulating hormone (FSH) in the hypophysis. The FSH stimulates the ovary to produce oestrogens which in turn stimulate the liver to produce vitellines. The transport of vitellines is largely regulated by oestrogens, while the uptake of vitellines in the oocyte is most probably regulated by FSH.

Reptilian reproductive cycles are generally closely related to environmental factors. These factors include light intensity which influences the hypophysis and, thus, the production of FSH. High light intensities are essential for optimal reproduction. In many Squamata stimulation of the hypophysis occurs long before mating. The basis for successful reproduction is often a high light intensity in the summer preceding the spring when reproduction is to occur.

Further factors which may affect the reproductive cycle are:-

1. The animal – must be mature, in good bodily condition and healthy.

2. Space and habitat – essential for normal behaviour, as is freedom from stressors.

3. The usual seasonal variation in length of daylight. It appears that for most snakes the increase and decrease of day length of moderate climatic areas are a strong stimulant for normal sexual development and behaviour. In addition, these variations synchronise the sexual activities of both males and females.

4. Temperature – in some cases a difference of some degrees centigrade at the right moment stimulates copulation.

5. Moisture – this may also induce copulation because, in nature, rains generally precede a relative abundance of food.

6. Food availability may also stimulate sexual development and activity.

After light intensity, the most important factors are variation in length of daylight, temperature (especially high temperatures which are favourable) and availability of food (many snakes are facultative breeders and breeding may depend on the availability of food). Many species of snakes reproduce biennially. However, if sufficient food is available, they may breed every year. This phenomenon is also seen in lizards. Captive-bred, adequately fed lizards will often reproduce much earlier in captivity than in the wild. Moisture in nature is a factor mainly related to the availability of food (Duvall *et al*, 1982). Rain in itself may trigger copulation.

During the normal ovarian cycle, pregnancy and deposition of the clutch of eggs, the postovulatory follicles are transformed into corpora lutea. The granulosa cells, which originally surround the egg follicle, hypertrophy and are transformed into lutein cells. In general, the corpora lutea are more persistent in viviparous than in oviparous Squamata: they produce progesterone and other steroids. In Squamata the continuation of pregnancy is only partly influenced by the presence of progesterone. In addition, hormones from the hypophysis are of importance. At parturition the level of progesterone in the blood plasma drops abruptly, probably making the uterus susceptible to oxytocin and arginine-vasotocin. The eggs or young are then produced. Reproductive strategies differ distinctly even in related species of Squamata. It is not known what evolutionary factors have led to such differences.

Pseudopregnancy

Pseudopregnancy is known to occur in snakes, especially in the Madagascan boa (*Sanzinia madagascariensis*). Following mating there is an apparently normal start to pregnancy. After some weeks or even shortly before parturition, the female changes her behaviour to that of a non-pregnant female. *Post-mortem* examination of two reptiles which died of causes which could not influence their hormonal status, showed early signs of resorption of egg follicles; there were no eggs in the oviduct.

Abnormal functioning of the ovaries

Under abnormal environmental circumstances changes may occur in the normal cycle and in the ovaries. The author's experiences with garter snakes (*Thamnophis radix* var. *haydeni*) indicate that under a strict regimen of 12 hours light/12 hours darkness over a period of two years at constant temperatures, erratic development of follicles occurred. Small primary vitellinogenic follicles were present; some secondary follicles had developed, while at the same time others were in the process of resorption (as could be seen from their haemorrhagic appearance). In addition, the vitelline found was the pale type characteristic of primary vitellinogenic follicles, whereas, as mentioned earlier, normal well developed secondary vitellinogenic follicles in this species have a light yellowish yolk. Under these circumstances no young were bred.

Atresia or resorption of follicular cells

This abnormality occurs fairly frequently in the primary previtellogenic phase and has been studied by Betz (1963) and Guraya (1965). Recently, it has also been recognised in ovaries which showed secondary vitellinogenic follicles (Zwart *et al*, 1990). On histology, the theca is found to be infiltrated by lymphocytes and the membrane of the oocyte is degenerated. From the connective tissue surrounding the follicle, an ingrowth of blood vessels into the mass of vitelline occurs. Enormous numbers of macrophages are transported via the blood vessels and these can be seen as cuffings of the blood vessels infiltrating the vitelline. Histiocytes phagocytose the vitelline. Occasionally, calcium may be deposited in the central remnant of vitelline. Such deposits of calcium can be detected radiographically. Gradually the production of fibro-angioblastic tissue increases, while on the exterior the production of dense connective tissue starts. Finally, fibrous scar tissue results. Resorption of egg follicles can take a considerable period of time.

Clinical signs may have been seen in a Pacific boa (*Candoia carinata*). Local thickening of the body at the site of the ovaries developed over at least two years. At surgery one ovary was found to have three, and the other four, follicles in the process of resorption.

Aseptic inflammation of the ovary

Aseptic inflammation of the ovary is a distinct, separate pathological feature (Zwart *et al*, 1990). It occurs fairly frequently in Squamata, particularly in older individuals, and is often combined with the resorption of egg follicles. The fat tissue of the ovary becomes inflamed. The process probably starts with a leakage of vitelline out of the follicle into the surrounding mesenchyma. Under the influence of the highly irritating vitelline, the mesenchymal fat tissue starts to degenerate. Lakes of fat are produced, which induce a granulomatous inflammation of the mesenchyma. The fat globules are surrounded by multinucleate giant cells and connective tissue. Heterophilic granulocytes invade. The inflammation spreads progressively in the fat tissue. This may lead to impressive diffuse scarring in which egg follicles are recognised only with difficulty. Occasionally, cholesterol clefts are found in the inflamed area. A fully developed case was found in an endangered Round Island boa (*Casarea dussumieri*). The spreading inflammation is likely to make the ovary non-functional. The initiating factor in this condition may be manipulation of the animal, for instance palpation of the follicles. This may cause rupture of the membrane of an egg follicle and subsequent irritation of the surrounding tissues.

Follicular rupture

Acute rupture of all developed egg follicles has been reported in a prairie rattlesnake (*Crotalus viridis*). This animal died suddenly and exhibited severe cramps of the abdominal musculature shortly before death. At *post-mortem* examination both ovaries were found to contain a number of fully developed egg follicles. The ovaries were transformed in a mass of vitelline, in which individual egg follicles could hardly be recognised. On histology it appeared that the egg follicles had ruptured and that vitelline had already been transported into the lymph vessels surrounding the intestinal tract.

DISEASES OF THE CLOACA

Cloacal prolapse

This can occur in all reptiles and is potentially a very serious problem. It can be caused by dystocia, protracted diarrhoea or cloacal calculi. In chelonians eggbinding (and osteodystrophy) may lead to cloacal prolapse, especially if the eggs are several years old and heavily calcified.

Cloacal calculi

Cloacal calculi consist of crystallised urates and, sometimes, other salts from the kidneys and may need to be broken up and removed while the animal is under general anaesthesia. The calculi can usually be diagnosed using an auroscope and speculum or, in larger reptiles, by cloacal examination with the aid of an arthroscope. The surgical treatment is described elsewhere (see "Surgery").

Cloacitis

This is a condition which can occur in all reptiles. It is perhaps more common in males following attempts at mating. Bacterial infections can occur and an isolated case of cloacal amoebiasis has been found in a *Thamnophis* sp.

Rough use of a metal probe for sexing snakes can cause either cloacitis (Cooper, 1981) or direct trauma to the pouch of the retracted hemipenis, with consequent secondary bacterial infection.

In cloacitis a discharge may be noticed and the cloacal lips will be inflamed and swollen. The patient will often be anorexic and lethargic. Diagnosis is based on the clinical signs and can be confirmed by cloacoscopy, using an arthroscope (Coppoolse and Zwart, 1985). Fluid floating through the arthroscope cleans the lens as well as the cloacal wall. A swab can be taken for culture and sensitivity testing to ascertain the appropriate antibiotic to use. Similar organisms to those isolated from stomatitis cases can be found, eg. *Pseudomonas* spp. and *Aeromonas* spp. (Cooper, 1981), with a similar narrow range of antimicrobial sensitivity. Cleaning should be performed several times daily with dilute povidone-iodine to remove any necrotic debris. Local infusion with an antibiotic paste, eg. Framomycin Anti-Scour Paste (C-Vet), is useful. Parenteral antibiotics will be required if the infection is severe; adequate hydration is important.

If a hemipenis or penis becomes necrotic it should be surgically amputated (see "Surgery"). This does not affect the passage of urine, which is not dependent on the hemipenis or penis, because there is only a sulcus and no urethra.

Tumors

Various tumors have been reported affecting the cloacal region of snakes and have been reviewed by Jacobson (1981). Few specific clinical signs are noted. Granulomas and entamoebiosis of the cloaca are possible differential diagnoses.

REFERENCES

BETZ, A. (1963). The ovarian histology of the diamond-backed water snake, *Natrix rhombifera*, during the reproductive cycle. *Journal of Morphology* **133**, 245.

COOPER, J.E. (1981). Bacteria. In: *Diseases of the Reptilia, Vol. 1.* (Eds. J.E. Cooper and O.F. Jackson). Academic Press, London.

COPPOOLSE, K.J. and ZWART, P. (1985). Cloacoscopy in reptiles. *Veterinary Quarterly* **7**, 243.

DUVALL, D., GUILLETTE, L.J. and JONES, R.E. (1982). Environmental control of reptilian reproductive cycles. In: *Biology of the Reptilia, Vol. 13.* (Eds. C. Gans and F.H. Pough). Academic Press, London.

FRANK, W. (1978). Blutharnsäurewerte und viszerale gicht bei reptilien. *Der praktische Tierarzt* **59**, 115.

FRANK, W. (1985). Infektionen (invasionen) mit parasiten (parasitosen). In: *Heimtierkrankheiten.* (Eds. E. Isenbügel and W. Frank). Ulmer, Stuttgart.

GURAYA, S.S. (1965). A histochemical study of follicular atresia in the snake ovary. *Journal of Morphology* **135**, 151.

IPPEN, R. and SCHRÖDER, H-D. (1977). Zu den Erkrankungen der Reptilien. *Verhandlungsbericht des 19th Internationalen Symposiums über die Erkrankungen der Zoo und Wildtiere.* Akademie-Verlag, Berlin.

JACOBSON, E.R. (1981). Neoplastic diseases. In: *Diseases of the Reptilia, Vol. 2.* (Eds. J.E. Cooper and O.F. Jackson). Academic Press, London.

KEYMER, I.F. (1981). Protozoa. In: *Diseases of the Reptilia, Vol. 1.* (Eds. J.E. Cooper and O.F. Jackson). Academic Press, London.

MONTALI, R.J., BUSH, R.J. and SMELLER, J.M. (1979). The pathology of nephrotoxicity of gentamicin in snakes. *Veterinary Pathology* **16**, 108.

MOYLE, V. (1949). Nitrogenous excretion in chelonian reptiles. *Biochemical Journal* **44**, 581.

WALLACH, L.J., WHITE, F.H. and GORE, H.L. (1966). Isolation of *Edwardsiella tarda* from a sea lion and two alligators. *Journal of the American Veterinary Medical Association* **149**, 881.

ZWART, P. (1964). Intraepithelial protozoon, *Klossiella boae* n.sp., in the kidney of a boa constrictor. *Journal of Protozoology* **11**, 261.

ZWART, P. (1985a). Erkrankungen der Niere. In: *Handbuch der Zootierkrankheiten. Band I, Reptilien.* (Eds. R. Ippen, H-D. Schröder and K. Elze). Akademie-Verlag, Berlin.

ZWART, P. (1985b). Erkrankungen der weiblichen Geschlechtsorgane. In: *Handbuch der Zootierkrankheiten. Band I, Reptilien.* (Eds. R. Ippen, H-D. Schröder and K. Elze). Akademie-Verlag, Berlin.

ZWART, P. and TRUYENS, E.H.A. (1975). Hexamitiasis in tortoises. *Veterinary Parasitology* **1**, 175.

ZWART, P., COOPER, J.E. and IPPEN, R. (1990). Pathology of the ovaries of Squamata with special emphasis on vitelline-protein induced ovaritis. *Verhandlungsbericht des 32nd Internationalen Symposiums über die Erkrankungen der Zoo und Wildtiere.* Akademie-Verlag, Berlin.

CHAPTER TWELVE

NEUROLOGICAL DISEASES

Martin P C Lawton BVetMed CertVOphthal FRCVS

INTRODUCTION

When investigating reptiles with neurological problems, it is important, in addition to following the general principles of a neurological examination, to have an understanding of their dietary requirements, as nutrition may often be implicated either as the primary cause or as a contributing factor (see "Nutritional Diseases").

All reptiles are ectothermic (poikilothermic) and, therefore, their metabolism is affected and, to a large extent, regulated by the external temperature at which they are being kept; reptiles may even show different clinical signs at different temperatures. Reflexes may vary at different temperatures because of variation in the speed of conduction of impulses through the nervous system (Rosenberg, 1977). Neurological examination of a reptile which has been out of its vivarium and, therefore, not at its preferred body temperature (PBT), is not recommended.

Some reptiles will hibernate if the external temperature falls below their activity range. During hibernation their responses are diminished or become virtually absent. However, even a tortoise that is in hibernation will move a forelimb in response to gentle stroking, albeit slowly (Jackson, 1981).

An inexperienced clinician may have difficulty in deciding whether or not a reptile is alive or "brain dead". An assessment can be made by returning the reptile to its PBT and noticing whether movement and reflexes return to normal.

Neurological problems of snakes, lizards, tortoises and crocodilians will be covered in this chapter.

ANATOMY

The reptilian brain has enlarged cerebral hemispheres, optic lobes and a cerebellum. The cephalic flexure (bending of the brain) is present, as are the cranial nerves, the XIth (accessory) and the XIIth (hypoglossal) (Davies, 1981).

The reptilian spinal cord is different from mammals in that it extends to the tail tip (Davies, 1981). Unlike mammals, the spinal cord retains a considerable amount of autonomy from the brain by possessing locomotor control centres (Davies, 1981). Snakes are considered to be spinal animals (Brazenor and Kaye, 1953) responding to stimuli more by reflex than by intuition or learning, and it is for this reason that spinal fractures carry a better prognosis than in mammals (see later).

SNAKES

Neurological examination

Neurological signs seen in snakes are generally associated with a loss of righting reflex, convulsions, inability to constrict, inability to strike at prey (with resulting cachexia) or lack of muscle tone

(Lawton, 1989). The lack of limbs should in no way discourage a full neurological examination of a snake. One of the most important reflexes to test is the righting reflex. This is most easily observed by turning the snake on its back in the same way as testing the righting reflex of a mammal. Snakes are only too ready to right themselves quickly, attempting to turn first the head then the rest of the body in a rolling motion. When there is partial paralysis, eg. as might be caused by a fracture of the spine, it is possible to demonstrate the paralysis by testing the righting reflex; the snake will be able to right the cranial portion of its body fairly quickly but the body caudal to the lesion may remain inverted (see Figure 1). The cloaca may also gape in cases of total transection of the spinal cord.

Figure 1
A paralysed snake trying to right itself. The body distal to the lesion remains inverted and the cloaca is gaping.

Next, it is important to assess how the snake moves. If placed on a table a snake will move across the surface. If placed on a sheet of perspex or glass, it is possible to view the snake from the underside of the glass and see how the gastropedges grip against the surface.

It is important to handle the snake: to feel how it moves across one's hands and arms and to assess muscle tone by the ability of the snake to grip on to the arm. Captured snakes tend to demonstrate better muscular tone than captive-bred snakes.

By holding the caudal part of the body and allowing the head and neck to hang, it is possible to see if the snake is able to rise in a straight line without jerking or twisting into unnatural positions. This is particularly useful in assessing water snakes with early thiamine deficiency.

The panniculus reflex may be demonstrated in reptiles, using a hypodermic needle and carefully pricking the skin to see if there is any "twitching" in response (Griffiths, 1989).

Radiographs, both lateral and dorsoventral views, should be considered a routine part of the neurological examination of a snake.

Convulsions

Convulsions, twitching or coma are commonly seen in snakes (Cooper and Jackson, 1981; Marcus, 1981; Lawrence, 1985a). Snakes with convulsions should be investigated to rule out causes such as dietary problems, septicaemia, colic, poisoning (eg. dichlorvos), dehydration, electrolyte imbalance and hypoglycaemia (especially in cachectic cases), and spinal cord damage. Convulsions encountered with terminal septicaemia are most commonly associated with respiratory disease, especially pneumonia. Micro- or macro-abscesses may also occur in the brain as a result of septicaemic spread. Cooper (1981) reported that reptiles with septicaemia may show no premonitory signs and may be found dead; alternatively, they can exhibit clinical signs ranging from lethargy through incoordination to convulsions.

Where a specific diagnosis of the cause of the convulsions is made, appropriate treatment should be carried out. In general, treatment of a convulsing snake should include:-

1. Fluid therapy, usually best achieved by the intracoelomic (intraperitoneal) route.

2. Prophylactic antibiotics.

3. Sedatives/anaesthesia if the convulsions are severe.

4. Gentle handling.
5. Modifying the temperature.

Frye (1991) stated that one of the most common findings on histopathological examination of snakes with nervous signs is mononuclear perivascular infiltration or "cuffs" around blood vessels in the brain substance: this is particularly so in viral encephalitic conditions. In some snakes peculiar myelopathy may be seen which resembles that caused by transection of the spinal cord. In a similar case of degeneration of the spinal cord in a reticulated python (*Python reticulatus*) showing neurological signs, viral-type inclusion bodies were found (Cooper and Lawton, unpublished data). In all these cases the cranial half of the affected snakes retained normal muscular tone, while the caudal half became flaccid and unresponsive to external stimuli.

Fractures

Fractures of the spine are usually due to trauma (Frye, 1981), generally resulting in paralysis distal to the fracture site. In very severe cases a distended or gaping cloaca may be noted. The fracture may be obvious due to distortion of the body contour, or it may only be diagnosed on radiography (see Figure 2) after a neurological examination where lack of response to stimuli with needle pricks, tail pinching or pressure on the body are seen (Cooper and Jackson, 1981).

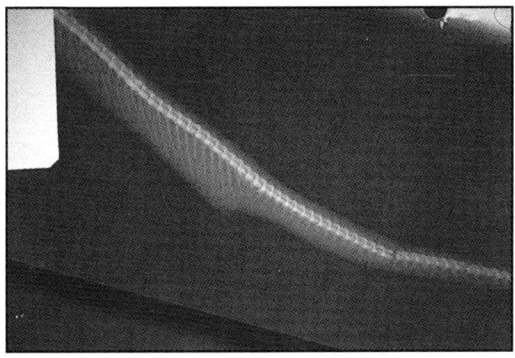

Figure 2
Radiograph of a fractured spine.

Diagnosis is ultimately by radiography. The prognosis is often more favourable than in mammals. With conservative therapy, such as force-feeding, diuretics and corticosteroids, there is a reasonable prognosis (Russo, 1985a). External casts can be used to stabilise the fracture site (Peavy, 1977; Frye, 1981). If there is severe displacement, an attempt to reduce the fracture under general anaesthesia can be made. It is advisable to use material such as Hexcelite which has the advantage over plaster of Paris of being lightweight and not likely to disintegrate or soften when the snake bathes. Hexcelite casts can be removed in order to examine the damaged area or to facilitate ecdysis, after which they can be easily replaced.

Thiamine (vitamin B_1) deficiency

As snakes are usually fed on whole prey, dietary problems are uncommon. However, fish-eating snakes, such as garter snakes (*Thamnophis* spp.) and dice snakes (*Natrix* spp.), are the exception. While in captivity these snakes tend to be fed on a variety of frozen white fish, eg. whitebait. Freezing not only reduces the amount of vitamins available but also increases thiaminase activity, which can result in a decrease of available vitamin B_1 (Frye, 1979; Lawrence, 1985a).

The presenting clinical signs depend on the severity of the deficiency which in turn is related to the snakes' diet. If the snake has been fed a frozen fish diet, supplemented with the occasional earthworm, frog, pinkie (baby hairless mouse) or mineral supplement, the presenting signs may not be as severe or as quick to develop as a snake that has been fed solely on frozen fish.

The clinical signs may vary from a slight "twitchiness" through incoordination or apparent blindness and an inability to strike correctly at prey, to torticollis (see Figure 3) or, eventually, to convulsions and death (Lawton, 1989).

Deficiency of thiamine results in cerebrocorticonecrosis (CCN) together with peripheral neuritis and, sometimes, damage to the heart resulting in cardiomyopathy.

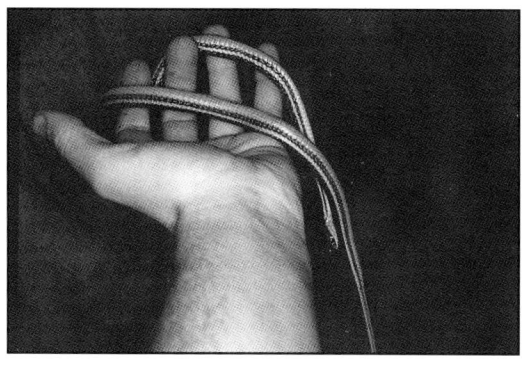

Figure 3
Torticollis in a garter snake (*Thamnophis* sp.).

Recovery in response to treatment with oral thiamine at 25mg/kg on a daily basis is dramatic. Such a response to treatment is considered diagnostic.

Prevention is by:-

1. Providing a more suitable diet.
2. Reducing the amount of thiaminase activity in the food supplied.
3. Addition of a thiamine supplement to the diet.

Garter snakes can be encouraged to eat earthworms, guppies and tadpoles as well as pinkies or cat food. If the snake is initially reluctant to eat these, one can try smearing whitebait over the pinkies or cat food and this may well entice it to eat. After a while these additional sources of food should be readily accepted.

If frozen fish is still to be offered as the main source of food, thiaminase activity may be reduced by heating the fish in boiling water for one minute in order to denature the thiaminase protein. In addition to this, it is advisable to give thiamine supplementation directly on to the fish, eg. by using a proprietary supplement.

Parasites

Parasites may cause neurological signs due to damage to the central nervous system when the snake is the intermediate host and the parasite has encysted or become encapsulated.

Frye (1991) described acanthamoebic meningoencephalitis, where snakes showed spasmodic opisthotonic contracture of the cervical musculature resulting in a "star gazing" posture. Frye (1991) considered this "star gazing" to be pathognomonic of acanthamoebiosis. However, this author has seen "star gazing" in terminal septicaemic snakes prior to death and Cooper and Jackson (1981) reported it following a blow to the head. A definitive diagnosis should, therefore, only be made if amoebae are found in cerebrospinal fluid.

Toxicity

Convulsions in snakes may be associated with organophosphorus poisoning, eg. from dichlorvos treatment for external parasites. This is particularly likely if snakes are fed in a tank where there are dichlorvos strips (Lawton, 1989). Treatment is mainly supportive and includes partial cooling of the snake to reduce the severity of the convulsions and lessen the risk of damage at this time. It is advisable to use atropine by injection, although Cooper and Jackson (1981) reported no success with this agent. However, Cooper (personal communication) has had one apparent success since. Hartmann's solution or Lectade (SmithKline Beecham) should be given by stomach tube to maintain adequate renal function and prevent dehydration. It is important that any remaining dichlorvos is removed from the tank.

Reptiles can prove sensitive to drugs, particularly antibiotics, where the half-life is affected by the ambient temperature. Toxicity due to drugs may result in incoordination and, in severe cases, convulsions or death. Acutely poisoned reptiles are usually found dead (Cooper and Jackson, 1981). Antibiotics that have been indicated as causing neurological problems are metronidazole, gentamicin, kanamycin, neomycin, streptomycin and polymyxin B (Jackson, 1976; Holt, 1981; Marcus, 1981; Lawrence, 1985a).

Aminoglycosides are particularly toxic to reptiles, especially when used in conjunction with anaesthetics. This is probably due to neuromuscular blockade. Gentamicin was found to be more nephrotoxic at higher temperatures (Hodge, 1978). At a reduced temperature and, therefore, a reduced metabolic rate, active transport of gentamicin is lowered and the intracellular concentrations are reduced. When dealing with a suspected toxicity problem, it may, therefore, be advantageous to reduce the ambient temperature in addition to giving supportive therapy and fluids. Sedatives or light anaesthesia may be used to control convulsions.

A reptile which eats a *Bufo* toad may be affected by the bufotoxin; most are presented dead. If not, treatment may be attempted using propranolol hydrochloride (5ml/kg i/v) together with oral and oesophageal lavage under anaesthesia (Cooper and Jackson, 1981).

Other substances which may affect snakes include iodine tincture, iodoform, lime, sulphur, nicotine salts (including nicotine from smokers' fingers, especially in the case of young hatchling snakes), naphthalene, paraffin, ether, chloroform, alcohol, paint solvents and lactic base (Lawrence, 1985a). Jacobson (1988) stated that wood shavings with a high resinous content, eg. cedar shavings, may cause ataxia within a week; this disappears when the snake is placed on newspaper and the wood shavings removed.

Gout

Dehydration or renal failure causing gout can result in convulsions and incoordination in snakes. This may be due to tophi forming around the heart and major blood vessels or directly in the brain. If blood chemistry indicates that the uric acid level is elevated, treatment with fluid therapy and allopurinol (20mg/kg) may be attempted. The prognosis is guarded.

Infections

Terminal septicaemia, as already mentioned, may result in convulsions. Other infections include viruses, eg. paramyxovirus or encephalomyopathy virus. Jacobson (1988) reported a viral encephalitis (boid inclusion disease) of boid snakes manifested as ataxia and head tilt. This was suspected as the cause of ataxia and flaccidity in a colony of boids in the UK (Cooper and Lawton, unpublished data). Pathological examination showed marked spinal cord degeneration. Electron microscopy demonstrated viral inclusion bodies.

Spinal abscesses

Abscesses may or may not be associated with neurological signs, depending on how deep the infection has spread. Radiography may show osteolysis of the vertebrae (see Figure 4). If more than one area is affected, secondary spread from another abscess should be suspected.

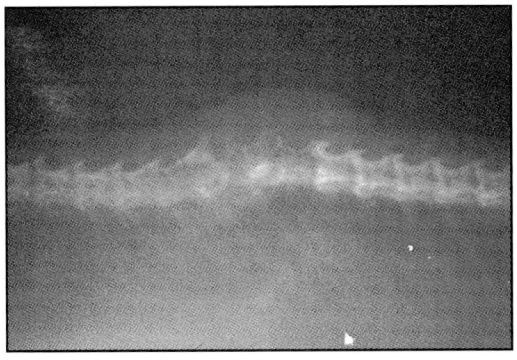

Figure 4
Osteolysis of vertebrae associated with a spinal abscess.

Under anaesthesia it is possible to open the abscess and take a swab for culture (aerobic and anaerobic) and sensitivity testing. Successful treatment with antibiotics should be followed by recalcification of the affected vertebrae.

LIZARDS

Lizards, probably more than any other reptiles, are susceptible to neurological problems associated with inadequate diets.

Calcium: phosphorus: vitamin D_3 imbalance

A lizard which is fed an insectivorous or herbivorous diet low in calcium but high in phosphorus is likely to develop nutritional osteodystrophy. The most common presenting signs in lizards are those of neurological deficit (Redisch, 1977; Russo, 1985b; Fowler, 1986; Bovin and Stauber, 1990). Severe continuous lack of calcium can result in lowered blood levels, which if less than 1.4mmol/l may cause hypocalcaemic tetany, seen initially as a bilateral or unilateral twitching of the muscles. In severe cases this can lead to paraplegia or heart failure.

Hypocalcaemic tetany responds dramatically to intravenous injections of calcium (2ml 10% calcium solution/kg).

True osteodystrophy can present as lameness due to the presence of a pathological fracture of a limb, to mechanical interference to locomotion caused by the inflammatory changes of the surrounding tissues (Lawrence, 1985b), or to collapse of the spinal cord (see Figure 5).

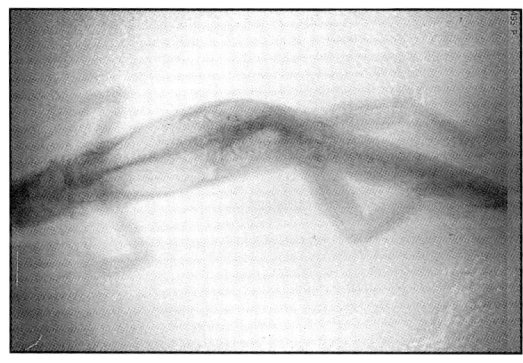

Figure 5
An iguana (*Iguana iguana*) with collapse of its spinal cord.

Nutritional osteodystrophy is generally treated by dietary supplementation and increasing the amount of calcium and vitamin D_3 in the food. The author has found baby food, eg. Milupa (fruit or vegetable varieties), together with Nutrobal (Vetark), a suitable food for lizards whose jaws are so soft that they are unable to eat more conventional food. Force-feeding affected lizards with this food and supplement eventually brings about an improvement (Lawton, unpublished data).

Fractures

Any lizard which is presented lame, especially if related to swelling in one of its limbs, should be radiographed to rule out pathological fractures associated with nutritional osteodystrophy. If necessary, blood samples for relative and absolute levels of calcium and phosphorus should be taken. Treatment of secondary pathological fractures of limbs should only be by external fixation (Lawton, 1989) as internal fixation can result in tearing or bending of the bones and failure of the implant to grip the fragments. The use of a lightweight cast made from Hexcelite, plaster of Paris, modified splints or dressings has been reported (Redisch, 1977; Frye, 1981).

Fractured spines, either simple or compression fractures, are usually osteodystrophic (Frye, 1991). The prognosis must always be guarded but, as in snakes, nursing and confinement may well result in improvement (Lawton, 1989).

Vitamin B deficiencies

B vitamin complex deficiencies can occur in lizards and cause neurological problems. Biotin deficiency may be seen, particularly in monitor lizards (*Varanus* spp.) fed raw eggs (Frye, 1979; Frye, 1991), as eggs contain an antivitamine (avidin) which can induce this deficiency. The presenting clinical sign in lizards is general muscle weakness and, as in thiamine deficiency in snakes, the condition is usually diagnosed on the response to supplementation with vitamin B complexes, especially biotin. Lawrence (1985b) advised routine use of B vitamin complexes in cases where no other cause of paresis could be found.

CHELONIANS

Neurological problems in chelonians include unilateral or bilateral paresis, circling, inability to hold up the head and overextension of the head, together with an inability to feed properly (Lawton, 1989).

Chelonians placed upside down right themselves by kicking and neck movements. This proves a useful aid for testing both the righting reflex and movement of the appendages.

A normal terrestrial chelonian should be able to support itself with the plastron clear of the ground. When retracted into the shell, the limbs and head should demonstrate good muscle tone and not be flaccid or easy to withdraw.

Paresis

Tortoises, especially females, presented with hindlimb paralysis should always be radiographed. Occasionally, the paralysis may be due to trauma to the limb, eg. an overexuberant male biting the limb and causing restrictive movement. Rubbing on the limb may also cause signs of paresis. Compression damage of the carapace may result in damage to the underlying spinal cord and subsequent paresis. Such injuries, caused by lawnmowers, have been seen in the American box-tortoise (*Terrapene carolina*) (Lawton, unpublished data).

Hindlimb paresis can be associated with pressure in the abdomen due to eggbinding and is diagnosed on radiography. These tortoises may show excessive wearing of the hindlimb claws and, occasionally, the plastron, indicative that they have been digging to try and lay eggs. Others may show only anorexia and an inability or unwillingness to move.

Treatment is by injection with a 10% calcium solution (0.5ml/kg i/m daily for three days), increasing the ambient temperature and giving warm baths. On the fourth day oxytocin (up to a maximum of 3iu i/m) is given, continuing with the increased temperature and baths. If this is not successful in removing the number of eggs that have been counted on the radiograph, after a days rest the procedure may be repeated but with several days of calcium injection before more oxytocin (up to 6iu) is given. If this is unsuccessful coeliotomy should be considered (Millichamp *et al*, 1983).

Freeze damage

A tortoise which is walking around in circles may have neurological damage associated with freezing episodes suffered during hibernation. Such signs may be due to toxaemia, especially associated with liver damage, eg. in an obese tortoise (fatty liver syndrome), or secondary bacterial spread and micro-abscesses. Damage to the central nervous system often results in the head being held over to one side (see Figure 6); this may become permanent and lead to walking in circles. Treatment is mainly supportive and the prognosis is guarded.

Where brain damage has occurred, diagnostic tests are of limited value. There may be differences in the pupillary light reflex, but these are unreliable (Cooper and Jackson, 1981). Reduced sensitivity to stimuli may be detectable or, conversely, the reptile may exhibit excitability or hypersensitivity. Administration of thiamine may be useful.

Blindness is frequently noted after freezing episodes and may be associated with hyphaema, unilateral or bilateral vitreal haze, bilateral lenticular opacities, or true retinal damage (Lawton and Stoakes, 1989). Retinal damage is observed on ophthalmoscopic examination as a general lack of reflectivity and normal structure of the retina, which instead of being the normal bright red, green and brown colour, tends to be a dull grey with a lack of cellular structure. Tortoises affected by freezing episodes may show various signs, partly depending on the changes that have occurred in their eyes; all are likely to be presented as neurological problems. A blind tortoise will not feed by itself and will be reluctant to move; it will walk in circles or hold its head abnormally high.

The retinal damage responds well to supplementation with high doses of vitamin A orally, but it may take up to 18 months for sight to return. Even mild lenticular opacities may disappear with supportive therapy over a long period of time. In the meantime it is essential for the tortoise to be handfed until it is able to eat. These tortoises show a lack of menace response. They are found to be on or above the mean of the Jackson's ratio (see "Examination and Diagnostic Techniques"), thus distinguishing them from tortoises which have post-hibernation anorexia.

Overextension of the head backwards, lack of movement or gasping is often associated with septicaemia or respiratory problems and these should be considered in the differential diagnosis.

Figure 6
The head being held over to one side in a *Testudo graeca*.

Fractures

Fractured limbs occasionally occur in tortoises. Treatment is by replacing the limb within the shell and using epoxyresin to immobilise it for a period of three months. This generally results in healing.

Osteolysis

An inability to use a limb may be associated with infection. Radiography may show osteolysis of the limb bones. Successful treatment with an appropriate antimicrobial agent, based on bacterial culture and sensitivity tests, should bring about improvement and normal use of the limb followed by remineralisation. If antibiotic treatment is unsuccessful, the only feasible treatment may be amputation in order to prevent secondary septicaemic spread.

Toxicity

Tortoises are reported to have a marked tolerance to poisonous plants (Frye, 1981), although acute poisoning and death due to oxalate toxicity may result from ingestion of large amounts of rhubarb.

Marcus (1981) reported that paint on turtles' shells is potentially dangerous and he advised that paint or paint solvents, which may also prove toxic, should never be used.

Thiamine deficiency

Thiamine deficiency may occur in aquatic chelonians which are fed on frozen white fish, similar to that described earlier under snakes. Frye (1991) stated that the most striking clinical sign is sinking of the eye into the orbit; this can, however, occur with weight loss due to other diseases, especially renal or liver failure.

CROCODILIANS

Osteodystrophy associated with incorrect diet may lead to limb deformities, muscular weakness and locomotory difficulties (Kuehn, 1974).

Thiamine deficiency may cause neurological signs similar to those already described (see Figure 7).

Figure 7
A caiman (*Caiman crocodilus*) with thiamine deficiency showing neurological signs.

Hypoglycaemia has been reported as a cause of neurological problems in crocodilians (Wallach, 1971; Frye, 1979; Frye, 1986). This is a stress-induced hypoglycaemia which results in muscle tremors, loss of righting reflex, mydriasis and reduction of the metabolic rate. Treatment is with glucose (3g/kg orally) and removal of the stressor (Frye, 1986).

REFERENCES

BOVIN, G.B. and STAUBER, E. (1990). Nutritional osteodystrophy in iguanas. *Canine Practice* **15 (1)**, 37.

BRAZENOR, C.W. and KAYE, G. (1953). Anesthesia for reptiles. *Copeia* **3**, 165.

COOPER, J.E. (1981). Bacteria. In: *Diseases of the Reptilia, Vol. 1.* (Eds. J.E. Cooper and O.F. Jackson). Academic Press, London.

COOPER, J.E. and JACKSON, O.F. (1981). Miscellaneous diseases. In: *Diseases of the Reptilia, Vol. 2.* (Eds. J.E. Cooper and O.F. Jackson). Academic Press, London.

DAVIES, P.M.C. (1981). Anatomy and physiology. In: *Diseases of the Reptilia, Vol. 1.* (Eds. J.E. Cooper and O.F. Jackson). Academic Press, London.

FOWLER, M.E. (1986). Metabolic bone disease. In: *Zoo and Wild Animal Medicine.* 2nd Edn. (Ed. M.E. Fowler). W.B. Saunders, Philadelphia.

FRYE, F.L. (1979). Reptile medicine and husbandry. *The Veterinary Clinics of North America: Symposium on Non-domestic Pet Medicine.* (Ed. W.J. Boever). **9 (3)**, 415.

FRYE, F.L. (1981). Traumatic and physical diseases. In: *Diseases of the Reptilia, Vol. 2.* (Eds. J.E. Cooper and O.F. Jackson). Academic Press, London.

FRYE, F.L. (1986). Feeding and nutritional diseases. In: *Zoo and Wild Animal Medicine.* 2nd Edn. (Ed. M.E. Fowler). W.B. Saunders, Philadelphia.

FRYE, F.L. (1991). *Biomedical and Surgical Aspects of Captive Reptile Husbandry.* 2nd Edn. Krieger, Malabar.

GRIFFITHS, I.R. (1989). Neurological examination of the limbs and body. In: *Manual of Small Animal Neurology.* (Ed. S.J. Wheeler). BSAVA, Cheltenham.

HODGE, M.K. (1978). The effect of acclimation temperature on gentamicin nephrotoxicity in the Florida broad-banded water snake (*Natrix fasciata*). In: *American Society of Zoo Veterinarians Annual Proceedings.*

HOLT, P.E. (1981). Drugs and dosages. In: *Diseases of the Reptilia, Vol. 2.* (Eds. J.E. Cooper and O.F. Jackson). Academic Press, London.

JACKSON, O.F. (1976). *Manual of Care and Treatment of Childrens and Exotic Pets.* (Ed. A.F. Cowie). BSAVA, Cheltenham.

JACKSON, O.F. (1981). Clinical aspects of diagnosis and treatment. In: *Diseases of the Reptilia, Vol. 2.* (Eds. J.E. Cooper and O.F. Jackson). Academic Press, London.

JACOBSON, E.R. (1988). The evaluation of the reptile patient. In: *Exotic Animals.* (Eds. E.R. Jacobson and G.V. Kollias). Churchill Livingstone, New York.

KUEHN, G. (1974). Crocodilian nutritional deficiencies. *Journal of Zoo Animal Medicine* **5**, 25.

LAWRENCE, K. (1985a). Snakes. In: *Manual of Exotic Pets.* Revised Edn. (Eds. J.E. Cooper, M.F. Hutchison, O.F. Jackson and R.J. Maurice). BSAVA, Cheltenham.

LAWRENCE, K. (1985b). Lizards. In: *Manual of Exotic Pets.* Revised Edn. (Eds. J.E. Cooper, M.F. Hutchison, O.F. Jackson and R.J. Maurice). BSAVA, Cheltenham.

LAWTON, M.P.C. (1989). Neurological problems of exotic species. In: *Manual of Small Animal Neurology.* (Ed. S.J. Wheeler). BSAVA, Cheltenham.

LAWTON, M.P.C. and STOAKES, L.C. (1989). Post-hibernation blindness in tortoises (*Testudo* spp.). In: *Third International Colloquium on Pathology of Reptiles and Amphibians.* (Ed. E.R. Jacobson). Orlando, Florida.

MARCUS, L.C. (1981). *Veterinary Biology and Medicine of Captive Amphibians and Reptiles.* Lea and Febiger, Philadelphia.

MILLICHAMP, N.J., LAWRENCE, K., JACOBSON, E.R., JACKSON, O.F. and BELL, D.A. (1983). Egg retention in snakes. *Journal of the American Veterinary Medical Association* **183 (11)**, 1213.

PEAVY, G.M. (1977). A non-surgical technique for stabilising multiple spinal fractures in a gopher snake. *Veterinary Medicine/Small Animal Clinician* **72**, 1055.

REDISCH, R.I. (1977). Management of leg fractures in the iguana. *Veterinary Medicine/Small Animal Clinician* **72**, 1487.

ROSENBERG, M.E. (1977). Temperature and nervous conduction in the tortoise. In: *The Proceedings of the Physiological Society* **April 1977.**

RUSSO, E.A. (1985a). Anorexia and spinal fracture in a boa constrictor. *Avian and Exotic Practice* **2 (3)**, 7.

RUSSO, E.A. (1985b). Nutritional osteodystrophy in an iguana. *Avian and Exotic Practice* **2 (2)**, 14.

WALLACH, J.D. (1971). Environmental and nutritional diseases of captive reptiles. *Journal of the American Veterinary Medical Association* **159 (11)**, 1632.

CHAPTER THIRTEEN

NUTRITIONAL DISEASES

Peter W Scott MSc BVSc MRCVS MIBiol ACIArb

INTRODUCTION

Various factors are involved in nutritional disease; they can usually be broadly categorised under:-

- anorexia/cachexia
- inappropriate temperature and conditions
- inappropriate food
- imbalanced food
- overfeeding.

Because of the basic unfamiliarity of reptiles it is better to begin with normality and consider how captivity changes the normal conditions and, in some cases, leads to nutritional disease.

Perhaps the most common problem in reptiles is 'straightforward' anorexia. This will often arise where owners are unaware of the normal types of food required and the conditions under which the reptile is likely to accept the food. The majority of reptiles are at least partially carnivorous, more so as juveniles. A few are specialised herbivores as adults, eg. some tortoises and the iguanas.

The anatomy of the gastrointestinal tract is relatively simple; snakes have a long fusiform stomach whilst the stomach of tortoises and lizards is more sac-like. Crocodilians have a two-chambered system similar to that of birds but reversed, ie. the gizzard (often containing stones) comes before the digestive section, no doubt relating to the type of food consumed. An enlarged modified caecum is found in chelonians, crocodilians and some lizards.

One needs to check that owners are keeping their reptiles within their preferred optimum temperature zone (POTZ). The POTZ is important as this helps the reptile maintain its preferred body temperature (PBT). Reptile enzyme systems also depend on the POTZ for normal function. For the majority of reptiles, 27° – 30°C is within this range but this is not true for all species. Many breeders simulate nature by inducing a winter by allowing an environmental temperature drop of 3° – 4°C. This, however, is for the specialist rather than the beginner who may well already be having problems feeding his/her reptiles.

Reptiles are ectothermic; they use external heat sources to control their own physiology. Despite this they are surprisingly sensitive to temperature; the desert dwelling earless lizard *(Holbrookia maculata)* is reported to be capable of maintaining its body temperature within a 3°C range all day by controlled basking (Wallach, 1971). The temperature of the food is very important; some reptiles can detect local variations of less than 0.1°C. When providing heat, it is important to consider whether the reptile is a thigmotherm, such as a python, requiring contact heat, or is a heliotherm, such as a chameleon, requiring a heat source under which it can bask.

Energy metabolism is extremely important. Two types of storage exist: as glycogen in the liver or as fat in fat bodies. It is these stores which are used during winter dormancy. The young of all species have problems accumulating stores since their needs for growth are high; it is these animals which have problems when cold periods are extended.

Light intensity levels are also important in feeding; simulation of a natural change in photoperiod is often needed to trigger normal seasonal activities. The standard incandescent light bulbs used as focal heat sources do not provide a full natural spectrum of light and the addition of full spectrum fluorescent tubes (Triton, Thorn) may stimulate poor feeders.

BACKGROUND

The detailed nutritional requirements of reptiles have not been well quantified but a few factors have been clarified. Parallels from other species are valid and useful in the absence of any hard and fast data.

Relatively few nutritional studies have been conducted on reptiles, although there is evidence that two species of garter snakes *(Thamnophis radix* and *T. sirtalis sirtalis)* synthesise their own vitamin C in the kidney and that this supply may be adequate even under mild stress (Vosburgh et al, 1982). Any disorder compromising kidney function may reduce this ability. Despite these data, hypovitaminosis C has been considered as being involved in the pathogenesis of ulcerative stomatitis. Certainly, vitamin C should be part of the treatment of this condition. Another study looked at the niacin requirements of the bull snake *(Pituophis melanoleucus sayi)* and its influence on susceptibility to necrotic enteritis (Bartkiewicz et al, 1982). This study was inconclusive, as no deficiency signs developed in snakes fed a totally niacin deficient diet for 132 days.

Some work on wild green turtles (*Chelonia mydas)* indicated a natural diet of marine angiosperms, primarily *Caulerpa* spp., and algae, plus occasional seagrass and jelly fish (Garnett and Murray, 1981). The *Caulerpa* spp., when analysed, had a crude protein level of less than 5% dry matter, up to 60% ash and approximately 20% insoluble carbohydrates. Seagrasses are similar in composition. These animals may derive 15% of their energy intake from cellulose breakdown and fermentation in their pseudocaecum. This situation is not dissimilar from the terrestrial situation and it is difficult to escape the conclusion that most of the nutritional problems which occur do so because, in captivity, diets are 'too good', encouraging too rapid growth which in turn reveals any inadequacies.

Garter snakes *(Thamnophis* spp.) and other semi-aquatic species may require vitamin B_1 (thiamine) supplementation when fed on fish, as many marine fish contain thiaminase (see "Neurological Diseases"). Vitamin E deficiencies may also affect garter snakes due to poor storage or rancidity of the fish.

Extra supplementation with other vitamins may be necessary at times of stress. This concept has developed from work which suggests a protective role for vitamin C in maintaining healthy mucous membranes. Vitamins A and E have both been used to maximise the immune response in the face of disease. The converse situation, ie. the role of sub-clinical deficiency, is still unclear. High level supplementation with vitamin A may improve or even promote ecdysis (sloughing) and so improve the rate of healing of skin lesions.

DRINKING

Some reptiles drink readily and others not at all. Some, eg. the Australian moloch or thorny lizard (*Moloch horridus*), obtain water by capillary conduction from the surface of the skin to the corners of the mouth. *Anolis* spp. and the Chamaeleonidae lick dew: therefore, it is necessary to spray their vivarium twice daily. Others, eg. spiny lizards (*Sceloporus cyanogenys*), are said only to 'register' available water by its movement, such as dripping or ripples.

FEEDING STRATEGIES AND PRACTICAL FEEDING

Feeding strategies are very important. Reptiles have all evolved a 'normal' method of catching prey which, in many cases, is such an important behavioural activity that without the appropriate foreplay they will not eat. The social life of reptiles is also influenced by feeding behaviour; easy prey availability can cause changes in normal behaviour. An important aspect of this is the type of energy budget which supports the feeding activity.

The major methods of food acquisition are:-

The majority of small lizards fall into the first group, ie. eurythermic, feeding over a wide temperature range, while larger lizards and most snakes fall into the last group, ie. stenothermic, feeding over a narrow temperature band (for the species).

The large constrictors are thigmotactic (requiring contact heat sources). They prefer to be in a 'tight spot', often requiring to be fed in the dark in a small box where they can constrict the prey, feel secure while they swallow it and sit relatively immobile. Also, in captivity a white laboratory rodent may not be recognised as food; in such cases gerbils are often taken with no problem.

Frozen mixed vegetables are a suitable source of nutrients for iguanas and tortoises. Trout pellets (lower protein maintenance pellets – 40% protein, <10% fat) have been used successfully as a staple for terrapins and as a part of the diet for larger chelonians. Moistened trout pellets top dressed with supplements are also useful for feeding to crickets and mealworms prior to their use as live food.

Table 1 summarises some of the preferred foods of Families of reptiles which may be encountered. This should, however, only be regarded as a starting point before more specialised information is sought.

Table 1
Preferred foods of some Families of reptiles.

Group	Families	Preferred foods
Snakes	boas, pythons, rat snakes, gopher snakes, bull snakes and vipers	warm blooded prey, eg. rodents and birds
	garter and water snakes	fish, frogs, earthworms and slugs
	indigo and king snakes	endothermic or ectothermic prey
	ring-neck and brown snakes	salamanders, earthworms small snakes and lizards
	egg-eating snakes	eggs
	racers and vine snakes	lizards (need not be live)
	king cobras	snakes (need not be live)

Group	Families	Preferred foods
Lizards	horned lizards	prefer ants, some will take small crickets and/or mealworms
	night and 'worm' lizards	termites and/or ants eggs
	green iguanas	frozen mixed vegetables, dandelions, crickets, pinkies, eggs and small amounts of dog food
	tegus and heloderms	raw eggs, chopped lean meat and pinkies
	monitors	as tegus plus larger rodents, birds and freshwater fish
	fence lizards, skinks alligator lizards, anoles and chameleons	appropriate sized insects, fruit flies, through to large crickets
	specialised species, eg. marine iguanas caiman lizards	marine algae or kelp, shelled molluscs
Chelonians	turtles and terrapins	earthworms, small whole fish, pinkies and green leafy vegetation (especially aquatic)
	tortoises	flowers, succulents, grass, cucumber, frozen mixed vegetables, fresh fruit and small amounts of dog food

based on Frye (1991).

The following is a guide to the frequency of feeding reptiles:-

smaller snakes and lizards — once or twice a week

young of large pythons and boas — 3 times weekly (larger specimens may fast for weeks; some specimens which are particularly inactive may wait even longer before taking food)

iguanas and aquatic chelonians — 2 – 3 times a week

The feeding behaviour of temperate chelonians is in most cases aimed at a rapid period of feeding during the warmest part of the day when they are most active; their beak simply cuts food and there is no chewing. The large stomach permits this 'gathering' and the long gut permits bacterial action on cellulose. Gut passage and enzyme activity are temperature-dependent and so basking is important for digestion.

Perceived wisdom cautions against the use of sand to avoid impactions, yet Aldabran tortoises *(Megalochelys gigantea)* have been seen to eat sand and small stones. The various island chelonians are reported to eat anything which stands still. European species are herbivores with omnivorous tendencies, eating a little animal protein when this is available in the wild. In captivity it is often too available and causes problems (see later). Box-tortoises (*Terrapene* spp.), especially young, and terrapins are primarily insectivorous and carnivorous, eating insects, snails, fish etc. Movement is a stimulus to chase and capture food; others, such as the snapping turtle (*Chelydra serpentina*) and the matamata (*Chelus fimbriatus*), employ a "sit and wait" policy.

Feeding tortoises (particularly *Testudo* spp.)

A balanced diet can be provided as follows:-

1. Most household fruits and vegetables provide the basis for the diet. Root vegetables are best grated; rhubarb and spinach should not be fed.

2. Grasses and weeds are popular and cheap; they will even be eaten when partially sun dried.

3. Animal protein is important, eg. tinned dog and cat food, sprats, shrimps and snails, eggs, liver and cheese. A small amount twice weekly is probably sufficient for adults, although young tortoises seem to benefit when they receive it on alternate feeding days.

4. Sprouted seeds are useful as supplements; on a dry matter basis most have a protein level of 8 – 12%. Lentils are as high as 25%; although a useful source of some minerals, they are low in calcium.

5. Calcium is needed, eg. cuttlebone, crushed egg shell or balanced supplement. The calcium content of many vegetables can be increased by liming the vegetable patch prior to and during growth.

Use of supplements

With all reptiles the use of appropriate supplements is important. Cultured species of insects, rodents or birds may not contain the vitamins and minerals of their wild equivalent; they are fed 'pelleted complete diets', are often starved before shipping or killing, and then often stored by freezing. Ross and Marzec (1990) disputed this view, at least regarding the constrictors, which, they reported, may be fed successfully on laboratory animals, although to cover all eventualities they recommended varying food items. Cooper (personal communication) also disputed this viewpoint. Certainly, examples can be sited to support either view; problems undoubtedly occur with crocodilians which grow rapidly on cultured foods. One should consider that 'a rat may not necessarily be a rat' from a nutritional standpoint, and that a day-old chick was an egg yesterday (with much of its calcium left behind in the shell when it became a chick). Cultured food items are generally preferred to avoid disease transmission.

The calcium/phosphorus ratio of mealworms *(Tenebrio molitor)* fed on bran is poor (in the region of 1:22); the phytates in bran and their effect on calcium are well reported in standard nutritional textbooks.

Mealworms normally have a very poor calcium/phosphorus ratio of 1:3 or even 1:14 (see Table 3). Supplementing their diet can make them a better balanced food source. The feeding to mealworms of a vitamin/mineral supplement (as the sole food for the last 24 hours) with acceptable palatability and a calcium/phosphorus ratio of greater than 20:1, can be expected to produce a ratio of 1.2 – 1.5:1. Nutrobal (Vetark) has a ratio of 46:1.

Crickets can be supplemented by putting a few in a plastic bag, adding the supplement as a dusting powder then lightly shaking the bag. Putting the bag in a refrigerator for about 10 – 15 minutes before a final shake will slow the crickets down and allow the reptiles to capture their prey before the supplement falls off.

Calcium lactate can be used to supplement drinking water at a rate of 1 teaspoonful (approx 3.5g) per litre when only a calcium supplement is required. In addition, cuttlefish bone is acceptable to many species, especially chelonians.

Supplementation of aquatic reptiles is more difficult. Animals can be trained to feed in a separate container where they will take their food more quickly. Vitamin/mineral mixes can be added to home-made foods held together with gelatin or agar, or placed inside 'nuggets' of food.

Basic information on supplements

Table 2
Veterinary supplements commonly used with reptiles.

	Vionate (Ciba Geigy) /g	**Collo-Cal D (C-Vet) /ml**	**Nutrobal (Vetark) /g**	**ACE-High (Vetark) /g**
Vitamin A (IU)	220	–	500	2530
Vitamin C (mg)	–	–	2.5	250
Vitamin E (IU)	0.12	–	20	122
Vitamin D_3 (IU)	22	70	150	20
Calcium (mg)	94.5	0.5	208	9.9
Phosphorus (mg)	63.6	–	4.5	4.9
Ca/P	1.48:1	no P	46:1	2:1

The recommended dose of Vionate is 1/8th level teaspoonful per 0.5kg bodyweight (approximately 0.5g/kg); for Nutrobal and ACE-High the recommended doses are 0.1g/kg/day.

Human preparations

1. Calcium lactate tablets = 300mg (1mmol Ca^{2+}).
2. Calcium gluconate tablets = 600mg (1.35 mmol Ca^{2+}).
3. Dicalcium phosphate = $CaHPO_4.2H_2O$; 23% calcium and 16% phosphorus is regarded as fairly inert and not a good source of calcium. It is used more as an excipient and antacid.
4. Human 'food products' in the UK (and possibly elsewhere in the EC) are labelled as containing all of their constituent vitamin D as 'mg cholecalciferol', ie. as vitamin D_3, as a standard unit rather than the IU of whichever form was present. Many products contain D_2 rather than D_3 which may not be effective in a number of exotic species, such as reptiles.
5. Cod-liver oil can be a good source of D_3 (210µg/100ml, ie. 8,400 IU/100ml) and retinol (18,000µg/100ml). It is, however, variable and potentially affected by rancidity.

SPECIFIC NUTRITIONAL PROBLEMS

The majority of nutritional problems become apparent during growth. The growth rate of reptiles is basically protein-dependent; a high protein level encourages fast growth. Unless the energy level is sufficient the protein may be utilised as an energy source and problems due to high blood levels of uric acid and renal deposition of urates may be seen. In addition to a correct protein/energy ratio, the vitamin and mineral levels need to be correct for the particular rate of growth. Little is known as yet of specific protein/energy ratios. It is very likely that these will interact with temperature requirements producing a very complex pattern.

Disorders of calcium metabolism probably comprise the single most important area of reptile nutrition and probably of captive reptile medicine as a whole. As in other areas of animal husbandry, deficiencies are particularly common where intensification occurs and where young animals are being reared, eg. as in crocodile farming.

Post hibernational anorexia (PHA) in Mediterranean tortoises

PHA has been investigated and reported by Lawrence (1987). The commonest presenting situation was an anorexic tortoise often with a low 'Jackson ratio'; this may also have been accompanied by various secondary problems such as mouth-rot or "runny nose syndrome". Blood urea levels were elevated (in severe cases over 150mmol/l), as were plasma proteins (over 100g/l), and the blood glucose level was depressed (below 1mmol/l). Signs of dehydration were usual. The tortoises did not eat until treatment had raised the blood glucose levels to 3.2mmol/l and lowered the blood urea levels substantially. Normal tortoises show a dramatic rise in blood glucose levels on waking from hibernation (to 20mmol/l). Tortoises with PHA appear not to show this rise which is thought to trigger (and presumably fuel) initial foraging and eating behaviour. In PHA catabolism begins, increasing plasma proteins and waste urea and uric acid.

Lawrence (1987) suggested that PHA cases not presented to a veterinary surgeon for appropriate treatment within seven weeks of waking, or those with a blood urea level of more than 200mmol/l, should carry a guarded prognosis.

Anorexia/cachexia in snakes

This is a common problem in constricting snakes, such as royal pythons *(Python regius)* and emerald tree boas *(Corallus caninus)*. In adults 'normal' fasting periods can be very long; gravid pythons may fast for 6 – 8 weeks and gravid boas for up to 9 months (Ross and Marzec, 1990). In general, with growing constrictors a three month fast is sufficient reason for veterinary intervention. The metabolic consequences are shown in Figure 1. Investigation involves reviewing the husbandry system, checking the temperature, lighting type and pattern, and feeding history, and noting whether or not hide boxes are provided for the more shy individuals. Without correction of the initial cause there is little point in rehydration and nutritional support.

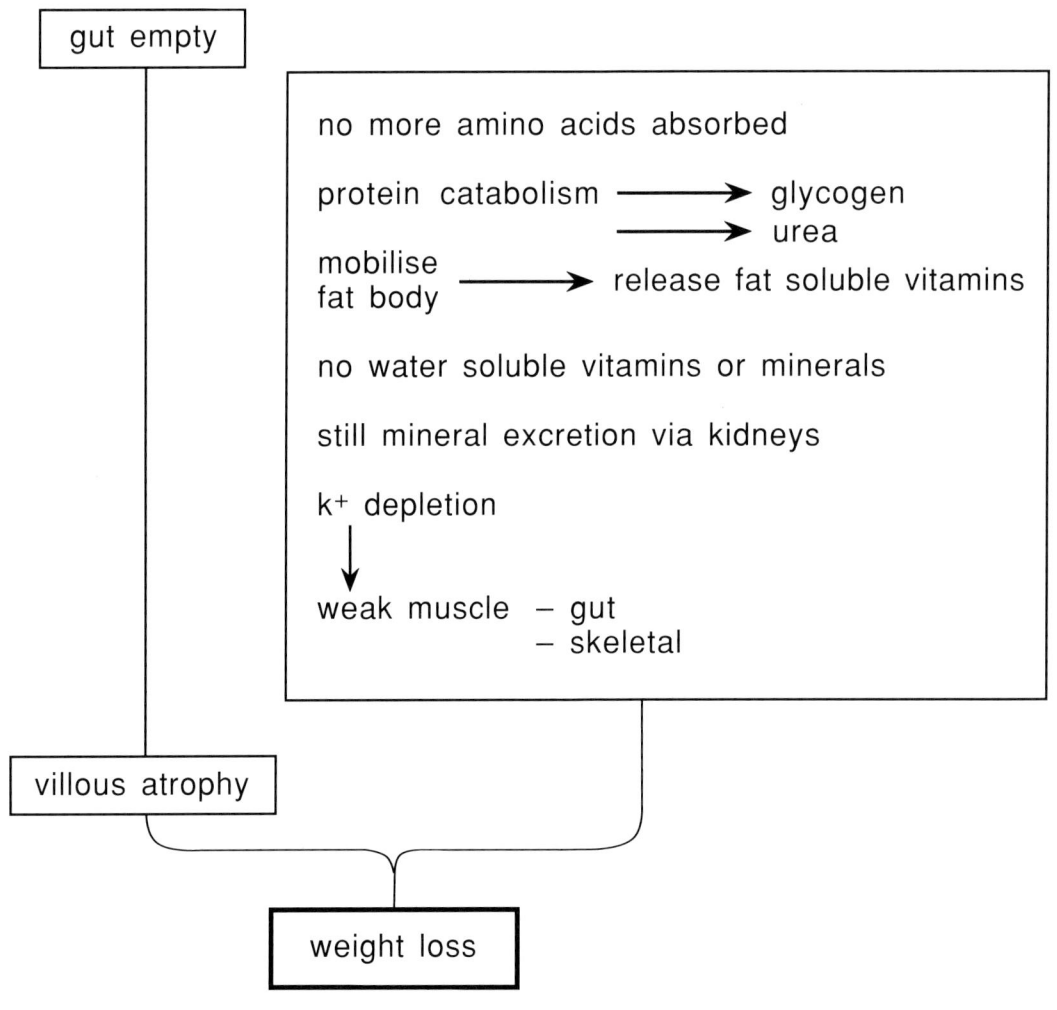

Figure 1
Metabolic consequences of prolonged anorexia in snakes.

Early cases of anorexia may respond simply to correction of the husbandry problem. In general, forced feeding with rodents is to be avoided. It may be successful but regurgitation may occur, if not immediately, then later on the way home. Advanced cases in which forced feeding has been attempted may result in regurgitation and further dehydration, or even death due to stress.

The preferred treatment is intracoelomic fluids – up to a maximum of 4% bodyweight. However, 2 – 3% may be safer (Jarchow, 1988). After four days, tube feeding can begin using high energy feeds with a low protein level, eg. Nutrical (Sanofi) or Hycal (SmithKline Beecham). A suitable probiotic, eg. Avipro, Vetark, can be included to help recolonise the inactive gut. Amino acid imbalances can be treated using injectable complexes such as Duphalyte (Solvay-Duphar Veterinary). Royal pythons may refuse every attempt to trigger natural feeding and the keeper must be warned that he or she may need to tube feed with liquidised cat food for years, possibly for the rest of the reptile's life (Lawton, 1991)

NUTRITIONAL EXCESSES

Obesity

Obesity in reptiles is a common problem. Acute deaths may be seen in large snakes due to "fatty liver" syndrome, and in chelonians excess fat deposits may physically restrict lung space.

Constipation may occur, particularly if heavily furred or feathered prey is supplied.

Mealworms should always be considered as a concentrated food source and the amounts fed should be limited; reptiles in the wild constantly hunt and eat. In captivity, unless diets are watched closely, reptiles can become obese.

Terrapins fed a diet consisting of bacon are reported as having cholesterol deposits in the cornea and, presumably, elsewhere. Confirmation of the diagnosis can be made by measuring serum levels – normally <6mmol/l (Lawton, unpublished data).

Protein

A high protein diet, with or without pre-existing dehydration, is considered to be the major cause of gout in reptiles. Uric acid is the major product of protein and purine metabolism in most reptiles and rapidly becomes a problem in the dehydrated reptile. Treatment consists of rehydration and reduction of protein intake and possible use of allopurinol (15 – 20mg/kg) (Jackson and Cooper, 1981; Lawton, 1991).

Gout is well recognised in commercially farmed alligators *(Alligator mississippiensis)*, primarily due to overfeeding a high protein diet when temperatures are below $20° - 25°C$ (when anabolism is less efficient) (Foggin, 1987). Signs usually begin with hindlimb paralysis, progressing to total paralysis. Secondary kidney infections may be seen in more chronic cases. Plasma levels of uric acid in normal alligators are less than 244μmol/l; in individuals with gout they may be as high as 4,164μmol/l (Coulson and Hernandez, 1964).

Vitamin D_3

Hypervitaminosis D_3 can occur, especially in iguanas, due to excess supplementation, usually when ultra-violet light is also being used. A dose of 200,000 IU/kg caused severe problems in a lizard (Lindt, 1968), whilst 100 IU/kg is reported to have caused problems in young iguanas (Zwart, 1980). Radiography may demonstrate calcification of the great vessels.

Nitrates

Secondary iodine deficiency is reportedly seen in herbivorous lizards and chelonians (especially the larger species) due to the feeding of green forage material grown using high supplementation with nitrates (especially kale, broccoli, Brussel sprouts and cabbage which store nitrate and are goitrogenic). Soya-bean sprouts also have some goitrogenic activity (Frye, 1991). Lethargy and swelling at the thoracic inlet are reported clinically. Treatment requires supplementation with T_4

and diet correction; the use of very high doses of iodine (as potassium iodide) to overcome blockage of the 'thyroid trap' may also be considered. Routine doses of sodium iodide are also used in prevention.

NUTRITIONAL DEFICIENCIES

Vitamin A

Hypovitaminosis A is relatively common in reptiles; the acute form is seen particularly in young red-eared terrapins *(Trachemys scripta elegans)* and box-turtles (*Terrapene* spp.) but the chronic form may be seen in all chelonians (Elkan and Zwart, 1967).

The major presenting sign is swollen eyelids, due to epithelial metaplasia and granulocyte infiltration of the lacrimal glands.

It is important to recognise the systemic nature of this disease and why it may carry a very poor prognosis. There is accompanying metaplasia of the cuboidal cells of the pancreas, kidneys, ureters and bladder. A number of animals will show inguinal and axillary swelling due to oedema caused by kidney failure (Lawton, 1989). Fatty degeneration of the liver is seen in severe cases. Chronic hypovitaminosis A results in hyperkeratosis of the 'beak' and cornea; affected animals may also be more susceptible to infections, such as 'runny nose syndrome'.

Diseased animals have been reported as having liver vitamin A levels of 9 – 19 IU/g of liver. Normal vitamin A levels have been recorded as being over 1,000 IU/g in monitor lizards and vipers (Elkan and Zwart, 1967).

Similar signs to those described in chelonians are described for Nile crocodiles *(Crocodylus niloticus)* with vitamin A deficiency (Foggin, 1987).

Zwart (1986) used high levels of vitamin A supplementation (100,000 IU/kg) when treating skin diseases in order to encourage sloughing.

Hypovitaminosis B_1

This manifests as a neurological condition (see "Neurological Diseases"), particularly in garter snakes (*Thamnophis* spp.) fed on seafish high in thiaminase activity. Opisthotonus is described by Zwart and van Ham (1980) in garter snakes fed cod, which actually contains relatively little thiaminase. They suggested cooking suspect fish in 1cm cubes for 5 minutes at 80°C and then adding 20mg B_1/kg of food. Treatment consisted of parenteral vitamin B_1 (80mg/kg) and correction of the diet. Supplements, such as BSP vitamin drops (Vetark), can be added to the water and freshwater fish, such as trout fry, can be used as a diet rather than deep frozen whole fish.

Biotin deficiency in large lizards

Monitor lizards (*Varanus* spp.), Gila monsters (*Heloderma suspectum*) and Mexican beaded lizards (*Heloderma horridum*) all relish raw eggs. In the wild these would only be part of a diet and would often be embryonated. In captivity, if fed large numbers of raw eggs (which contain the anti-vitamine avidin), they may show signs of biotin deficiency. This appears as muscular weakness, although one might also expect skin lesions.

Hypovitaminosis B_{12}

Whether this condition exists as a clinical entity is uncertain. Vitamin B_{12} is used commonly (and often successfully) as a dietary stimulant in anorexic specimens. Large doses orally (Cytacon Liquid, Duncan Flockhart) seem more effective than injections. However, the rationale for this is unclear.

Hypovitaminosis C

Hypovitaminosis C may be involved in infectious stomatitis; certainly, extrapolation from other species would support this view. Intramuscular vitamin C (10 – 20mg/kg/day) is a rational part of the routine treatment (Frye, 1991). Its use routinely in oral supplements could also be valuable where stress may be depleting levels of vitamin C more rapidly than normal. Levels of 25mg for a small patient up to several grams for large patients can be supplemented by using vitamin C on its own or as part of a specialist multivitamin preparation, eg. ACE-High, Vetark. Hypervitaminosis C has not been reported.

Hypovitaminosis E

Hypovitaminosis E has been seen in crocodilians fed mackerel, terrapins fed oily fish, such as whitebait, and in some snakes fed on overweight laboratory rodents. Steatitis is seen, with hard fatty pads around the legs in chelonians and linear ulcerations of the cloaca in crocodilians (Frye, 1991). Often the only presenting sign is anorexia followed by death. The greater susceptibility of embryos to vitamin E deficiency is shown in crocodilians (Lance, 1987), where a sub-clinical deficiency may produce no overt signs in adults but may dramatically increase fetal mortality.

Hypovitaminosis K

This has been reported in crocodilians (Frye, 1991). Affected animals had gingival bleeding without petechiation and bleeding from tooth sockets as the deciduous teeth were lost. Oral supplements are required. This condition, as with others, may occur when reptiles are fed a captive 'artificial' diet rather than a 'natural' diet.

Sodium chloride

It may be necessary to supplement the diet of captive marine species of reptile with sodium chloride. These have active lacrimal salt glands which continue to excrete salt even when the reptiles are kept in fresh water and fed a 'normal' diet without high salt levels. Some attention may also be required in desert reptiles which often show a salt deposit around the nostrils.

Hypoglycaemia in crocodilians

High stocking levels of crocodilians seem to cause social stress and low liver glycogen levels, which at the time of a feeding frenzy may lead to hypoglycaemic shock. Mydriasis is typically seen early in the development of the problem. An attack may follow with tremors, incoordination, opisthotonus and drowning. This is also seen at times when a new animal is introduced to a group. Alligators show a seasonal fluctuation in normal glucose levels from 5.6mmol/l in summer to 2.8mmol/l in winter (Wallach, 1971). The relationship between this and the reported lactic acidosis on capture (Seymour et al, 1987) is uncertain, but the inter-relationship between feeding, season and capture needs to be considered. Treatment using oral glucose at a dose of 3g/kg is suggested by Wallach (1971).

OTHER NUTRITIONAL PROBLEMS

A variety of injuries and other problems have been associated with the act of feeding. These include:-

1. Venomous snakes may strike others.

2. Crocodilians and terrapins demonstrate a feeding frenzy and may injure cage mates.

3. Live feeding has been the subject of much debate but there are potential hazards which need to be considered. Live food in a reptile cage which is not eaten quickly will itself need feeding; rat bite injury to large constrictors is not uncommon. Released crickets and cockroaches are reported to have nibbled satiated snakes.

4. Banana needs to be fed cautiously as it tends to become sticky around the corners of the mouth and may lead to infections. It can also cause constipation in chelonians.

5. Cabbages and kale are goitrogenic (see earlier).

6. Stomach tubing of tortoises with milk based foods, such as some Milupas or Complan, may cause problems of fermentation and colic, leading to enteritis. Lawton (personal communication) suggested that colic may be successfully treated using spasmolytics combined with probiotics.

CALCIUM, PHOSPHORUS AND VITAMIN D$_3$

A number of conditions can be grouped together under the heading nutritional secondary osteodystrophy, eg. osteomalacia, osteoporosis, rickets, osteodystrophia fibrosa and nutritional secondary hyperparathyroidism.

The Ca:P ratio needed by reptiles is approximately 1.5 – 2.0:1, the higher calcium levels being required by juveniles or breeding females. As a consequence most captive diets need some supplementation. This should be in the form of calcium without phosphorus (bone flour is unsuitable). Pure calcium carbonate seems unpalatable; powdered cuttlefish bone, egg shell or the specially formulated supplements may be preferred.

In absolute terms red-eared terrapins *(Trachemys scripta elegans)* have been shown to require 2% dietary calcium on a dry matter basis (Kass *et al*, 1982).

The use of tinned dog and cat foods as protein sources is advised as these are already supplemented at approximately the correct level. They can, however, only be used in small quantities as they may cause too rapid growth, outstripping mineral supply, and some reptiles will not eat them.

Over and above the calcium supplementation, vitamin levels need to be optimal for healthy growth. There is little definite evidence as to which form of vitamin D is antirachitic in reptiles, although the D$_3$ pathway has been demonstrated in lizards and turtles. A more detailed discussion of metabolic bone disease in exotic species can be found in Fowler (1986).

Vitamin D$_3$ supplementation at a level of 50 – 100 IU/kg on alternate days is the suggested maximum. The intestinal tract is not responsive to calcium unless primed by vitamin D$_3$; when blood levels are high the excess calcium is stored in the bones or endolymphatic sacs, or is excreted via the kidneys.

DISORDERS OF CALCIUM AND POTASSIUM METABOLISM

Shell growth in chelonians

The calcium and phosphorus content of the majority of the vegetables, fruits and salad items used as diets for captive herbivores is poor compared with that of the grasses, mixed shrubbery, cactuses etc eaten in the wild. Problems occur when protein and energy levels encourage rapid growth, outstripping the available calcium and phosphorus. Tortoises fed excessively on lettuce will grow more slowly but the calcium and phosphorus levels are lower still. Shredded lettuce is an ideal food item within a diet to dust with vitamin/mineral preparations.

Nutritional secondary osteodystrophy is very easy to produce in young growing chelonians by feeding deficient diets. Two manifestations are seen:-

1. Soft shell.

2. Lumpy shell, pyramiding.

Although still uncertain, it seems likely that these are degrees of the same problem. The former is commoner in the turtle (including terrapins) and may be a true rickets-type hypovitaminosis D$_3$, and the latter is commoner in tortoises where calcium deficiency may be more significant. Most turtles are fed solely on meat/fish (or the dried invertebrates sold as turtle food) and grow rapidly, whilst tortoises may be fed exclusively on vegetables and grow very slowly. Chelonians with soft shells will often become grossly deformed as the limb muscles which attach to the inside of the carapace pull the limb in and produce a camel saddle appearance. Limbs are weak and the claws and beak may become overgrown.

Table 3
Calcium/phosphorus content of various foods.

Food	Dry matter %	Protein %	Fat %	Energy kcal/g	Calcium %	Phosphorus %	Ca:P ratio	Ref.
Mealworms	38	47	35		0.23	0.71	0.32	1
	42.2	52.8	35	6.53	0.06	0.53	0.11	3
	38.1	54.6	31.4	5.35	0.07	0.71	0.1	3
Locusts	31.2	61.7	19.4		0.1	0.75	0.13	1
Crickets	38.2	55.3	30.2		0.23	0.74	0.31	1
Earthworms	22	49.9	5.8		0.59	0.85	0.69	1
Beef muscle	26	20.3	4.6	1.23	0.007	0.18	0.04	2
Chicken muscle	25.6	20.5	4.3	1.21	0.01	0.20	0.05	2
Ox liver	31.4	21.1	7.8	1.61	0.006	0.36	0.016	2
Cod fillet	17.9	17.4	0.7	0.76	0.016	0.17	0.09	2
Egg, whole	25.2	12.3	10.9	1.47	0.052	0.22	0.023	2
Day-old chick		15.3	4.4	1.04	0.44	0.4	1.1	4
Day-old chick		17.17	6.7	1.66	0.38	0.28	1.36	4
Mice								
1 – 2 days					1.6	1.8	0.88	5
7 days					1.43	1.29	1.1	5
adult		19.86	8.81	2.07	0.84	0.61	1.37	4
Rat		21.6	7.6	1.99	0.69	0.51	1.35	4
Bluegrass (lawn)	33	2.4	1.2	1.58	0.1	0.09	1.1	4
Alfalfa hay		15.5	37.1	3.94	1.29	0.21	6.14	4
Iceberg lettuce		1.2	2.5	0.14	0.035	0.026	1.34	4
Lettuce	4.1	1	0.4	0.12	0.023	0.027	0.85	2
Oranges	13.9	0.8	trace	0.35	0.041	0.024	1.71	2
Grapes (white)	20.7	0.6	trace	0.63	0.019	0.022	0.86	2
Bananas	29.3	1.1	0.3	0.79	0.007	0.028	0.25	2
Carrots (old)	10.1	0.7	trace	0.23	0.048	0.021	2.29	2
Tomato	6.6	0.9	trace	0.14	0.013	0.021	0.62	2

References: 1 = Zwart, 1980; 2 = Paul and Southgate, 1988; 3 = Frye, 1991;
4 = Fowler, 1986; 5 = Allen and Oftedal, 1982.

Even providing young active tortoises with a good diet which allows rapid growth can lead to secondary deficiency, ie. there may be sufficient calcium for slow growth but not for the rapid growth that occurs on a high protein diet; hence the need for supplementation. Over enthusiastic calcium supplementation of chelonians may result in increased shell density and an overall thicker shell.

Figure 2 shows the growth rate over 12 months of a group of juvenile *Testudo graeca*. J1, J2, J5 and J7 were from one clutch, J9 from a second and P2 from a third clutch. Final lengths and Jackson's ratios for the animals are shown alongside.

The basic diet consisted of moderate amounts of cabbage, cress, grass and mixed weeds with small amounts of carrot, banana, apple, tomato, and orange. All the tortoises received 2 – 3g of Vionate (Ciba Geigy) daily sprinkled on the feed. J2 and P2 received brown bread once a week for the first 6 months; after this all of the tortoises received brown bread. Protein sources were alternated daily, using tinned dog food, spratts, liquidised chicks and heart or liver. J1 and J2 both received an animal protein source four days a week (J1 was fed four days/week and J2 was fed six days/week). They also received extra calcium in the form of ground cuttlefish bone. J5 received no animal protein for the first six months. J7 received mainly grass and mixed weeds with little cultivated food and no animal protein. Its counterpart (P2) received animal protein throughout (Reid, 1986).

Those tortoises which received animal protein grew most rapidly and had the better shell shape and final Jackson's ratios. Continual monitoring and correction is vital to achieve 'normal' shell structure.

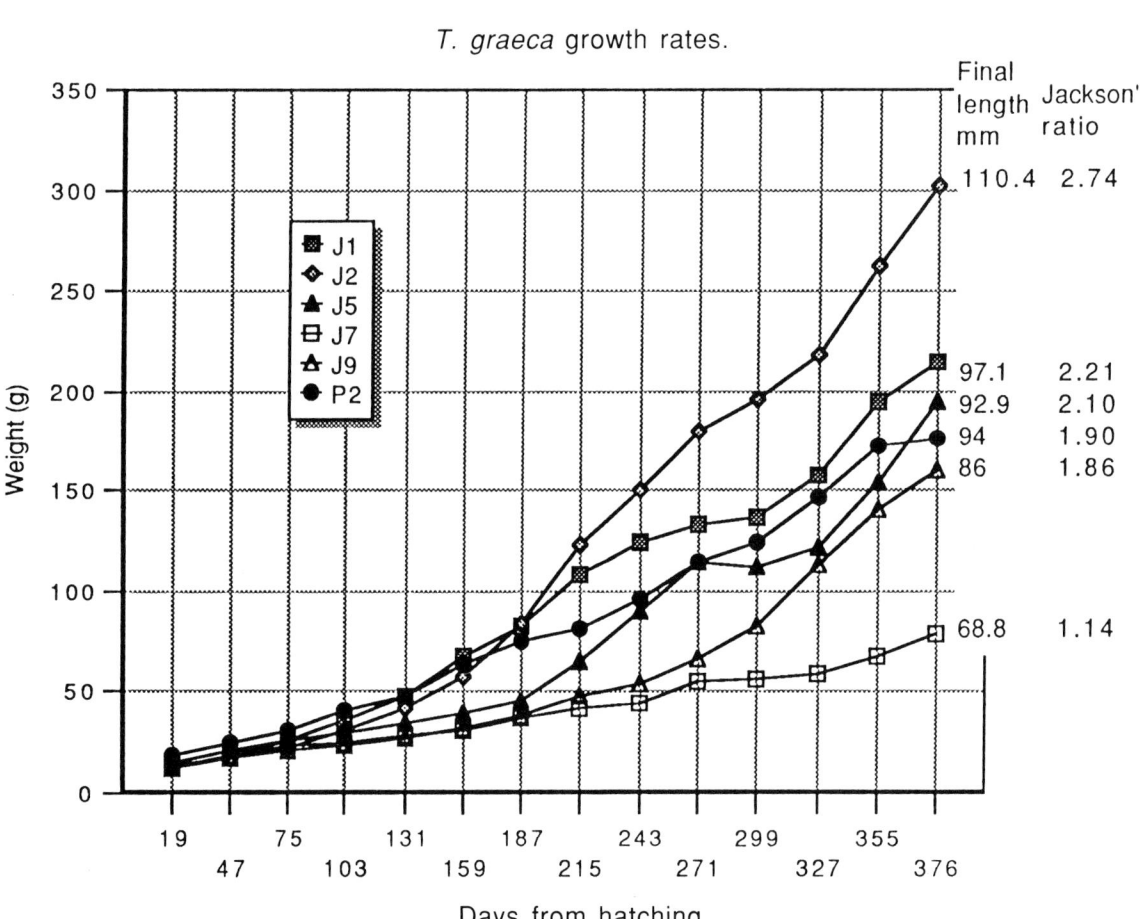

Figure 2

T. graeca growth rates.

Source of data:– Reid (1986).

Nutritional osteodystrophy in iguanas

Nutritionally induced osteodystrophy is probably the commonest disease of iguanas *(Iguana iguana)*, affected animals being dull and lethargic with a decreased appetite. Activity is usually reduced, often to the point that the hindlimbs barely move, and muscles may be atrophied. The animals appear closer to the ground and may present with 'spontaneous' fractures. Tetany may also be seen (Boivin and Stauber, 1990). In females the onset of egg laying may precipitate the appearance of signs, despite the eggs being normally soft-shelled (Scott, unpublished data). Undermineralisation is relatively obvious on radiography and blood calcium levels will be below the normal 2.3 mmol/l (Lawton, 1991).

Treatment of acute hypocalcaemia involves rapid correction of the blood levels using 1% calcium gluconate (0.5ml of 20% in 9.5ml water) at a rate of 1 – 2ml per 100g body weight, ie. 100 – 200mg calcium gluconate/kg (Fowler,1986). Repeated doses may be required to achieve remission of signs. If the intravenous route is used the calcium should be given slowly.

Dietary calcium levels need correction with high calcium supplements, eg. Nutrobal (Vetark). This can be given at a rate of 0.1g/kg per day over a period of 6 – 8 weeks or until the problem is corrected. The dose is then adjusted to perhaps alternate days. Such supplements are designed for routine long-term supplementation of reptiles. Monitoring of serum calcium levels is advisable. Radiography to monitor progress is also valuable. A calcium level of 2% (on a dry matter basis) seems to be required by iguanas. Vionate (1g/kg per day) plus extra calcium in mealworms, combined with the use of ultra-violet light is also described for long-term maintenance (Boivin and Stauber, 1990). The use of ultra-violet light to convert 7-dehydrocholesterol into the active form of vitamin D is probably important and needs to be taken into account when assessing the level of dietary supplementation required.

Treatment of associated fractures depends on the level of undermineralisation of the bone; severely undermineralised bone needs prior supplementation before intramedullary fixation. In the interim, external supports, eg. Hexcelite, fitted like a jacket may be useful (see "Surgery").

Other problems

In both chelonians and lizards disorders of calcium and phosphorus metabolism at any time can cause skeletal abnormalities, which in the pelvis may lead to eggbinding, compounding the direct effects of hypocalcaemia on the uterus. Langenwerf (1980) associated low calcium diets in breeding stock with poorly mineralised embryos which did not hatch or which died soon after hatching. Dietary correction appeared to eliminate the problem. Kuehn (1974) reported traumatic deaths of crocodilians due to damage to weakened and soft bones. He pointed out the absurdity of the argument which implies suitability of a diet simply because an animal has remained alive on it.

REFERENCES

ALLEN, M.E. and OFTEDAL, O.Y. (1982). Calcium and phosphorus levels in live prey. In: *Proceedings of the North East Section of the American Association of Zoos, Parks and Aquaria*.

BARTKIEWICZ, S.E., ULLREY, D.E., TRAPP, A.L. and KU, P.K. (1982). A preliminary study of niacin needs of the bull snake. *Journal of Zoo Animal Medicine* **13,** 5.

BOIVIN, G.P. and STAUBER, E. (1990). Nutritional osteodystrophy in iguanas. *Canine Practice* **15 (1)**, 37

BUSTARD, R. (1972). *Sea Turtles: their Natural History and Conservation.* Collins, Glasgow.

COULSON, R.A. and HERNANDEZ, T. (1964). *Biochemistry of the Alligator.* Louisiana State Press, Baton Rouge.

ELKAN, E. and ZWART, P. (1967). The ocular disease of young terrapins caused by vitamin A deficiency. *Pathologia Veterinaria* **4,** 201.

FRYE, F.L. (1991). *Biomedical and Surgical Aspects of Captive Reptile Husbandry.* 2nd Edn. Krieger, Malabar

FOGGIN, C.M. (1987). Diseases and disease control on crocodile farms in Zimbabwe. In:*Wildlife Management: Crocodiles and Alligators.* (Eds G.J.W. Webb, S.C. Manolis and P.J. Whitehead). Surrey Beatty, Chipping Norton, New South Wales.

FOWLER, M.E. (1986). Metabolic bone disease. In: *Zoo and Wild Animal Medicine.* 2nd Edn. (Ed. M.E. Fowler). W.B. Saunders, Philadelphia.

GARNETT, S.T. and MURRAY, R.M. (1981). Farm management and nutrition of the green turtle *(Chelonia mydas).* In: *Proceedings of the Melbourne Herpetological Symposium.* (Eds. C.B. Banks and A.A. Martin). Zoological Board of Victoria, Melbourne.

JACKSON, O.F. and COOPER, J.E. (1981). Nutritional diseases. In: *Diseases of the Reptilia, Vol. 2.* (Eds. J.E. Cooper and O.F. Jackson). Academic Press, London.

JARCHOW, J.L. (1988). Hospital care of the reptile patient. In: *Exotic Animals.* (Eds. E.R. Jacobson and G.V. Kollias). Churchill Livingstone, New York.

KASS, R.E., ULLREY, D.E. and TRAPP, A.L. (1982). A study of calcium requirements of the red-eared slider turtle (*Pseudemys scripta elegans*). *Journal of Zoo Animal Medicine* **13**, 62.

KUEHN, G. (1974). Crocodilian nutritional deficiencies. *Journal of Wildlife Diseases* **5 (4)**, 25.

LANCE, V.A. (1987). Hormonal control of reproduction. In: *Wildlife Management: Crocodiles and Alligators.* (Eds. G.J.W. Webb, S.C. Manolis and P.J. Whitehead). Surrey Beatty, Chipping Norton, New South Wales.

LANGENWERF, B.A.W.A. (1980). The successful breeding of lizards from temperate regions. In: *The Care and Breeding of Captive Reptiles.* (Eds. S. Townson, N.J. Millichamp, D.G.D. Lucas and A.J. Millwood). British Herpetological Society, London.

LAWRENCE, K. (1987). Post hibernational anorexia in captive Mediterranean tortoises *(Testudo graeca and T.hermanni). Veterinary Record* **120,** 87.

LAWTON, M.P.C. (1989). Health problems associated with feeding. *Testudo* **3 (1)**, 75.

LAWTON, M.P.C. (1991). Lizards and snakes. In: *Manual of Exotic Pets.* New Edn. (Eds. P.H. Beynon and J.E. Cooper). BSAVA, Cheltenham.

LINDT, S. (1968). Kalzinose durch hypervitaminose D bei verschidenen tieren. *Wiener Tierärztliche Monatsschrift* **55**, 148.

PAUL, A.A. and SOUTHGATE, D.A.T. (1988). *The Composition of Foods.* MRC Special Report No. 297. HMSO, London.

REID, D. (1986). Notes on the dietary regimes of juvenile spur-thighed tortoises (*Testudo graeca*) and a comparison of their growth over 12 months. *Journal of the Association for the Study of Reptilia and Amphibia* **1 (3)**, 2.

ROSS, R.A. and MARZEC, G. (1990). *The Reproductive Husbandry of Pythons and Boas.* Institute for Herpetological Research, Stanford.

ROSSKOPF, W.J. and WOERPEL, R.W. (1981). Rat bite injury in a pet snake. *Modern Veterinary Practice* **62 (11)**, 871.

SEYMOUR, R.G., WEBB, G.J.W., BENNETT, A.F. and BRADFORD, D.F. (1987). Effect of capture on the physiology of *Crocodylus porosus.* In: *Wildlife Management: Crocodiles and Alligators.* (Eds. G.J.W. Webb, S.C. Manolis and P.J. Whitehead). Surrey Beatty, Chipping Norton, New South Wales.

VOSBURGH, K.M., BRADY, P.S. and ULLREY, D.E. (1982). Ascorbic acid requirements of garter snakes: plains *(Thamnophis radix)* and eastern *(T.sirtalis sirtalis). Journal of Zoo Animal Medicine* **13**, 38.

WALLACH, J.D. (1971). Environmental and nutritional diseases of captive reptiles. *Journal of the American Veterinary Medical Association* **159 (11)**, 1632.

ZWART, P. (1980). Nutrition and nutritional disturbances in reptiles. In: *Proceedings of the European Herpetological Symposium.* Cotswold Wildlife Park, Burford.

ZWART, P. (1986). Infectious diseases of reptiles. In: *Zoo and Wild Animal Medicine.* 2nd Edn. (Ed. M.E. Fowler). W.B. Saunders, Philadelphia.

ZWART, P. and van HAM, B. (1980). Keeping, breeding and raising garter snakes *(Thamnophis radix).* In: *The Care and Breeding of Captive Reptiles.* (Eds. S. Townson, N.J. Millichamp, D.G.D. Lucas and A.J. Millwood). British Herpetological Society, London.

CHAPTER FOURTEEN

MISCELLANEOUS

Martin P C Lawton BVetMed CertVOphthal FRCVS

There are a number of conditions that do not readily fall into the systems approach that is adopted in this manual. This may be due to their multifactorial nature, both in cause and effect, or simply that it is not possible to place them elsewhere. The result is a miscellany of conditions which are unrelated and often poorly understood.

AGEING

Reptiles tend to live longer in captivity than in the wild. They are protected from predators and offered an abundance of food. Improved husbandry and biological knowledge has led to the successful breeding, rearing and, ultimately, the survival of captive reptiles. As a consequence of the potentially increased life-span of captive reptiles, problems associated with ageing and degenerative disease are more commonly encountered (Cosgrove and Anderson, 1984).

Arteriosclerosis (see "Cardiovascular System") and **gout** (see "Urogenital System") are the more commonly seen problems associated with ageing. Less commonly seen is **amyloidosis**, where amorphous proteins are deposited in various organs and elsewhere in the body. In one report 30 out of 52 captive Hermann's tortoises (*Testudo hermanni*) were found to have amyloidosis at *post-mortem* examination (Trautwein and Pruksaraj, 1967), although other authors have found this condition to be a rarer occurrence.

The significance of amyloidosis in reptiles is still unclear (Cosgrove and Anderson, 1984), although Frye (1991) suspected that it is linked to the presence of highly antigenic substances in the circulation. The clinical signs will depend on the organs affected and the amount of amyloid deposited. Single or multi-organ failure may be encountered but usually this condition is diagnosed *post mortem*.

CANNIBALISM

Cannibalism can be a problem where reptiles are kept in captivity. It is usually associated with overcrowding or failure to supervise when feeding.

The only species that is known routinely to eat other snakes is the black-headed python (*Aspidites melanocephalus*), although the Papuan python (*Liasis papuanus*) is thought to be cannibalistic (Ross and Marzec, 1990). Generally, cannibalism in snakes occurs by accident when two snakes compete for the same prey food. If unsupervised, it is possible that one snake (usually the larger one) may ingest the other (usually the smaller one).

Lizards are highly territorial; some species may attack and eat other lizards in the same cage.

Terrapins (*Trachemys scripta elegans*) fed in a small tank may go into a feeding frenzy and attack and possibly eat smaller species in the tank. A terrapin with open wounds following trauma, may often be bullied and bitten by the others in the tank; this may result in the death of the terrapin.

EUTHANASIA

There are times when euthanasia must be carried out. Such occasions include the necessity to alleviate suffering, to cull due to overbreeding and overcrowding, and to aid in the investigation of colony disease problems. Reptiles are still used as food sources in certain parts of the world and a humane method of euthanasia is required. Cooper *et al* (1989) reviewed the options available for the euthanasia of amphibians and reptiles.

As part of the humane approach to euthanasia, it is important that the reptile should not be handled incorrectly or with excessive force. Cooper *et al* (1989) stated that there is a need for further research into the euthanasia of the lower vertebrates in order to find a method that is both humane and causes the least stress and pain to the animal.

Intravenous barbiturates are the method of choice. This technique is quick, resulting in death within a couple of minutes. Details of intravenous injection techniques can be found in "Laboratory Investigations" and "Anaesthesia". In species which are aggressive or difficult to handle, prior sedation or anaesthesia by intramuscular injection of a suitable agent, or gaseous induction may be useful.

Intraperitoneal or intramuscular injection of barbiturate is effective but very slow - several hours (Lawton, unpublished data). Intraperitoneal injection should not be used if a *post-mortem* examination is planned. Intracardiac injection can be used in Squamata but the lizard or snake should ideally be sedated or even anaesthetised prior to this technique.

Inhalation of volatile agents within an anaesthetic chamber is useful for terrestrial species and, especially, for poisonous snakes. The concentration of volatile agent must be high enough and the exposure to the agent long enough to ensure the death of the reptile. Care must also be taken to avoid contact of the reptile with the volatile fluid. Euthanasia of aquatic species should never be attempted in this way, as they are able to revert to anaerobic respiration, the length of time required to achieve euthanasia makes this method impracticable.

Injection of a volatile agent directly into the lung space has been used as a method of euthanasia, although this should only be done in anaesthetised reptiles (Cooper *et al*, 1989).

A **captive bolt** or **free bullet** may be used in larger species, particularly crocodilians, monitor lizards and giant chelonians (Cooper *et al*, 1989). In mature species of crocodilians, the use of a heavy calibre firearm is advised. There is often a thin area of the skull where a single layer of bone exists between the brain and the skin; this is often located midline and posterior to the eyes.

In an emergency the head of the reptile can be struck with such force as to destroy the brain (Cooper *et al*, 1989).

Decapitation should not be used as a sole method of euthanasia, unless the brain is destroyed by pithing immediately afterwards (Cooper *et al*, 1984). The assumption that decapitation results in unconsciousness followed by rapid deterioration of the nervous system is disputed: the brain remains viable for up to an hour after decapitation (Cooper *et al*, 1984).

Hypothermia has been claimed as being both a suitable and humane method for the euthanasia of small reptiles (Frye, 1984a), although this is strongly contested. Whilst hypothermia makes reptiles torpid and may facilitate handling, freezing is considered painful due to the formation of ice crystals on the skin and in the tissues (Cooper *et al*, 1989). The exception to this is dropping small reptiles directly into liquid nitrogen; this is considered to cause instantaneous death (Hillman, 1978).

ENDOCRINOLOGICAL DISORDERS

There is little reported evidence of endocrinological disorders in reptiles. With the increase in captive breeding more work is required on gonadal hormone levels and, in particular, the effect of environmental conditions on their production, especially light and temperature. Ross and Marzec

(1990) found that a number of Boidae required a wide array of changes in environmental conditions in order to achieve breeding. These conditions, even when supplied, may take years to have their effect and allow the snakes to "reset" their biological clocks.

Several changes have been reported in the thyroid gland, including developmental cysts, abscessation, granulomas, involution, hypertrophy and hyperplasia (Cooper and Jackson, 1981). Goitrogenic effects of feeding kale, cabbage and other similar foods to chelonians has resulted in goitre, which is often characterised by swelling of the ventral neck region. Treatment is by addition of iodine to the diet and other dietary changes (Frye, 1991). Snakes with an underactive thyroid may shed more frequently, often starting to do so before the new skin is properly formed (Lawton, unpublished data). Frye (1981) suggested that in some circumstances dysecdysis and frequent shedding may occur with both hypothyroidism and hyperthyroidism. Although normal blood levels for thyroid hormones are not available, a comparison can be made between the snake under investigation and another snake of the same species (or at least genus) and the same sex.

Hyperplastic pancreatic regeneration has been reported in captive snakes, while it is considered very rare in wild snakes (Cosgrove and Anderson, 1984). The pathological changes involve necrosis of pancreatic tissue which is replaced by atypical adenomatous areas. Although the cause of this condition is unknown, it may be part of the maladaptation syndrome (see later).

IMMUNOLOGICAL DISORDERS

Cooper and Jackson (1981) questioned whether or not reptiles were susceptible to immunological disorders such as anaphylactic shock, immune complex diseases or even allergic reactions.

Little work has been done on immunological disorders (including immunosuppresion) in reptiles, although it is often postulated that they occur but often with little evidence to support this.

Reptiles are capable of immunological responses, both cellular and humoral. Kollias (1984) gave a review of the immunologic capabilities of reptiles.

Little is recorded on the use of or responses to vaccination in reptiles, although autogenous vaccines have been used and found to be helpful (Addison and Jacobson, 1974).

SHOCK

Shock is well recognised in mammals and birds and is characterised by physiological changes, especially of the cardiovascular system, as well as endocrinological changes. Based on clinical experience, shock is thought to occur in reptiles (Cooper and Jackson, 1981). Affected reptiles show similar clinical signs to those seen in mammals, especially signs associated with the cardiovascular system. Treatment should be aimed at improving the circulatory volume by the use of fluid therapy and even corticosteroids, as well as maintaining the core body temperature by keeping the reptile at its preferred body temperature (PBT).

Hypoglycaemic "shock" of alligators is discussed in "Neurological Diseases".

STRESS

Cooper and Jackson (1981) stated that many deaths in reptiles are attributed to "stress" but without any supporting evidence. Selye (1936) was the first to postulate the concept of stress. Stressors, eg. pain or changes in temperature and environment, result in physiological modifications to cardiovascular, renal and intestinal systems, as well as an increased production of their corticosteroids and changes in the haematological profiles.

Stress over a short time may not cause any lasting problems. However, should the stress continue, it may threaten the well being of the animal.

MALADAPTATION SYNDROME

Maladaption syndrome may be associated with prolonged stress. This syndrome is most commonly seen in captured reptiles which fail to adapt to life in captivity, and is often associated with unsuitable accommodation, temperature and/or food. Affected reptiles stop feeding or never start to feed. They will lose weight and be prone to infections associated with opportunist bacteria. Death is usually associated with emaciation and dehydration (Cooper and Jackson, 1981). *Post-mortem* examination confirms inanition with lack of fat deposits and often secondary infections. A suspected case of maladaptation should be carefully investigated. A full clinical examination should be undertaken to rule out pathological causes. A good case history should highlight problems with the environment which could result in failure to adapt. Treatment is aimed at improving the environment and providing therapeutic and supportive therapy.

Starvation is a common problem in reptiles (Frye, 1984b) and is the main presenting clinical sign of maladaptation syndrome, although parasitic and other pathological problems must be ruled out.

DISEASES OF UNKNOWN AETIOLOGY

A thorough investigation (as described in other chapters in this manual), especially *post-mortem* examination and the submitting of samples for further evaluation, should lead to a diagnosis of the majority of diseases encountered in reptiles. Some conditions, especially viral, are often difficult to confirm; it is even more difficult to prove that viral inclusions seen on histopathology or electron microscopy are capable of causing disease. More research is needed into reptilian diseases to increase our understanding of their pathogenesis and aetiology.

REFERENCES

ADDISON, B. and JACOBSON, E.R. (1974). An autogenous bacterin for a chronic mouth infection in a reticulated python. *Journal of Zoo Animal Medicine* **5**, 10.

COOPER, J.E., EWBANK, R., PLATT, C. and WARWICK, C. (1989). Eds. *Euthanasia of Amphibians and Reptiles*. Universities Federation for Animal Welfare, Potters Bar.

COOPER, J.E., EWBANK, R. and ROSENBERG, M.E. (1984). Euthanasia of tortoises. *Veterinary Record* **114**, 635.

COOPER, J.E. and JACKSON, O.F. (1981). Miscellaneous diseases. In: *Diseases of the Reptilia, Vol. 2.* (Eds. J.E. Cooper and O.F. Jackson). Academic Press, London.

COSGROVE, G.E. and ANDERSON, M.P. (1984). Aging and degenerative diseases. In: *Diseases of Amphibians and Reptiles*. (Eds. G.L. Hoff, F.L. Frye and E.R. Jacobson). Plenum Press, New York.

FRYE, F.L. (1981). Traumatic and physical diseases. In: *Diseases of the Reptilia, Vol. 2.* (Eds. J.E. Cooper and O.F. Jackson). Academic Press, London.

FRYE, F.L. (1984a). Euthanasia, necropsy techniques and comparative histology of reptiles. In: *Diseases of Amphibians and Reptiles*. (Eds. G.L. Hoff, F.L. Frye and E.R. Jacobson). Plenum Press, New York.

FRYE, F.L. (1984b). Nutritional disorders in reptiles. In: *Diseases of Amphibians and Reptiles*. (Eds. G.L. Hoff, F.L. Frye and E.R. Jacobson). Plenum Press, New York.

FRYE, F.L. (1991). *Biomedical and Surgical Aspects of Captive Reptile Husbandry*. 2nd Edn. Krieger, Malabar.

HILLMAN, H. (1978). Humane killing of animals for medical experiments. *World Medical Journal* **25 (5)**, 68.

KOLLIAS, G.V. (1984). Immunologic aspects of infectious diseases. In: *Diseases of Amphibians and Reptiles*. (Eds. G.L. Hoff, F.L. Frye and E.R. Jacobson). Plenum Press, New York.

ROSS, R.A. and MARZEC, G. (1990). *The Reproductive Husbandry of Pythons and Boas*. Institute for Herpetological Research, Stanford.

SELYE, H. (1936). A syndrome produced by diverse nocuous agents. *Nature* **138**, 32.

TRAUTWEIN, C. and PRUKSARAJ, D. (1967). Uber amyloidose bei Schildkroten. *Deutsche Teirarztl Wochenschr* **74**, 184.

CHAPTER FIFTEEN

OPHTHALMOLOGY

Martin P C Lawton BVetMed CertVOphthal FRCVS

INTRODUCTION

Jackson (1981) stated that little use appears to have been made of the ophthalmoscope in reptiles. Ensley *et al* (1978) claimed that "ocular disorders in snakes are not a common finding" and reported three individual cases of ophthalmic problems in snakes. However, one of these cases, in an indigo snake (*Drymarchon corais*), was incorrectly diagnosed as a "proptosis": what they described was an enlargement of the sub-spectacular space and a protruding spectacle. With the growing interest in ophthalmology in all species, the importance of an ophthalmological examination as part of the clinical assessment and investigation of the reptilian patient is now becoming well recognised.

EXAMINATION

The general approach to the ophthalmological examination of reptiles should differ in no way from that performed in mammals or birds.

In order to examine the whole of the lens and fundus mydriasis is useful but, unlike mammals, reptiles have striated musculature in the iris which is under voluntary control (Walls, 1942). Mydriatics, therefore, tend not to work successfully, even after repeated applications. Mydriasis is best achieved under general anaesthesia, although Millichamp and Jacobson (1986) advised the use of D-tubocurare in larger species of lizards or crocodilians. The tubocurare (0.05 − 1ml) is injected via a 27G − 30G needle into the anterior chamber, giving mydriasis from 30 minutes to several hours.

Although direct light response is rapid, a consensual light reflex cannot be elicited (Millichamp *et al*, 1983) and should not be confused with a neurological deficit.

The other major difference in the examination of mammalian and reptilian eyes is one of practicality. Care has to be taken in examining species which can prove dangerous, eg. lizards that bite, poisonous snakes or crocodilians, and for this reason anaesthesia or sedation may be required. The use of indirect ophthalmoscopy, with either a 30 diopter lens for the larger species or a 90 diopter lens for a very small eye, has the advantage of keeping the animal's head at arm's length from the veterinary surgeon's face. Likewise, all the advantages of indirect ophthalmoscopy described by Slatter (1981) are just as true for reptiles as they are for any other animals, ie. a large area of fundus may be examined at any one time, although this image is upside down and inverted (see Figure 1). Indirect ophthalmoscopy also allows a three dimensional assessment and, therefore, better visualisation of lesions of the lens, fundus and vitreous, and even the conus papillaris in lizards.

Some degree of magnification, ideally a slit lamp biomicroscope or a plus 10 or plus 15 lens on an ophthalmoscope, is recommended for the examination of the adnexa. This is especially useful when dealing with the smaller species where, without magnification, lesions would be missed.

Figure 1
Indirect ophthalmoscopic study of the fundus of a tortoise (*Testudo* sp.).

ANATOMY AND PHYSIOLOGY

The eye of the snake is the exception in the Class Reptilia and requires special attention. This will, therefore, be dealt with separately. It is thought that the first snakes that evolved lived underground with the result that their eyes regressed (Walls, 1942). Subsequently, the eye developed all over again.

In general, the reptilian eye is very similar to that of mammals, although there are differences. The iridocorneal angle has some similarities to that of mammals, although it is less well developed (Millichamp *et al*, 1983). There is no Descemet's membrane of the cornea (Duke-Elder, 1958) and the retina is avascular (anangiotic). One of the major anatomical differences is the presence of ossicles in the sclera: these maintain the shape and size of the eye. In lizards there are approximately 14 scleral ossicles located behind the corneoscleral limbus adjacent to the ciliary body. The chelonian eye has these ossicles arranged so as to form a cup of cartilage (Walls, 1942). These structures are important as they play a major part in accommodation by changing the shape of the eye and, therefore, altering the distance between the cornea and the fundus. The ossicles require decalcification before histopathological examination can be performed (Millichamp and Jacobson, 1986).

The eye of ophidians has no ossicles; the sclera is composed entirely of tendinous connective tissue and resembles more closely the mammalian eye where accommodation is brought about by changes in lenticular shape and not by the shape or size of the eye. However, in the snake pressure changes within the eye result in the lens moving forward and backwards (Davies, 1981).

Eyelids

Except for ophidians and some lizards, most reptiles have mobile eyelids and many species also have a third eyelid. Some species of lizards have a transparent lower eyelid where the scales are reduced or absent; this allows vision when the eyelids are closed, but provides protection from sand or grit (Duke-Elder, 1958).

In true chameleons the eyelids are fused except for a small central aperture (Marcus, 1981). This stage is taken one step further in ophidians, some geckos and the ocellated skink *(Ablepharus* sp.), where there is no palpebral fissure, the upper and lower eyelids being fused. These fused eyelids form a transparent membrane known as the spectacle, brille or eye-cap. There are three types of spectacles described by Walls (1942). The type found in Squamata is the tertiary spectacle, which embryologically is mainly composed of the lower eyelid. It is a horny, dry, transparent eye scale which, by microsilicone injection, has been demonstrated to be highly vascular (Millichamp *et al*, 1983).

In chelonians the lower eyelid is larger and more mobile than the upper eyelid, which is almost fixed.

In crocodilians the upper eyelid is the more mobile and can move to cover the eye.

Tear film

All reptiles produce tears, although there are variations in which tear glands are present or more dominant. In some species, eg. sea turtles (*Chelonia mydas*), the Harderian gland may be modified as an osmoregulatory organ and have the ability to secrete salt.

Even snakes, which have no lacrimal gland, produce tears from the Harderian gland (Millichamp and Jacobson, 1986). These tears form an area of lubrication between the cornea and the overlying spectacle, thus allowing the free movement of the globe underneath the spectacle. As in mammals, a nasolacrimal system operates in snakes in order to drain the oily fluid from the sub-spectacular space into the mouth near to the vomeronasal organ. Unlike mammals, however, should there be any damage to the nasolacrimal system tears cannot overflow the eyelid margins; this results in a buildup of fluid in the sub-spectacular space (see later).

Some chelonians, eg. *Testudo* spp., do not have a nasolacrimal system; if tears overflow the eyelid margins they are lost by spillage or evaporate from the face.

Blood supply

Lizards have a similar blood supply to birds, where a large vascular network protrudes into the vitreous, thus providing nutrients and oxygen but also removing metabolic waste. In birds this structure is known as the pecten, while in reptiles it is known as the conus papillaris (Duke-Elder, 1958) (see Figure 2).

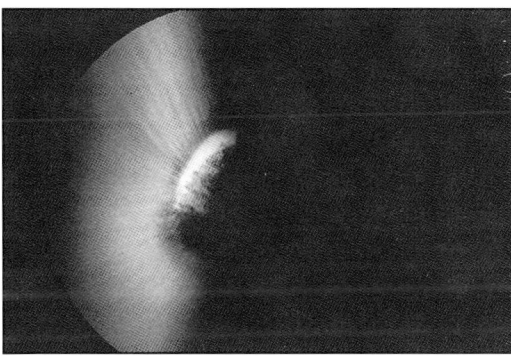

Figure 2
The fundus of a river monitor (*Varanus* sp.) showing conus papillaris.

Some lizards have a modified blood supply to the eye. The horned lizard (*Phrynosoma cornutum*) has the ability to rupture blood vessels in the medial angle of the eye and squirt blood as a defensive mechanism (Evans, 1986).

Chelonians have a true avascular (anangiotic) retina and no conus papillaris.

The retina of the snake is also avascular, nutrients being supplied by the membrana vasculosa retinae, a branching array of vessels from the choroid, which runs into the posterior vitreous near the optic disc.

In crocodilians the conus papillaris is functionless and reduced to a glial pad with 1 – 2 capillaries found on the adult disc which scarcely protrudes into the vitreous (Walls, 1942; Duke-Elder, 1958). The crocodilian retina seems to have lost any need for nutritive provision other than that provided by the choroid.

Third eye

The parietal eye (third eye) is the sole remaining median eye from the original paired visual organs on the roof of the head of provertebrates (Walls, 1942) and is found in a number of Squamata (Sauria and Sphenodontia) (see Figure 3). This is a primitive structure which, on histological examination, has been shown to be an eye, having a neurological input and primitive retina. Indeed, in the Tuatara (*Sphenodon punctatus*) there is a lens as well as the fairly fine grained retina (Walls, 1942).

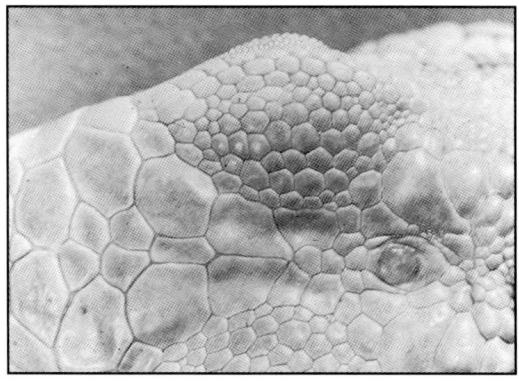

Figure 3
An iguana showing the parietal (third) eye.
(Courtesy of O.F. Jackson)

The parietal eye is located just below a hole in the parietal bone (Davies, 1981); in the Tuatara the overlying scales are transparent (Duke-Elder, 1958). There is a nerve connecting the parietal eye to the epithalamus, entering the brain at the base of the pineal body. There is, therefore, a relationship between the parietal eye, the pineal body and the habenular nucleus (Evans, 1986) and this is thought to play a part in hormone (especially gonadal) production and thermoregulation. For this reason Frye (1991) referred to the parietal eye as a dosimeter.

Iris

All New World non-venomous snakes, except the boa constrictor *(Boa constrictor constrictor)*, have round pupils, while pit vipers *(Crotalidae)* have vertical slit pupils (Marcus, 1981). Coral snakes (*Micrurus* spp.) also have round pupils.

The pupils of diurnal lizards are usually round and relatively immobile, while nocturnal lizards, eg. geckos, have a slit pupil which can almost completely close in bright light, leaving only a row of stenopoeic openings down its length which, acting together, produce an image of considerable clarity without any dioptric mechanisms or accommodative adjustments (Duke-Elder, 1958). When not fully closed this may give the appearance of two or more pupils.

The iris of the common box-tortoise (*Terrapene carolina*) is peculiar in that it shows sexual dimorphism, the male iris being red and the female iris brown (Duke-Elder, 1958).

Posterior segment

The retina is avascular, usually with nerve fibres radiating uniformly outwards from the disc, which is obscured in Families having a conus.

Crocodilians possess a tapetum (Duke-Elder, 1958; Marcus, 1981; Millichamp *et al*, 1983; Evans, 1986) which produces an orange to reddish reflex glow when a light is shone on to the eyes, as it is rich in guanine crystals. This can be used at night to the crocodilian's detriment: a torch light easily discloses the animal's presence to the hunter.

The ophidian retina is usually grey mottled with white or red spots and with semi-opaque nerve fibres radiating uniformly from the optic disc. Over the surface of the retina, in the vitreous, is the membrana vasculosa retinae (Duke-Elder, 1958).

DISEASES OF THE EYELIDS

Trauma

Trauma to eyelids may occur in all species. This may be as a result of a fight at the time of feeding, eg. terrapins fed in a small tank may go into a feeding frenzy.

Frye (1972) described traumatic burns to eyelids from a heatlamp; this led to closure of the palpebral fissure.

Treatment of traumatised eyelids, irrespective of the cause, involves cleaning with a povidone-iodine preparation and topical antibiosis where indicated. If the damage is such as to affect the continuity of the eyelid margin, then, under general anaesthesia, blepharoplasty, as used for mammals (Slatter, 1981), should be performed to repair the defect.

Damage to the spectacles of snakes may occur as one of the hazards of feeding live prey. Treatment requires topical antiseptics and antibiosis and, where possible, the spectacle should be repaired under anaesthesia using 8/0 or 10/0 suture material.

Retained spectacle

In snakes the spectacle is shed as part of the normal shedding cycle. If this fails to occur then a retained spectacle results.

Retained spectacles are a common finding in snakes (Cooper, 1975; Cooper and Jackson, 1981). If not removed, subsequent shedding will also be inhibited and the spectacles will buildup to form a thickened mass of dead skin which will affect the snake's vision (see Figure 4) and may affect its willingness or ability to feed. Retained spectacles, if left, may become secondarily infected and can result in blindness (Zwart et al, 1973). Even a single retained spectacle will have an effect, the surface appearing irregular or even dented. The more retained layers, the easier it is to diagnose the retained spectacle. The majority of retained spectacles are associated with environmental conditions (Lawton, unpublished data). Other factors, such as inadequate nutrition or systemic disease, may also be contributory (Millichamp and Jacobson, 1986).

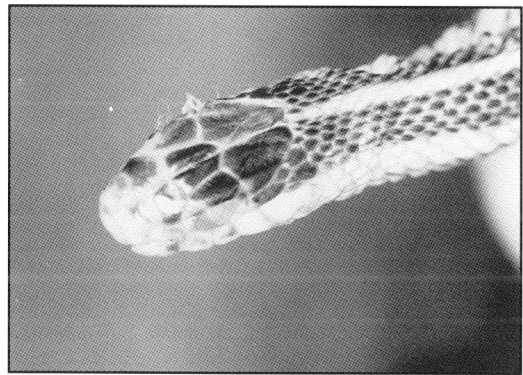

Figure 4
A retained spectacle in a garter snake (*Thamnophis* sp.).

Figure 5
Removing a retained spectacle by rubbing at the lateral canthus.

Treatment involves gentle soaking of the retained spectacle to aid its removal. This is best done by use of a wetted cotton bud, rubbing from the medial and lateral canthi towards the centre of the eye (see Figure 5). If the spectacle fails to come away with this technique, the use of artificial tears, eg. hypromellose, over a number of days will help to soften the spectacle before its removal. Although Frye (1981) advised the use of forceps or other instruments to remove a retained spectacle, the author considers these techniques ill-advised and liable to cause damage to the underlying spectacle. If gentle rubbing or the use of hypromellose will not allow removal of the retained spectacle, it is better to leave it, if necessary, until the next slough. Damage to the underlying spectacle resulting from an attempt to remove a retained spectacle can have a deleterious effect on the health of the eye (see later).

Avulsed spectacle

Damage to the eyelids in snakes is more serious than in other reptiles. The whole spectacle may be avulsed if an inexperienced herpetologist uses sellotape in an attempt to remove what is considered to be a retained spectacle. The spectacle may also be lost by other forms of trauma or infection. If totally lost this can lead to desiccation of the cornea, as the eyelids in all species are important in maintaining the continuity of the tear film over the cornea.

Topical antibiosis and artificial tears are seldom sufficient to save the eye in the case of total avulsion of the spectacle. Treatment may be attempted using a cut-down, soft contact lens or by performing an oral mucosa transposition over the eye (Lawton, unpublished data).

Parasites

It is not uncommon to find parasites, especially mites or ticks, around the adnexae (Jacobson, 1988). In snakes these may lie in the fossa between the periocular scales of the face and the origins of the spectacle and may be difficult to see. Parasites may be physically removed by flushing and appropriate treatment taken to eliminate the parasitic burden (see "Therapeutics").

Blepharoedema

Oedema of the eyelids (blepharoedema) is a common finding in reptiles and is often associated with vitamin A deficiency (see "Nutritional Diseases").

Neoplasia

A number of superficial neoplasms may affect the eyelids (see "Integument").

Green sea turtles (*Chelonia mydas*) may develop tumours of the eyelids, associated with a herpesvirus: these can be fibromas, papillomas or fibropapillomas (Millichamp et al, 1983; Millichamp and Jacobson, 1986). Green lizards (*Lacerta* spp.) may also develop viral papillomas around the eyelids (see Figure 6). Poxvirus may cause focal raised papules of the eyelids in the spectacled caiman (*Caiman crocodilus*) (Millichamp et al, 1983).

If any of these tumours interfere with the normal function of the eyelids or interfere with vision, surgical removal may be attempted, though they may recur.

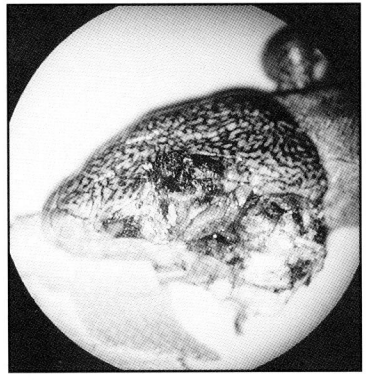

Figure 6
Papillomas around the eyelids of a green lizard (*Lacerta* sp.).

OPHTHALMIC FOREIGN BODIES

Foreign bodies in the conjunctival fornix may cause blepharospasm; general anaesthesia or nerve blocks may be required in order to examine the eye properly.

Tortoises (*Testudo* spp.) hibernated in hay or straw may acquire seeds as a foreign body. Lizards kept on peat or sand may occasionally get such material into their eyes. Discomfort caused by the foreign body reaction prompts them to rub their faces even more into the substrate and exacerbate the problem.

Foreign bodies that are untreated or penetrate the cornea can result in panophthalmitis (Frye, 1991).

Treatment involves the removal of the foreign body by grasping it with fine forceps, or flushing using Hartmann's solution via a soft intravenous cannula. The cornea should be checked with fluoroscein for ulceration or abrasions and treated accordingly.

TEARS

Keratitis sicca

Reptiles can suffer from a lack of production of tears in a similar way to mammals. The most common cause of keratitis sicca is changes in the lacrimal and Harderian glands associated with vitamin A deficiency (Elkan and Zwart, 1967). There is also epithelial metaplasia and excessive keratosis, not just of the conjunctiva but of the cornea. Providing the deficiency has not been left too long, there is often some response to vitamin A supplementation and lacrimation may eventually return.

Keratitis sicca in reptiles is seen as a lustless cornea, an absence of any fluid and, usually, a mucoid tacky discharge. If this drying of the cornea is chronic, it will lead to hyperkeratosis and pigmentation (see Figure 7). A definitive diagnosis is made using Schirmer tear test strips, if necessary cut down to size.

Treatment is similar to that employed in mammals, ie. using artificial tear preparations. If the xerophthalmia is due to vitamin A deficiency, administration of vitamin A on a weekly basis is advised.

Tear staining syndrome

Excess tear staining is common in some reptiles, eg. *Testudo* spp., due to the lack of a functional nasolacrimal system. The tears spill naturally over the eyelids and down the side of the face and eventually evaporate. This is normal and requires no treatment.

Blockage of the nasolacrimal system

Should the nasolacrimal system need to be investigated, fluorescein may be used in a similar way to mammals. Ophidians require the injection of a fluorescein suspension through the spectacle using a 27G or 30G needle (Millichamp *et al*, 1986; Millichamp, 1988).

Blockage of the nasolacrimal system is more commonly encountered as a clinical problem in snakes. Any fluid not draining down the nasolacrimal ducts will cause an enlargement of the sub-spectacular space and distortion or bulging of the spectacle (see Figure 8). This should not be confused with the normal process of ecdysis, where fluid is present between the old and new skin layers of the spectacle, resulting in a bluish opacity of the spectacle. Chronic blockage of the nasolacrimal system leads to the accumulation of fluid and a bulging of the spectacle, but enophthalmus of the globe itself.

The blockage may be due to infection, pressure, fibrosis or congenital abnormality. The distention should consist of a clear fluid and should not be confused with a sub-spectacular abscess (see

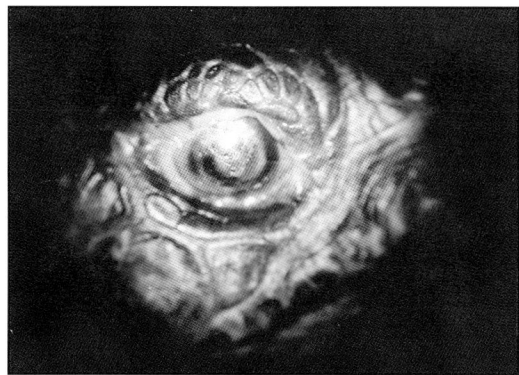

Figure 7
Keratitis sica in a
Testudo sp.

Figure 8
A bulging sub-spectacular space (clear fluid)
in a royal python (*Python regius*).

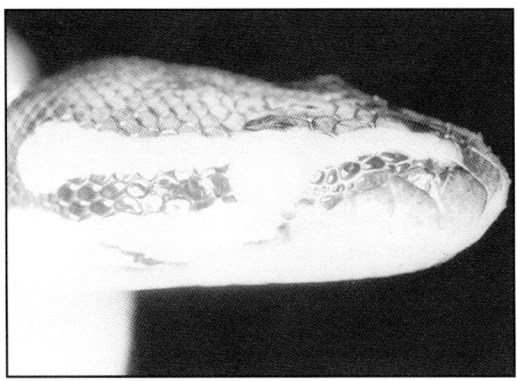

Figure 9
Sub-spectacular abscess.

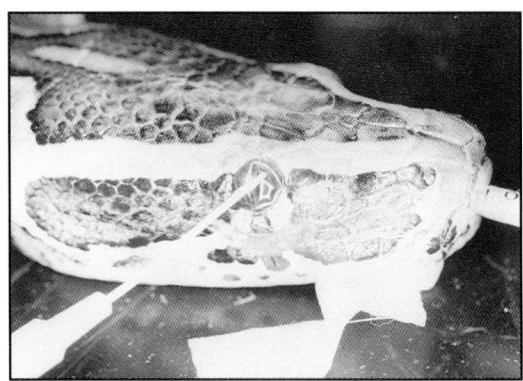

Figure 10
A small wedge is cut in the lateral canthus and the sub-spectacular space is flushed with Hartmann's solution.

Conjunctivitis). It is advisable to investigate the patency of the nasolacrimal duct by injecting a small amount of fluorescein through the lateral canthus of the spectacle into the sub-spectacular space, as mentioned earlier. Chronic increase in pressure can damage the eye.

Total blockage which cannot be readily unblocked may require the formation of a surgical fistula and conjunctivoralostomy as described by Millichamp et al (1986). Under general anaesthesia an incision is made through the spectacle and an 18G needle is passed between the inferior fornix of the sub-spectacular space and the roof of the mouth, emerging between the palatine and maxillary teeth. Sialistic tubing (0.635mm diameter) is threaded through the needle to maintain the patency. The tubing is left *in situ* for one month and then removed.

CONJUNCTIVITIS

Conjunctivitis is a common problem in all reptiles. However, due to the nature of the inflammatory response and lack of lysosomes from the heterophils, a mucopurulent discharge is seldom seen. An infectious conjunctivitis usually results in a caseous plaque which is retained in or on the eye. This can cause a foreign body reaction. When presented with any reptile (other than ophidians) with a closed eye, one should flush out the conjunctival fornix with Hartmann's solution using a fine soft catheter (24G) or explore using wetted endodontic paper points in order to remove the caseous plaque. Antibiosis, using tobramycin or gentamicin drops, should be applied four to six times daily.

Reptiles suffering from vitamin A deficiency may have a similar caseous mass in the conjunctival fornix. This is associated with desquamated cells in combination with the xerophthalmia due to the squamous cell metaplasia of the lacrimal glands and not due to infection. The plaque should be removed in a similar fashion to that described above. If it is not removed, a reptile, even when successfully treated with vitamin A, will still not open its eye. This is the commonest cause of apparent lack of response to treatment in terrapins.

Snakes do suffer from conjunctivitis but, because of the spectacle, this is referred to as a sub-spectacular abscess. The presenting sign is a cloudy or white spectacle which may or may not be distorted (see Figure 9). Sub-spectacular abscesses may occur as a result of stomatitis and infection ascending the nasolacrimal duct (Millichamp et al, 1986), or through haematogenous spread of systemic infection (Cooper, 1981; Millichamp and Jacobson, 1986; Millichamp et al, 1986), although often only one eye is affected.

Treatment involves cutting a small wedge through the spectacle at the lateral canthus (see Figure 10). A swab for culture and sensitivity testing should be taken. The sub-spectacular space should be flushed with Hartmann's solution and a suitable antibiotic solution, such as tobramycin or gentamicin. This should be repeated on at least a daily basis. It is advisable to give systemic antibiotics as well.

DISEASES OF THE CORNEA

Keratitis

Infection of the cornea is especially noted in chelonians, where a lesion involving an infectious agent may result in a white plaque appearing on the cornea (see Figure 11). Under general anaesthesia it is possible to debride this plaque; culture and sensitivity testing should be undertaken. The bacteria involved are varied and may include *Moraxella* spp. (Lawton, unpublished data), *Pseudomonas* spp. or *Aeromonas* spp. (Cooper *et al*, 1980). This type of keratitis in chelonians appears to be contagious and should be considered a colony problem.

Treatment with a suitable antibiotic, such as tobramycin or gentamicin topically, will often prove successful.

Not all corneal opacities are associated with bacterial infections. Fungal keratitis may also be encountered in snakes. If left untreated this may result in a panophthalmitis and require enucleation of the affected eye (Zwart *et al*, 1973; Collette and Curry, 1978). Diagnosis is made by taking a scraping of the lesion and demonstrating the fungal hyphae. However, non-pathogenic, non-invasive fungi are probably present on the normal healthy cornea of all reptiles.

The long-term use of corticosteroids (in combination with or without antibiotics) can result in connective tissue break down. This reduces the mucopolysaccharides and causes degenerative changes in the basement membrane of the epithelium, allowing fungi to proliferate (Collette and Curry, 1978).

Treatment involves the topical use of miconazole (Conoderm 1%, C-Vet).

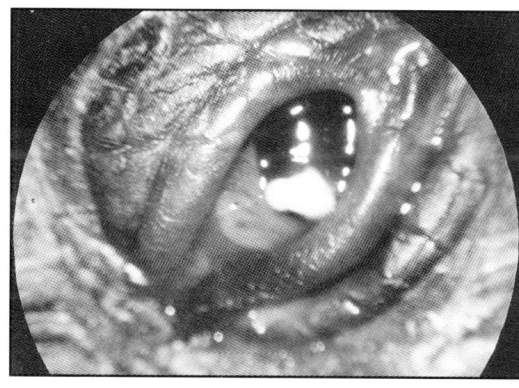

Figure 11
Small white plaque (keratitis) on the cornea of a *Testudo* sp.

Ulceration

Ulceration, especially associated with foreign bodies or trauma, is a common finding. If the damage to the cornea is severe, thought should be given to performing a third eyelid flap in a similar way to that employed in dogs and cats (see "Surgery"). Traumatic lacerations of the cornea should be cleaned and sutured with fine suture material (8/0 or 10/0) and treated with a topical antibiotic (Northway, 1970).

Arcus lipoides corneae

Arcus lipoides corneae is seen as a white area partially or totally around the cornea usually near the limbus (see Figure 12). Although perfectly normal in chelonians, arcus lipoides corneae may be misdiagnosed as a problem. It is part of the normal ageing process and is seen in the majority of adult *Testudo* spp. (Lawton, unpublished data).

Frye (1991) described this condition in monitor lizards (*Varanus* spp.), together with a reversible corneal opacity in salt water snakes and marine iguanas (*Amblyrhynchus* spp.) kept in fresh water.

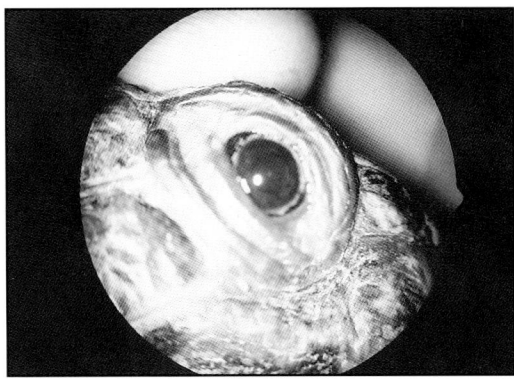

Figure 12
Arcus lipoides corneae
in a *Testudo* sp.

DISEASES OF THE ANTERIOR CHAMBER

Although Cooper and Jackson (1981) stated that intraocular abnormalities appear to be rare in reptiles, they are now being recognised as ophthalmic examination becomes a routine part of the clinical examination of reptiles.

Uveitis

Uveitis may be traumatic, bacterial, fungal, viral or due to neoplastic infiltration of tissues of the uveal tract (Frye, 1991). Clinical signs are the same as for mammals.

Topical treatment for uveitis is with antibiotics, steroids and cycloplegic drugs, although in reptiles with spectacles injection is required into the sub-spectacular space.

Hyphaema

Hyphaema may be associated with uveitis or be independent of other clinical signs. Hyphaema may be seen in chelonians following a freezing episode or trauma (Lawton, 1989). With time and conservative treatment this will usually disappear.

Hypopyon

Hypopyon has been associated with pneumonia in both chelonians (Tomson *et al*, 1986) and lizards (Bonney *et al*, 1978). The condition resolved after successful treatment of the systemic disease in the tortoise but was unsuccessful in the gecko.

DISEASES OF THE LENS

Cataracts

As in all species, cataracts may occur for a variety of reasons. They may be congenital, primary or secondary.

Lawton (1989) and Lawton and Stoakes (1989) reported that cataracts in chelonians may be associated with freezing episodes. The chelonian lens is extremely soft and almost fluid in consistency (Duke-Elder, 1958) and, therefore, may be particularly prone to freeze damage.

Reptiles should have their lenses examined to make sure that there are no opacities which could result in anorexia or behavioural abnormalities. Surgery may be performed on these cataracts. Technically this is difficult and Millichamp *et al* (1983) have questioned the advantages of such surgery.

Frye (1991) has reported the occurrence of nuclear sclerosis, although the author has yet to see this condition.

Luxation

Traumatic luxation, usually posteriorly, has been seen in *Testudo* spp. (Lawton, unpublished data).

FUNDUS

Retinal damage has been reported in *Testudo* spp. in association with vitamin A deficiency and following freezing episodes (Lawton, 1989). In certain cases treatment with vitamin A may result in regression of the retinal lesion and some retinal function may return.

In Tokay geckos (*Gekko gecko*) retinal degeneration has been reported (Schmidt and Toft, 1981), diagnosed ophthalmoscopicaly and confirmed on histopathology as loss of photoreceptors, disorganisation of nuclear layers and microphages full of debris and pigment. This was also found incidentally in another Tokay gecko with hypopyon (Bonney *et al*, 1978). Millichamp and Jacobson (1986) considered retinal degeneration a sporadic finding in most Families of reptiles.

Optic neuropathy in the Mexican wood turtle (*Rhinoclemmys pulcherrima*) is thought to be associated with thiamine deficiency (Frye, 1991) as is blindness in garter snakes (*Thamnophis* spp.) in long-standing cases.

ANOPHTHALMIA/MICROPHTHALMIA AND OTHER ABNORMALITIES

Anophthalmia is the complete absence of a globe, which in reality is very rare. Although certain reptiles may be presented with what appears to be anophthalmia, if histopathology is performed it is often shown that some ophthalmic tissue is present, ie. it is microphthalmia (Millichamp, personal communication).

Microphthalmia may occur in terrapins (*Trachemys* spp.), especially if the parents were fed on a vitamin A deficient diet (see Figure 13) (Lawton, unpublished data). Microphthalmia has also been seen in pythons (*Python molurus molurus*) (Millichamp, 1989) and red-headed rat snakes (*Elaphe moellendorffi*); histologically the eyes lacked lenses (Ensley *et al*, 1978). In both reports this was considered to be due to genetic or incubation factors.

Microphthalmia in lizards is associated with abnormalities of the interorbital septum (Bellairs, 1981).

Cyclopia has also been reported (Bellairs, 1981).

Cooper (1975) described a congenital **exophthalmia** in a rhinoceros viper (*Bitis nasicornis*) due to a developmental cyst. Such conditions may be more common than formerly believed and might increase in prevalence because of inbreeding in captivity.

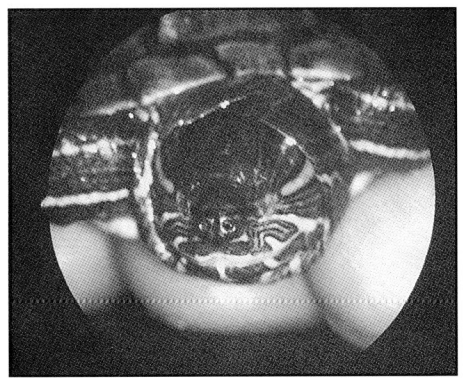

Figure 13
Microphthalmia in a terrapin
(*Trachemys scripta elegans*).

REFERENCES

BELLAIRS, A. d'A. (1981). Congenital and developmental diseases. In: *Diseases of the Reptilia, Vol. 2.* (Eds. J.E. Cooper and O.F. Jackson). Academic Press, London.

BONNEY, C.H., HARTFIEL, D.H. and SCHMIDT, R.W. (1978). *Klebsiella* pneumonia infection with secondary hypopyon in Tokay gecko lizards. *Journal of the American Veterinary Medical Association* **173 (9)**, 1115.

COLLETTE, B.E. and CURRY, O.H. (1978). Mycotic keratitis in a reticulated python. *Journal of the American Veterinary Medical Association* **173 (9),** 1117.

COOPER, J.E. (1975). Exophthalmia in a rhinoceros viper (*Bitis nasicornis*). *Veterinary Record* **97**, 130.

COOPER, J.E. (1981). Bacteria. In: *Diseases of the Reptilia, Vol. 1.* (Eds. J.E. Cooper and O.F. Jackson). Academic Press, London.

COOPER, J.E. and JACKSON, O.F. (1981). Miscellaneous diseases. In: *Diseases of the Reptilia, Vol. 2.* (Eds. J.E. Cooper and O.F. Jackson). Academic Press, London.

COOPER, J.E., McCLELLAND, M.H. and NEEDHAM, J.R. (1980). An eye infection in laboratory lizards associated with an *Aeromonas* sp. *Laboratory Animals* **14**, 149.

DAVIES, P.M.C. (1981). Anatomy and physiology. In: *Diseases of the Reptilia, Vol. 1.* (Eds. J.E. Cooper and O.F. Jackson). Academic Press, London.

DUKE-ELDER, S. (1958). The eyes of reptiles. In: *Systems of Ophthalmology. Vol. 1: the Eye in Evolution.* Kimpton, London.

ELKAN, E. and ZWART, P. (1967). The ocular diseases of young terrapins caused by Vitamin A deficiency. *Pathologica Veterinarian* **4 (3),** 201.

ENSLEY, T.K., ANDERSON, M.P. and BACON, J.P. (1978). Ophthalmic disorders in three snakes. *Journal of Zoo Animal Medicine* **9,** 57.

EVANS, H.E. (1986). Introduction and anatomy. In: *Zoo and Wild Animal Medicine.* 2nd Edn. (Ed. M.E. Fowler). W.B. Saunders, Philadelphia.

FRYE, F.L. (1972). Blepharoplasty in an iguana. *Veterinary Medicine/Small Animal Clinician* **67,** 1110.

FRYE, F.L. (1981). Traumatic and physical disease. In: *Diseases of the Reptilia, Vol. 2.* (Eds. J.E. Cooper and O.F. Jackson). Academic Press, London.

FRYE, F.L. (1991). *Biomedical and Surgical Aspects of Captive Reptile Husbandry.* 2nd Edn. Krieger, Malabar.

JACKSON, O.F. (1981). Clinical aspects of diagnosis and treatment. In: *Diseases of the Reptilia, Vol. 2.* (Eds. J.E. Cooper and O.F. Jackson). Academic Press, London.

JACOBSON, E.R. (1988). The evaluation of the reptile patient. In: *Exotic Animals.* (Eds. E.R. Jacobson and G.V. Kollias). Churchill Livingstone, New York.

LAWTON, M.P.C. (1989). Neurological problems of exotic species. In: *Manual of Small Animal Neurology.* (Ed. S.J. Wheeler). BSAVA, Cheltenham.

LAWTON, M.P.C. and STOAKES, L.C. (1989). Post hibernation blindness in tortoises (*Testudo* spp.). In: *Third International Colloquium on Pathology of Reptiles and Amphibians.* (Ed. E.R. Jacobson). Orlando, Florida.

MARCUS, L.C. (1981). *Veterinary Biology and Medicine of Captive Amphibians and Reptiles.* Lea and Febiger, Philadelphia.

MILLICHAMP, N.J. (1988). Surgical techniques in reptiles. In: *Exotic Animals.* (Eds. E.R. Jacobson and G.V. Kollias). Churchill Livingstone, New York.

MILLICHAMP, N.J. (1989). Congenital defects in captive-bred Indian pythons (*Python molurus molurus*). In: *Third International Colloquium on Pathology of Reptiles and Amphibians.* (Ed. E.R. Jacobson). Orlando, Florida.

MILLICHAMP, N.J. and JACOBSON, E.R. (1986). Ophthalmic diseases of reptiles. In: *Current Veterinary Therapy IX. Small Animal Practice.* (Ed. R.W. Kirk). W.B. Saunders, Philadelphia.

MILLICHAMP, N.J., JACOBSON, E.R. and DZIEZYC, J. (1986). Conjunctivorialostomy for treatment of an occluded lacrimal duct in a blood python. *Journal of the American Veterinary Medical Association* **189 (9),** 1136.

MILLICHAMP, N.J., JACOBSON, E.R. and WOLF, E.D. (1983). Diseases of the eye and ocular adnexae in reptiles. *Journal of the American Veterinary Medical Association* **183 (11),** 1205.

NORTHWAY, R.B. (1970). Repair of a fractured shell and lacerated cornea in a tortoise. *Veterinary Medicine/Small Animal Clinician* **65**, 944.

SCHMIDT, R.E. and TOFT, J.D. (1981). Ophthalmic lesions in animals from a zoological collection. *Journal of Wildlife Diseases* **17 (2),** 267.

SLATTER, D.H. (1981). *Fundamentals of Veterinary Ophthalmology.* W.B. Saunders, London.

TOMSON, F.N., McDONALD, S.E. and WOLF, E.D. (1986). Hypopyon in a tortoise. *Journal of the American Veterinary Medical Association* **169 (8),** 942.

WALLS, G.L. (1942). The vertebrate eye and its adaptive radiation. In: *Cranbrook Institute of Science, Bulletin No. 19.* Cranbrook.

ZWART, P., VERWER, M.A.J., De VRIES, G.A., HERMANDIEZ-NIJHOF, E.J. and De VRIES, H.W. (1973). Fungal infection of the eyelids of the snake *Epicrates chenchria maurus* : enucleation under halothane narcosis. *Journal of Small Animal Practice* **14**, 773.

CHAPTER SIXTEEN

ANAESTHESIA

Martin P C Lawton BVetMed CertVOphthal FRCVS

INTRODUCTION

Anaesthetic drugs play a major role in reptile medicine and surgery, not only for surgical procedures but also, often in lower doses, for sedation or immobilisation to facilitate clinical examination and investigative tests, such as endoscopy, radiography and blood sampling. Great advances have been made since Brazenor and Kaye (1953) described anaesthesia of Australian snakes and lizards as "an amusing by-way of anaesthesia".

The basic approach to anaesthesia is very similar to that in any other species, except one has to remember that reptiles are ectothermic (poikilothermic) and, therefore, the response to the drugs used will depend on the temperature before, during and after anaesthesia (Boever, 1979). Reptiles have different metabolic rates and, therefore, direct extrapolation from mammalian dosages should be used with great caution (Burke, 1986). It is advisable, wherever possible, for the reptile to be maintained at its preferred body temperature (PBT) both pre-operatively and in the recovery phase.

Reviews on anaesthesia of reptiles are provided by Calderwood (1971), Bonath (1978), Holt (1981), Jackson and Cooper (1981), Marcus (1981), Burke (1986), Carbo and Martin (1986), Sedgwick (1988), Bennet (1991) and Frye (1991).

An approach to anaesthetising reptiles is given in the flow chart (see Table 1).

ANAESTHETIC REQUIREMENTS

Whichever technique or combination of drugs is used, there are four criteria that are required of an anaesthetic.

1. **Restraint** - the degree of restraint will depend on the procedure to be performed. Minimal restraint is required for a general clinical examination and, therefore, it is unlikely that an anaesthetic is necessary, though occasionally sedation may be advantageous. For surgery it is essential that the animal is both immobilised and anaesthetised.

2. **Muscle relaxation** - the degree of muscle relaxation required is dependent upon the procedure to be performed. Substantial muscle relaxation is required for coeliotomy (exploratory laparotomy), while for clinical examination or removal of a superficial skin lesion it may not be necessary.

3. **Analgesia** - this is an important part of anaesthesia. Reptiles should never be considered incapable of feeling pain; they are highly sensitive to skin incisions (Sedgwick, 1980).

4. **Uncomplicated recovery** - no matter how good a surgeon or how intricate the surgery to be performed, if the animal does not recover afterwards the procedure is to no avail. It is for this reason that an anaesthetic which allows a more rapid and uncomplicated recovery should always be considered.

PRIOR TO ANAESTHESIA

Bodyweight. Prior to sedating or anaesthetising any reptile, it is important to know its exact bodyweight (mass). The weight is important not only for assessing postoperatively any fluid loss but also for calculating an accurate dose of the drugs that are to be given.

Premedication. Atropine as a premedicant is considered unnecessary as salivary excretions are not a problem in reptiles (Burke, 1986). Sedatives, such as metomidate, may prove helpful in snakes or lizards prior to induction.

Warmth. Reptiles should be maintained at their PBT both prior to anaesthesia and in the recovery stage. A reptile that is returned to a vivarium already at its PBT will recover more quickly than one put into a cold tank (Lawton, unpublished data), so a vivarium should be prepared prior to anaesthetising the patient. Kraner *et al* (1965) advised placing snakes in warm water at 35°C to hasten recovery but this author would not recommend such an approach.

Table 1
An approach to anaesthetising reptiles.

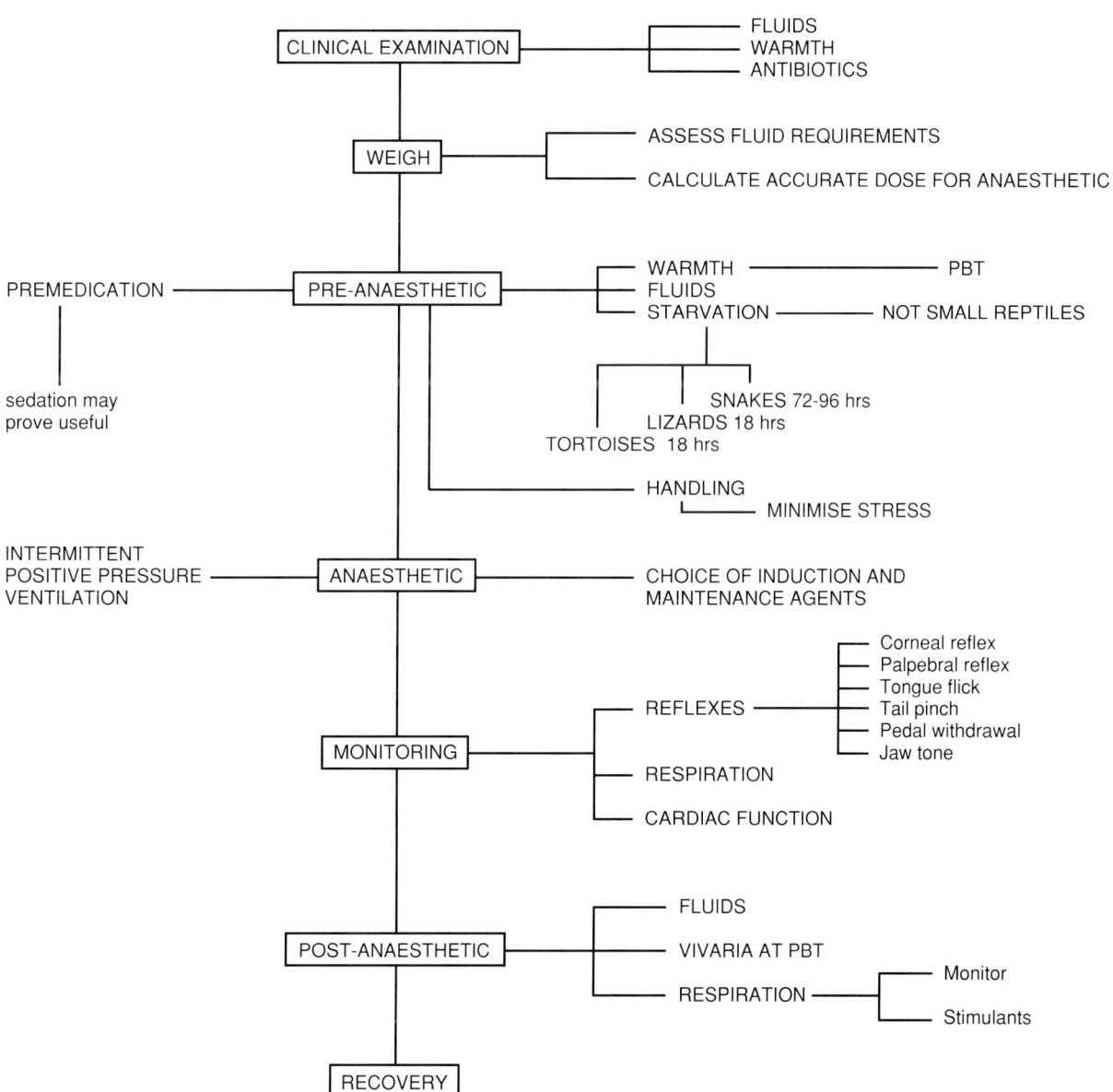

Fluid therapy. Dehydrated reptiles should be given fluids prior to being anaesthetised or undergoing surgery. If the type of surgery to be performed is one that might result in blood loss, prophylactic fluid therapy prior to surgery is advised. With certain drugs, such as ketamine, following which recovery may be prolonged, it is important to maintain a good fluid balance in the recovery stage.

Starvation. Prior to anaesthesia it is not necessary to starve smaller species, though one should avoid anaesthetising them while there is fresh food in the oesophagus or live insects in the stomach. Tortoises and lizards should be starved for a period of 18 hours but larger snakes, eg. boids, take longer to digest their food and may be starved for 72 – 96 hours.

Handling. As far as possible reptiles should not be stressed or bruised prior to, or at the time of, giving an anaesthetic. Correct handling is essential, especially when intravenous injections are being administered. In some cases it may be easier to sedate the animal prior to intravenous injection (see later).

ANAESTHETIC AND SEDATIVE AGENTS

Dose rates and routes of administration of the agents described in this chapter are given in Table 2.

INTRAVENOUS INJECTION

A summary of the sites for intravenous injection of reptiles is given in Table 3, and mention is also made elsewhere in this manual (see "Laboratory Investigations"). Figures 1, 2 and 3 show intravenous injection of a tortoise, lizard and snake respectively.

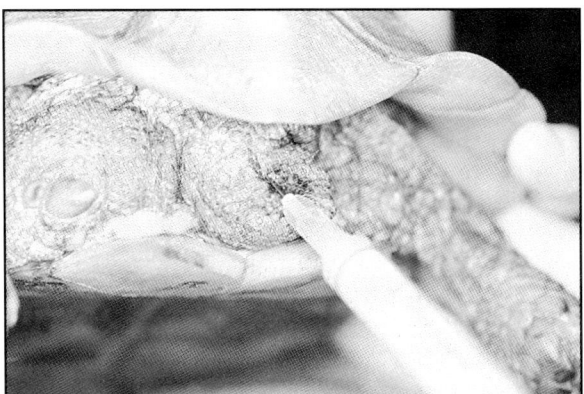

Figure 1
I/v injection of a tortoise (dorsal venous sinus).

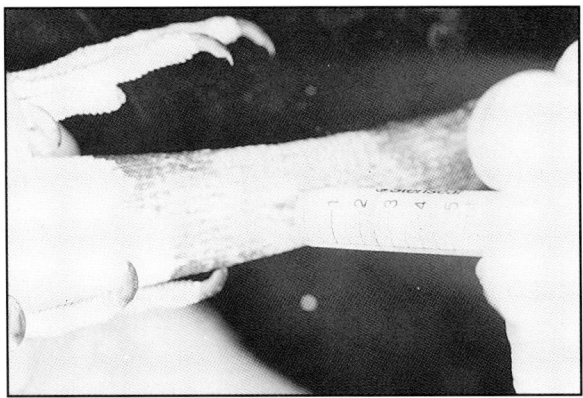

Figure 2
I/v injection of a lizard (ventral venous sinus).

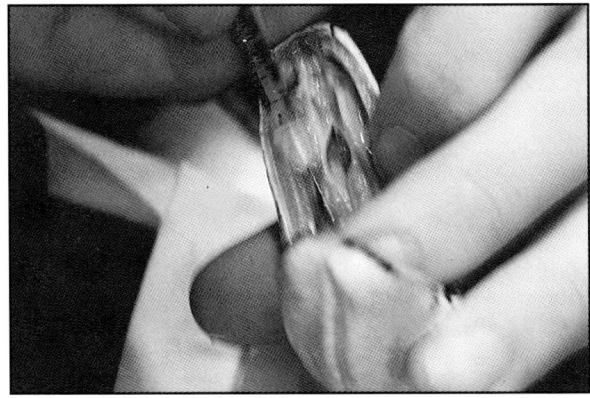

Figure 3
I/v injection of a snake (palatine vein).

INJECTABLE ANAESTHETICS

Pentobarbitone

Pentobarbitone has been used in reptiles by Betz (1962), Cooper (1971) and Jackson and Cooper (1981). The disadvantages are the long induction time (40 – 60 minutes), very long recovery (18 – 36 hours (Betz, 1962), up to three days (Jones, 1977)), respiratory depression and poor control over the depth of anaesthesia (Jackson and Cooper, 1981). This agent has no place in reptilian anaesthesia other than for euthanasia, when it is given by the intravenous or intraperitoneal route (see "Miscellaneous").

Table 2
Anaesthetic and sedative agents.

Drug	Trade Name	Dosage	Site
Alphaxalone/ alphadolone	Saffan, Pitman-Moore.	6 – 9mg/kg 9 – 15mg/kg	i/v i/m
Ketamine hydrochloride	Vetalar, Parke-Davis; Ketaset, Willows Francis.	20 – 100mg/kg	s/c i/m or i/p
Metomidate	Hypnodil, Janssen.	10 – 20mg/kg	i/m
Propofol	Rapinovet, Pitman-Moore; Diprivan, ICI.	Tortoises 14mg/kg Lizards 10mg/kg Snakes 10mg/kg	i/v
Pentobarbitone	Sagatal, Rhône Mérieux.	10 – 30mg/kg	i/m i/p
Thiopentone	Intraval Sodium, Rhône Mérieux.	20 – 30mg/kg of 2.5%	i/v i/p
Etorphine	M99, C-Vet; Immobilon, C-Vet.	0.3 – 2.75mg/kg	i/m
Tricaine methanesulphonate	MS222, Sandoz.	15 – 30mg/kg	i/p
Succinylcholine (NB. this is a muscle relaxant)	Scoline, Duncan Flockhart.	0.34 – 1mg/kg	i/m or i/p
Tiletamine hydrochloride	Telazole, Parke-Davis.	3.5 – 210mg/kg	i/m
Xylazine	Rompun, Bayer.	0.1 – 10mg/kg	i/m
Halothane	Halothane, Rhône Mérieux; Fluothane, Pitman-Moore.	1 – 4%	Inhalation
Isoflurane	Forane, Abbots.	1 – 6%	Inhalation
Methoxyflurane	Metofane, C-Vet.	1 – 4%	Inhalation

Table 3

Sites for intravenous injection.

Site	Technique	Comments
Chelonians		
Dorsal venous sinus (vena coccygea superior)	Located on the dorsal midline of the tail. The needle is angled at 45° cranially and advanced until bone is reached, then withdrawn slightly.	This is the vessel of choice (Samour et al, 1984)
Ramus coccyggeus lateralis pracipus	Used in giant tortoises; it is a deep vein lateral to the coccygeal vertebrae.	Richter et al, 1977.
Femoral venous plexus	Located on the medial aspect of the thigh. The needle is advanced until it strikes the femur and then angled slightly caudally and laterally.	Repeated attempts may be required to find this vein (Richter et al, 1977).
Cardiac puncture	Requires trephination of the plastron in adults.	Not advised as trephination should only be performed under anaesthesia.
Jugular veins	As per mammals.	Requires experience (Lawrence, 1985).
Lizards (saurians)		
Ventral venous sinus	Located on the ventral midline of the tail. The needle is angled at 45° cranially and advanced to the midline aiming between the coccygeal vertebrae.	This is the vessel of choice (Esra et al, 1975; Samour et al, 1984).
Ventral vena cava	Located in the midline of the abdominal wall.	This site is not advised as may result in large haematoma / haemorrhage as complications.
Cardiac puncture	May be approached from the ventral abdomen angling the needle cranially under the rib cage.	Potentially dangerous. Only advised in animals that are terminally ill or are to be culled (Calderwood, 1971; Sedgwick, 1980).

Table 3 (continued)

Sites for intravenous injection.

Site	Technique	Comments
Snakes (Serpentes)		
Ventral venous sinus	The needle is angled at 45° as described for saurians, although it is initially placed between paired caudal scales.	Care must be taken in males not to damage the hemipenes.
Palatine and sub-lingual veins	These vessels are readily seen when the mouth is open. Bending the needle is usually necessary.	Sampling is easier if the reptile is sedated with metomidate or lightly anaesthetised.
Cardiac puncture	The heart may be palpated ventrally about 1/5 to 1/4 the distance between rostrum and cloaca.	Potentially dangerous. Only recommended in animals that are terminally ill or are to be culled (Lawrence, 1985).
Jugular veins	As per mammals.	Requires experience.

Thiopentone

Thiopentone induces anaesthesia in 30–45 minutes (Jackson and Cooper, 1981). It must be given intravenously because an intraperitoneal injection causes severe peritonitis (Jackson and Cooper, 1981). Perivascular injection also causes irritation. Kraner et al (1965) required near lethal doses to achieve anaesthesia and reported heavy mortalities due to respiratory depression. Recovery takes up to six hours (Jackson and Cooper, 1981).

Metomidate

Metomidate is a very useful sedative for snakes (Holt, 1981). It facilitates clinical examination, endoscopy or blood sampling. This author has also found metomidate useful for dealing with snakes with severe stomatitis: it can be used safely on a daily basis for debriding, cleaning and treating the mouth.

Metomidate has a rapid onset: when given intramuscularly the snake is usually heavily sedated after 15–20 minutes. It should not, however, be considered an anaesthetic as it has no analgesic property (Burke, 1986). It is also useful for sedating snakes prior to an intravenous injection to induce anaesthesia.

Etorphine

Etorphine has been reported as useful in all reptiles (Wallach and Hoessle, 1970), especially crocodilians and terrapins, while Hinsch and Gandal (1969) reported very little effect in snakes. In crocodilians, etorphine produces 45–100 minutes of analgesia after 10–30 minutes induction (Burke, 1986). In red-eared terrapins (*Trachemys scripta elegans*) dosages up to 5mg/kg have been used (Jones, 1977). Because of its potential danger to humans, this author has not used etorphine and would not recommend its use other than in an emergency or, possibly, for dealing with large crocodiles in Africa. Safer drugs are available.

Tricaine methanesulphonate (MS222)

Klyde and Klein (1973) reported that MS222 at a dose rate of 50mg/kg by pleuroperitoneal injection failed to have a satisfactory effect in a caiman, while Brisbin (1966) reported satisfactory immobilisation at 90mg/kg by the same route but with an undesirably long recovery time of ten hours. MS222 has been used in snakes (Kaplan, 1969; Green and Precious, 1978); induction time was 12 – 14 minutes, giving 30 – 60 minutes of anaesthesia.

Succinylcholine

Jacobson (1988) considered succinylcholine to be the injectable agent of choice for crocodilians and for premedicating large tortoises prior to intubation for gaseous anaesthesia.

Klyde and Klein (1973) used succinylcholine (0.34 – 0.68mg/kg i/m) in crocodilians for restraint during transport. Ventilation equipment was available in case the patients became apnoeic, although the duration time of the drug was considered to be not more than the time the animal might normally spend under water. Brisbin (1966) also reported a wide range of tolerance at doses of 3 – 5mg/kg with no mortalities but recovery taking up to nine hours. The ethics of using this agent as the sole anaesthetic drug have to be considered critically with regard to the welfare of the animal and any possible legal implications for the anaesthetist. No analgesia is provided by this drug, just paralysis of muscles. Therefore, operating on a reptile so immobilised may be considered to be causing unnecessary suffering (see "Appendix Four – Legal Aspects").

Ketamine hydrochloride

Ketamine is very useful and safe, both by itself and in combination with other agents (Cooper, 1974). Ketamine hydrochloride is a phencyclidine analogue with approximately one fifth to one sixth of phencyclidine's activity (Beck, 1972). The dose rate varies depending on the reason for use, together with effect required. Glenn et al (1972) showed that the LD_{50} for Western diamond rattlesnakes (Crotalus atrox) was 154mg/kg and recommended endotracheal intubation and ventilation for doses above 110mg/kg. Cooper (1974) reported an inadvertent overdose of 227mg/kg; following manual ventilation for six hours respiration resumed and recovery ensued.

Ketamine has the advantage that it is efficacious if given subcutaneously, intramuscularly or intravenously. It is advisable to start at a lower dose rate of about 30mg/kg and then use incremental doses at 20 – 30 minute intervals until the desired state of anaesthesia is obtained (Burke, 1986). However, Sedgwick (1980) considers this approach to be irrational, the only result being prolonged incapacity of the animal. Doses of less than 50mg/kg usually produce tranquillisation or light anaesthesia, while doses above 50mg/kg prove satisfactory for surgery (Glenn et al, 1972 ; Cooper, 1974).

Ketamine is contraindicated in debilitated or dehydrated reptiles, especially those with hepatic or renal damage. The other major disadvantage of ketamine is that the recovery period is proportional to the dosage used; thus, with very high doses, such as 100mg/kg, recovery may take up to several days (Glenn et al, 1972; Jones, 1977). Some reptiles, particularly snakes, have been reported as becoming permanently aggressive after recovery from ketamine anaesthesia (Lawrence and Jackson, 1983).

Tiletamine hydrochloride

Tiletamine is neither available nor licensed at present in the UK. The dose range is 3.5 – 210mg/kg i/m (Holt, 1981). Tiletamine is a phencyclidine analogue with approximately half its potency, so it is 2 – 3 times more potent than ketamine hydrochloride (Beck, 1972). Tiletamine acts in a very similar way to ketamine; smaller dosages are required but a side effect is that it is highly convulsive (Beck, 1972). Studies have been carried out combining tiletamine (10 – 75mg/kg) with arylcycloalkylamine or zolazepam (both of which are diazepinones) in an attempt to prevent the convulsions and provide better muscle relaxation (Gray et al, 1974; Boever and Caputo, 1982). Boever and Caputo (1982) concluded that unless used as a part of balanced anaesthesia, tiletamine was only suitable as an anaesthetic in iguanas.

Xylazine

Xylazine at a dose of 0.1 – 1.25mg/kg i/m produces states varying from torpidity to anaesthesia (Frye, 1991). Immobilisation lasts from 45 minutes to 12 hours. Sedgwick (1988) recommended using xylazine (10mg/kg) and ketamine (50mg/kg) in combination and stated that it is an improvement over ketamine *per se*.

Alphaxalone/alphadolone

Alphaxalone/alphadolone is a very good anaesthetic in reptiles, especially when given intravenously. As in mammals, the intramuscular route is less predictable (Lawrence and Jackson, 1983) because fascial plains and fat may affect absorption. It has the disadvantage intramuscularly that fairly large volumes are required and the dose may have to be given at multiple sites in snakes.

Intravenously there is a fairly rapid response, allowing intubation within three minutes of the injection. However, induction by the intramuscular route is slower (25 – 40 minutes) (Lawrence and Jackson, 1983). Anaesthesia lasts 15 – 35 minutes with an average of 25 minutes (Lawrence and Jackson, 1983).

Recovery is more rapid than with ketamine, but may still take 1.5 – 4 hours with an average of 2.5 hours (Lawrence and Jackson, 1983).

Propofol

The desirable anaesthetic properties of propofol are its rapid smooth induction, minimal accumulation on repeated injections, relative freedom from excitatory side effects and rapid recovery with little apparent "hangover" effect (Glen and Hunter, 1984).

This author considers propofol to be the induction anaesthetic of choice for reptiles. It has an advantage over alphaxalone/alphadolone in that it is shorter acting – used by itself it gives approximately 20 minutes restraint and anaesthesia (Lawton, unpublished data) – while recovery postoperatively is rapid compared with alphaxalone/alphadolone. Another advantage is that propofol can be injected through a fine needle (27G) and so can be given by the intravenous route in very small reptiles, eg. garter snakes (*Thamnophis* spp.); alphaxalone/alphadolone is difficult to administer even through a 25G needle.

The disadvantage of propofol is that it has to be given intravenously whereas alphaxalone/alphadolone can be administered by other routes.

If a snake has been previously sedated with metomidate 20 minutes prior to induction, the dose rate of propofol required is only 5mg/kg i/v (Lawton, unpublished data). However, this prolongs the recovery time and is only advised in order to facilitate intravenous injection in this group of reptiles. Lizards and chelonians usually present less of a problem.

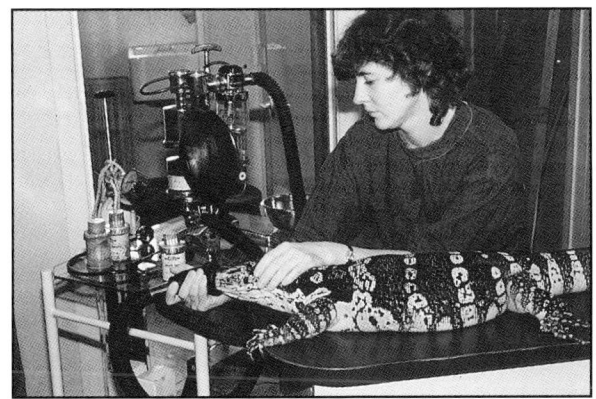

Figure 4
A lizard with its head in an anaesthetic mask.

ANAESTHETIC CIRCUITS

Open circuit

Face mask. Face masks can be used to some effect in iguanas and other large lizards, as many will bury their head in the mask and breathe and not try to escape (see Figure 4). They may also be used for snakes (Jackson, 1970), although more restraint is required to keep the head in the mask. Jackson (1970) stated that faster moving species take longer to anaesthetise. Although the reason for this was not suggested, it may be an effect of the temperature. In tortoises and terrapins the use of a face mask can prove to be very difficult and is not advised.

Anaesthetic chamber. An anaesthetic chamber can be used for most reptiles (Cooper, 1989). It is important that the animal is kept away from the volatile fluid. A chamber has the disadvantage that it is an inaccurate method unless an anaesthetic machine is attached which is delivering a known percentage of the anaesthetic gas. For health and safety reasons a scavenging system should be used. The advantages of an anaesthetic chamber are low cost, ease of induction and suitability for venomous species. The technique is more suitable for snakes (Hackenbrock and Finster, 1963), but can also be used for tortoises. It should not be used for terrapins as they are able to revert to total anaerobic metabolism for up to 27 hours (Calderwood, 1971) and induction may take a very long time.

Semi-closed circuits

Examples of semi-closed circuits are Ayre's T-piece, Bain Co-axil, McGill and Bethune. These circuits require the reptile to be intubated. Semi-closed breathing circuits rely on the animal breathing by itself and are wasteful of anaesthetic agents and gases.

Closed circle circuits

As it is very difficult to monitor respiration of reptiles it is better to practise intermittent positive pressure ventilation (Brazenor and Kaye, 1953; Gandal, 1968; Calderwood, 1971), or, as the Americans call it, "intermittent mandatory ventilation for reptiles" (Sedgwick, 1988).

Closed circuits make positive pressure ventilation easier. The rebreathing bag can be pressed once every 30 seconds throughout anaesthesia and into the recovery period until the reptile is obviously breathing by itself.

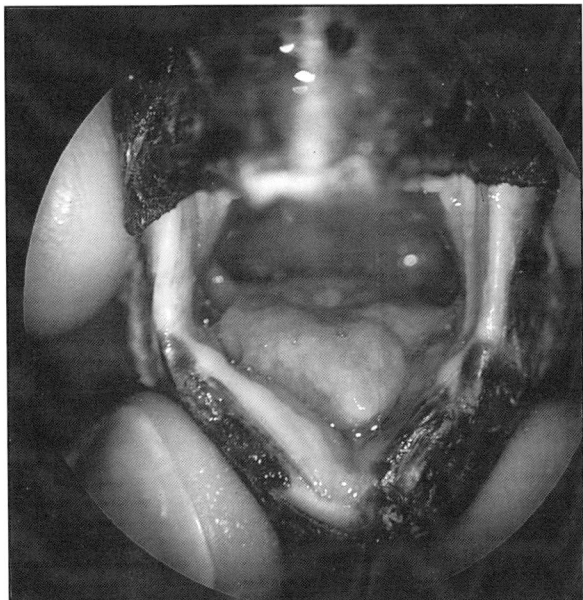

Figure 5
The glottis of a tortoise (*Testudo* sp.).

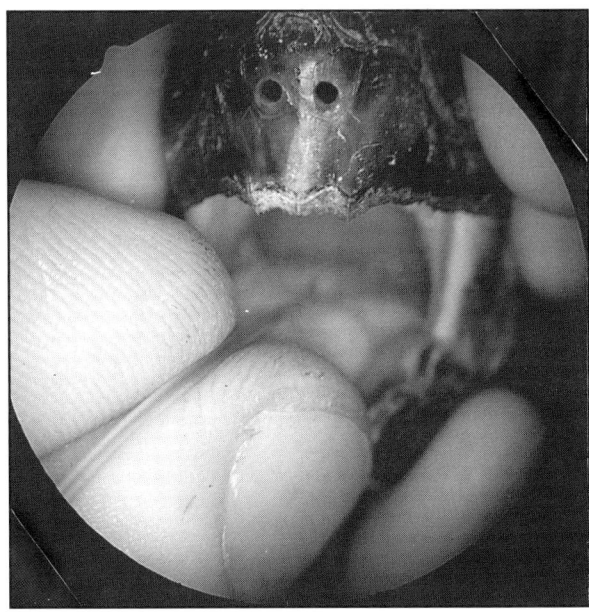

Figure 6
Intubation of the glottis using an introducer.

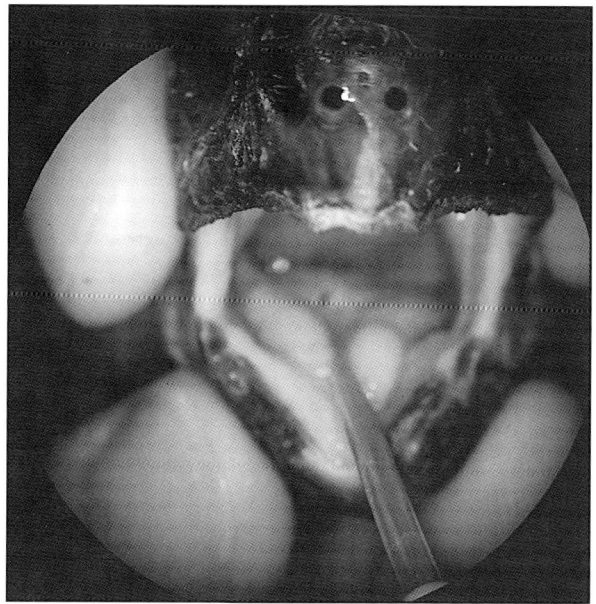 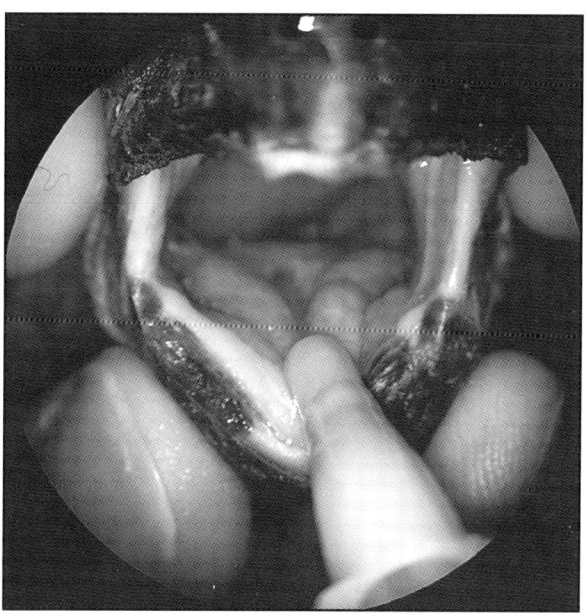

Figure 7
Passing a catheter
over the introducer.

Figure 8
"Endotracheal" tube
(catheter) in place.

INTUBATION

Intubation in all reptilian species is relatively easy. In lizards and snakes the glottis is readily visible. In tortoises the glottis is at the back of the large fleshy tongue and, therefore, they may be harder to intubate (see Figure 5). By use of intravenous cat or dog catheters and introducers it is possible to intubate even the smallest reptile (see Figures 6, 7 and 8).

Although Sedgwick (1980) stated that reptiles may be intubated with manual restraint only, it is this author's opinion that for welfare reasons they should be deeply sedated or anaesthetised before this is attempted.

INHALATION ANAESTHESIA

Five anaesthetic inhalation agents have been used as part of balanced anaesthesia in reptiles.

Ether. Although described by Brazenor and Kay (1953), ether is irritant and explosive; therefore, it should not be considered further.

Nitrous oxide. This agent may be used in combination with oxygen and a volatile anaesthetic. It has the advantage that the percentage of the volatile anaesthetic required is reduced as nitrous oxide provides good analgesia. Nitrous oxide also increases muscle relaxation.

Halothane. Halothane is the inhalation agent used for reptiles in most veterinary practices (Hackenbrock and Finster, 1963; Cooper, 1974). However, as many reptiles have underlying liver damage, its use may be contraindicated. 15 – 20% of the agent is metabolised in the body; thus, it takes a while for the anaesthetised animal to recover fully, especially if there is any hepatic damage.

Although Calderwood (1971) reported a difference in effective dosages of halothane between poisonous and non-poisonous snakes, the poisonous species requiring a higher percentage, this has yet to be confirmed.

With halothane, apnoea and cardiac arrest tend to occur simultaneously. Muscle relaxation is only moderate and there is marked respiratory depression. Postoperatively the reptile may need to be ventilated on oxygen for some time until breathing returns to normal.

Methoxyflurane. Although fatalities were reported using methoxyflurane in three elapids in a series involving seven snakes (Burke and Wall, 1970), the anaesthetic technique and recovery care left much to be desired. Others (Gandal, 1968; Lawton, unpublished data) have found it safe and reliable. The high solubility coefficient necessitates ventilation with pure oxygen prior to extubation, longer than for other volatile anaesthetics (Calderwood, 1971).

50% of the methoxyflurane is metabolised in the body; thus, there is a hang-over effect and a prolonged recovery period. Apnoea and cardiac arrest may occur simultaneously but methoxyflurane does have an advantage over halothane in that it gives excellent muscle relaxation and good analgesia. However, methoxyflurane can cause marked respiratory depression.

Isoflurane. This author considers isoflurane to be the volatile anaesthetic of choice, at present, for reptiles, although its cost may well be prohibitive.

Isoflurane has the advantage that only 0.3% is metabolised in the body; the blood gas partition coefficient is 1.4 at 37°C (halothane 2.3 and methoxyflurane 12) which means a very low solubility in blood, together with a lower fat solubility than methoxyflurane or halothane (Werner, 1987). It is almost entirely excreted by the lungs. Therefore, as soon as the isoflurane administration is stopped, the reptile starts to exhale it and begins to recover rapidly. Isoflurane has properties similar to methoxyflurane in that it provides excellent muscle relaxation and good analgesia. It has an advantage over halothane and methoxyflurane in that it produces less respiratory depression, and apnoea and cardiac arrest do not occur simultaneously.

MISCELLANEOUS TECHNIQUES

Electroanaesthesia. Using both low and high frequency waves via temporal electrodes, this technique is reported as giving anaesthesia from 35 – 75 minutes (Calderwood, 1971). Electroanaesthesia has a reported fast induction and rapid recovery time. The welfare aspect of this procedure is in doubt and it is surprising that it is not mentioned in the Royal College of Veterinary Surgeons Working Party Report on the Mutilation of Animals (1987).

Hypothermia. Lack of reaction in an animal subjected to hypothermia is often mistaken for analgesia (Bonath, 1978). Hypothermia does not stop nerve conduction in tortoises (Rosenberg, 1977) but reduces their speed; the sensory withdrawal reflexes are still present, although very much reduced. Hypothermia does not prove consistently effective in reptiles (Sedgwick, 1980) and this, together with the failure of hypothermia to provide true analgesia (Kraner et al, 1965), means that it should never be used or considered as an anaesthetic. In fact, cold can cause pain or be stressful (Burke, 1986; Frye, 1991). Hypothermia, however, may sometimes facilitate clinical examination. It can also reduce the effect of some toxic drugs or minimise the effects of hypoxia.

MONITORING ANAESTHESIA

Although Jackson (1970) described three distinguishable stages of anaesthesia and Bonath (1978) four stages, the depth and stage of anaesthesia in reptiles will depend upon:-

1. The anaesthetic used.
2. The analgesic property of that anaesthetic.
3. The relaxation required.
4. The procedure to be performed.

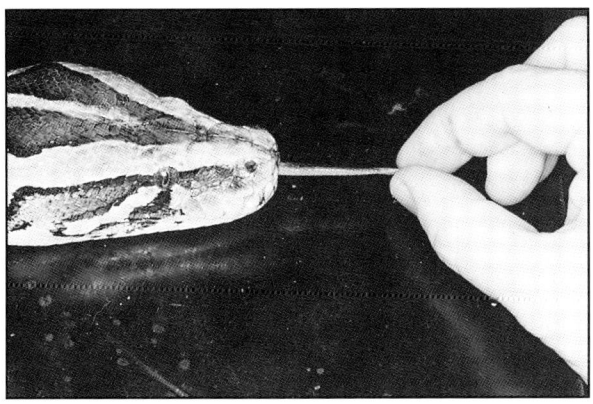 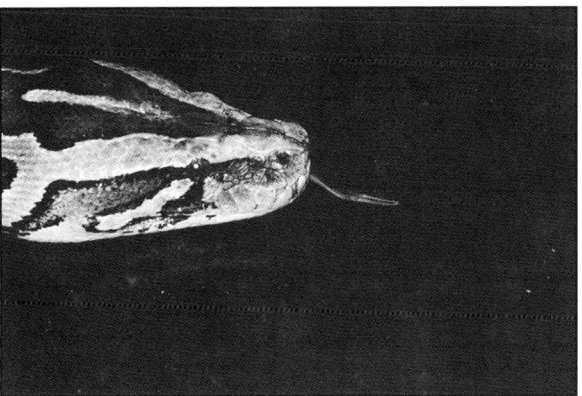

Figure 9
Tongue reflex of an anaesthetised snake (a).

Figure 10
Tongue reflex of an anaesthetised snake (b).

If a painful procedure is being carried out but the degree of analgesia offered by the anaesthetic choice is poor, then it will be necessary to deepen anaesthesia, eg. the depth of anaesthesia required with isoflurane may be less than that required when halothane is used.

Bodily functions can be monitored in the same way as in mammals to assess the depth of anaesthesia. The reflexes to be monitored are very similar to those of mammals. In lizards and chelonians the palpebral and corneal reflexes are useful. In snakes, as the eyelids are fused, no palpebral or corneal reflex may be elicited.

The tongue flick reflex is more useful in snakes but can also be used in lizards (see Figures 9 and 10). The tongue is pulled gently out of the mouth to see if it twitches or is withdrawn (Betz, 1962). With increased depth of anaesthesia the response is diminished or, finally, eliminated.

Assessment of jaw tone is also useful for judging if a reptile is lightly or moderately anaesthetised, although care has to be taken with some species due to their teeth.

Pinching the tail of an anaesthetised snake or lizard (Betz, 1962; Cooper, 1974), or the foot of a lizard or chelonian, will elicit a withdrawal response which will vary or disappear with depth of anaesthesia.

Respiration may be monitored in lightly anaesthetised reptiles using a respiratory monitor. This appears to work best in snakes and lizards, which will usually activate a thermocouple monitor, such as an IMP (Veterinary Instrumentation); tortoises appear unable to do so. Lizards tend to breathe spontaneously, though sporadically, even when under a moderate depth of anaesthesia, while snakes may appear not to breathe at all. The problem of assessing breathing of reptiles, together with assuring adequate ventilation, is the reason for recommending intermittent positive pressure ventilation during anaesthesia.

Heart rate is a good indicator of analgesia and response to pain and may also be used to monitor the depth of anaesthesia. Heart rate must be monitored via an oesophageal stethoscope, Doppler or ECG. Although it may prove difficult to record the heart electrical waves and all that is being measured is muscular movement or respiration (Cooper, unpublished data), it is reassuring in demonstrating that the reptile is still alive.

Above all else, one must remember that the temperature at which the reptile is kept will have a substantial influence on the depth and duration of anaesthesia and response to stimuli (Bonath, 1978). The duration of anaesthesia may be increased for a given dose by lowering the temperature; a drop of 6°C may double the half-life of some drugs (Lawrence, 1983).

POSTOPERATIVE CARE OF REPTILES

Following surgery or investigation, the anaesthetised reptile should be returned to a vivarium at the species' preferred body temperature (PBT).

It is important to monitor respiration during the recovery period, especially following the use of ketamine, halothane or methoxyflurane, because, although reptiles are relatively tolerant of hypoxia, it may still kill them. The use of respiratory stimulants/anaesthetic reversal agents, such as, doxapram (Dopram, Willows Francis) (0.25ml/kg i/v), is useful to stimulate spontaneous respiration and to reduce the amount of postoperative observation that is required.

Cooper (1974) considered that recovery had occurred when the righting reflex had returned to normal. In snakes this is assessed by placing the reptile on its back and observing if it moves on to the ventrum. In lizards and chelonians obvious return of the pedal reflexes is a good indicator of recovery.

Fluid therapy is essential prior to anaesthesia of any debilitated reptile. Slight dehydration can result in visceral gout due to the insolubility of uric acid in the blood and the ease with which it forms microcrystals and tophi in tissues. Therefore, it is good practice to give intraperitoneal, intravenous or oral fluids, up to 4% of bodyweight, in the post–anaesthetic period.

REFERENCES

BECK, C.C. (1972). Chemical restraint of exotic species. *Journal of Zoo Animal Medicine* **3 (3)**, 3.

BENNETT, R.A. (1991). A review of anesthesia and chemical restraint in reptiles. *Journal of Zoo and Wildlife Medicine* **22 (3)**, 282.

BETZ, T.W. (1962). Surgical anesthesia in reptiles with special reference to the water snake (*Natrix rhombifera*). *Copeia* **2**, 284.

BOEVER, W.J. (1979). The restraint of non-domestic pets. In: *Veterinary Clinics of North America: Small Animal Practice; Symposia on Non-domestic Pet Medicine.* (Ed. W.J. Boever). **9 (3)**, 391.

BOEVER, W.J. and CAPUTO, F. (1982). Tilazol (CI 744) as an anesthetic agent in reptiles. *Journal of Zoo Animal Medicine* **13**, 59.

BONATH, K. (1978). Halothane inhalation anaesthesia in reptiles and its clinical controls. *International Zoo Yearbook* **19**, 112.

BRAZENOR, C.W. and KAYE, G. (1953). Anesthesia for reptiles. *Copeia* **3**, 165.

BRISBIN, I.L. (1966). Reaction of the American alligator to several immobilising drugs. *Copeia* **1**, 129.

BURKE, T.J. (1986). Reptile anesthesia. In: *Zoo and Wild Animal Medicine.* 2nd Edn. (Ed. M.E. Fowler). W.B. Saunders, Philadelphia.

BURKE, T.J. and WALL, B.E. (1970). Anesthetic deaths in cobras (*Naja naja* and *Ophiophagus hannah*) with methoxyflurane. *Journal of the American Veterinary Medical Association* **157 (5)**, 620.

CALDERWOOD, H.W. (1971). Anesthesia for reptiles. *Journal of the American Veterinary Medical Association* **159 (11)**, 1618.

CARBO, L.M. and MARTIN, R.J.M. (1986). Anestesiaen los reptiles (Turtugas serpientes y cocodrilos). *Medicina Veterinaria* **3 (1)**, 51.

COOPER, J.E. (1971). Surgery on a captive iguana (*Iguana iguana*). *Journal of Zoo Animal Medicine* **2**, 29.

COOPER, J.E. (1974). Ketamine hydrochloride as an anaesthetic for East African reptiles. *Veterinary Record* **95**, 37.

COOPER, J.E. (1989). Exotic animal anaesthesia. In: *Manual of Anaesthesia for Small Animal Practice.* 3rd Edn. (Ed. A.D.R. Hilbery). BSAVA, Cheltenham.

ESRA, G.N., BENIRSCHKE, K. and GRINER, L.A. (1975). Blood collecting techniques in lizards. *Journal of the American Veterinary Medical Association* **167**, 555.

FRYE, F.L. (1991). *Biomedical and Surgical Aspects of Captive Reptile Husbandry.* 2nd Edn. Krieger, Malabar.

GANDAL, C.P. (1968). A practical anesthetic technique in snakes utilising methoxyflurane. *Journal of the American Animal Hospital Association* **4**, 258.

GLEN, J.B. and HUNTER, S.C. (1984). Diprivan: an update. *Journal of the Association of Veterinary Anaesthetists of Great Britain and Ireland* **12**, 40.

GLENN, J.L., STRAIGHT, R. and SNYDER, C.C. (1972). Clinical use of ketamine hydrochloride as an anesthetic agent for snakes. *American Journal of Veterinary Research* **33 (9)**, 1901.

GRAY, C.W., BUSH, M. and BECK, C.C. (1974). Clinical experience of CI-744 in chemical restraint and anesthesia of exotic specimens. *Journal of Zoo Animal Medicine* **5 (4)**, 12.

GREEN, C.J. and PRECIOUS, S. (1978). Reptilian anaesthesia. *Veterinary Record* **102**, 110.

HACKENBROCK, C.R. and FINSTER, M. (1963). Fluothane, a rapid and safe inhalation anesthetic for poisonous snakes. *Copeia* **2**, 440.

HINSCH, H. and GANDAL, C.P. (1969). The effects of etorphine (M99), oxymorphone hydrochloride and meperidine hydrochloride in reptiles. *Copeia* **2**, 404.

HOLT, P.E. (1981). Drugs and dosages. In: *Diseases of the Reptilia, Vol. 2.* (Eds. J.E. Cooper and O.F. Jackson). Academic Press, London.

JACKSON, O.F. (1970). Snake anaesthesia. *British Journal of Herpetology* **6**, 172.

JACKSON, O.F. and COOPER, J.E. (1981). Anaesthesia and surgery. In: *Diseases of the Reptilia, Vol. 2.* (Eds. J.E. Cooper and O.F. Jackson). Academic Press, London.

JACOBSON, E.R. (1988). The evaluation of the reptile patient. In: *Exotic Animals.* (Eds. E.R. Jacobson and G.V. Kollias). Churchill Livingstone, New York.

JONES, D.M. (1977). The sedation and anaesthesia of birds and reptiles. *Veterinary Record* **101**, 340.

KAPLAN, H.M. (1969). Anesthesia in amphibians and reptiles. *Federation Proceedings* **28 (4)**, 1541.

KLYDE, A.M. and KLEIN, L.V. (1973). Chemical restraint in three reptilian species. *Journal of Zoo Animal Medicine* **4 (1)**, 8.

KRANER, K.L., SILVERSTEIN, A.M. and PARSHALL, C.J. (1965). Surgical anesthesia in snakes. *Experimental Animal Anesthesiology.* USAF School of Aerospace Medical Centre.

LAWRENCE, K. (1983). The use of antibiotics in reptiles: a review. *Journal of Small Animal Practice* **24**, 741.

LAWRENCE, K. (1985). An introduction to haematology and blood chemistry of the Reptilia. In: *Reptiles: Breeding, Behaviour and Veterinary Aspects.* (Eds. S. Townson and K. Lawrence). British Herpetological Society, London.

LAWRENCE, K. and JACKSON, O.F. (1983). Alphaxalone/alphadolone anaesthesia in reptiles. *Veterinary Record* **12**, 26.

MARCUS, L.C. (1981). *Veterinary Biology and Medicine of Captive Amphibians and Reptiles.* Lea and Febiger, Philadelphia.

RICHTER, A.G., OLSEN, J., FLETCHER, K., BENIRSCHKE, K. and BOGART, M. (1977). Collecting blood from Galapagos tortoises and box-turtles. *Veterinary Medicine/Small Animal Clinician* **72**, 1376.

ROSENBERG, M.E. (1977). Temperature and nervous conduction in the tortoise. *Proceedings of the Physiological Society* **April 1977.**

ROYAL COLLEGE OF VETERINARY SURGEONS (1987). *Report of Working Party Established by RCVS Council to Consider the Mutilation of Animals.* RCVS, London.

SAMOUR, H.J., RISLEY, D., MARCH, T., SAVAGE, B., NIEVA, O. and JONES, D.M. (1984). Blood sampling techniques in reptiles. *Veterinary Record* **114**, 472.

SEDGWICK, C.J. (1980). Anesthesia of reptiles. In: *Current Veterinary Therapy VII. Small Animal Practice.* (Ed. R.W. Kirk). W.B. Saunders, Philadelphia.

SEDGWICK, C.J. (1988). Anesthesia for reptiles, birds, primates and small exotic mammals. In: *Manual of Small Animal Anesthesia.* (Ed. R.R. Paddleford). Churchill Livingstone, New York.

WALLACH, J.D. and HOESSLE, C. (1970). M99 as an immobilising agent in poikilotherms. *Veterinary Medicine/Small Animal Clinician* **65**, 163.

WERNER, R.E. (1987). Isoflurane anesthesia: a guide for practitioners. *Compendium on Continuing Education for the Practicing Veterinarian* (*North American Edn.*) **9 (6)**, 603.

CHAPTER SEVENTEEN

SURGERY

Martin P C Lawton BVetMed CertVOphthal FRCVS
Lynne C Stoakes BVetMed MRCVS

INTRODUCTION

The principles of surgery as applied to dogs, cats or other mammals apply equally to reptiles. In some circumstances obtaining access to the surgical site may prove a challenge, eg. chelonians, but the problem is never insurmountable.

An understanding of basic reptilian anatomy is desirable. There are major anatomical differences compared with mammals, birds and fish: the most obvious example is the single or paired ventral abdominal veins (ventral vena cavae); the surgeon should be aware of the position of these veins in reptile species before exploratory laparotomy is attempted. In snakes there is a constant relationship between the visceral organs and the ventral scales, often expressed as a percentage of the total length from the rostrum to the cloaca of the snake (Bragdon, 1953).

It is assumed that any surgery, no matter how minor, will be performed under local or general anaesthesia (see "Anaesthesia"); this may even be a legal requirement. Likewise, thought and consideration must be given to the relief of pain intra-operatively and postoperatively.

In view of the size of some species, magnification, as provided by a binocular loop or operating microscope, makes a surgical task easier and, in some cases, is essential. Likewise, fine ophthalmic instruments are desirable to minimise tissue damage, especially as visceral organs, eg. oviduct, may be very friable.

Haemorrhage is a significant complication, especially in smaller species. Prompt ligation or use of radiosurgery, in particular bipolar forceps (Surgitron, Ellmans), is advised.

PREPARATION OF THE SURGICAL SITE

As the majority of bacterial infections in reptiles are due to opportunist Gram-negative bacteria, correct cleaning and preparation of the surgical site is essential to prevent postoperative wound contamination.

Antiseptics such as hibitane (Hibiscrub, ICI), cetrimide and chlorhexidine (Savlon, ICI) and povidone-iodine are to be encouraged. There is good evidence that povidone-iodine is a superior antiseptic for cleaning and protection of the surgical wound. Surgical spirit can be applied after cleaning the wound with an antiseptic. However, care must be taken when surgical spirit is being used near the head or cloaca, or in aquatic species which have relatively permeable skin.

The reptilian patient can be positioned for surgery using sandbags and positioning pads or ties as for mammals. Drapes should be used. Op-site (Smith and Nephew) is transparent and, when used in smaller species, allows observation of the patient under general anaesthetic: in addition it will adhere to the shell of chelonians where towel clamps may be inappropriate (Bennett, 1989a).

BASIC PRINCIPLES OF SURGERY

The skin of reptiles is different from that of mammals and birds (see "Integument"). In most reptiles the skin may be incised with a scalpel blade, although chelonian shell and crocodilian dermal plates may need an oscillating or double-sided diamond saw to complete the incision.

Whatever the choice of instrument to make the incisions, the patient should be properly anaesthetised, as reptiles are highly sensitive to skin incisions and will react vigorously if there is insufficient analgesia (Sedgwick, 1980).

Chiodini et al (1982) described the basic anatomy of northern water snakes (*Natrix sipeton sipeton*) and the eastern milk snake (*Lampropeltis triangulum*). A review of the basic surgical anatomy of reptiles is given by Bennett (1989a).

Wound healing

Factors that influence wound healing include:-

1. **Environmental temperature.** This should be at the higher end of the optimum temperature range for each individual species; temperatures between 30°–36°C have been shown to promote healing in certain species (Smith et al, 1988; Bennett, 1989a). Hibernation should always be postponed until healing is complete.

2. **Orientation of the wound.** In snakes a craniocaudal wound will heal faster than a ventrodorsal wound (Bennett, 1989a).

3. **Good environmental hygiene.** The ideal substrate is paper towelling which can be discarded daily to prevent contamination of wounds by enteric bacteria. Bacterial contamination of wounds can lead to life-threatening septicaemia.

4. **Nutritional status.** Hypoproteinaemia will affect the healing ability of wounds; hypocalcaemia will affect the healing of bones.

Suturing

It is advisable to make incisions between scales rather than through them. This may result in a slightly ragged incision but allows best apposition for suturing and first intention healing. The sutures should also be placed between the scales.

Thin-skinned reptiles, especially snakes, have a tendency for the cut edges of a skin wound to roll inwards. A slightly everting suture, such as a horizontal mattress suture, overcomes this (Frye, 1980). There can be complications when snakes shed as they may avulse the sutures (Lawton, unpublished data) or suffer dysecdysis around the area of the old surgical wound (Cooper, 1971; Crawley et al, 1988). The use of mattress sutures as opposed to simple interrupted sutures in snakes also reduces the risk of dehiscence.

Reptilian heterophils lack many of the proteolytic enzymes that are responsible for absorption of "catgut", so it is advisable to use suture materials that are not dependent on this mode of absorption, eg. Vicryl (polyglactin 910), Dexon (polyglycolic acid) or PDS (polydioxanone) (Bennett, 1989a). In addition, Dexon and Vicryl have a lower infection potential than the more inflammatory sutures which produce an exudative foreign-body response and local tissue autolysis; this creates a protein-rich area which is capable of supporting bacterial growth and movement (Smeak and Wendelburg, 1989).

In crocodilians, chelonians and certain large lizards, where dermal plates or shell are involved, the use of stainless steel sutures is advised.

In aquatic species it is advantageous to make the wound watertight so that the reptile may bathe as normal. Use may be made of cyanoacrylic tissue glues (Isodent, Ellmans) or, if the shell is involved, cyanoacrylic plus resin or aquarists' glue.

The use of reconstructive surgery utilising skin flaps is more difficult in reptiles than in mammals due to the relative inelasticity of the skin and the presence of scales or dermal bone. It may be necessary, following trauma such as burns, or removal of large abscesses or growths, to leave the wound to heal by second intention healing and granulation. Crawley et al (1988) found it necessary to perform a gastrotomy to empty the stomach of its contents before undertaking closure of a traumatic coeliotomy and ventral skin trauma in a snake.

As healing is generally very slow in reptiles it is sensible to protect the wound so that it does not become contaminated; the use of surgical dressings (Op-site, Smith and Nephew) is recommended. These allow the wound to be monitored visually; they may also be removed to allow wound treatment and application of fresh dressings. In snakes with ventral or lateral wounds, the dressing provides good protection against contamination of the wound, even if suturing and first intention is the aim (Cooper, 1971).

Repair of the shell of chelonians

The repair of the chelonian shell, whether it is following a surgical incision or trauma, requires good apposition of the fragments to produce a good cosmetic and functional repair. Often the use of surgical staples or wire sutures to hold the pieces in apposition is advisable (Northway, 1970). Bridging any gaps by use of an epoxyresin and fibreglass has been described (Zeman et al, 1967; Frye, 1973; Holt, 1981). There is a fairly long period of polymerisation and temperature change, but the time can be reduced by using rapid curing epoxyresin, although more specialised veterinary products, eg. Technovite or Ellman's Beak Repair Kits, are recommended. When building up and repairing the deficit it is important that none of the repair material should be trapped between the edges of the shell being repaired. If this occurs it will delay or stop the healing process and result in a sequestration of the bone fragments. One technique the authors have found useful is to insert an antibiotic ointment into any gaps first and then to apply the resin over the top. It may be necessary, at a later stage, to replace and repair any covering, as the healing of chelonian shells is often prolonged and has been reported to take several years. In *Testudo* spp. hibernation following shell repair is not advised as this reduces metabolism and prolongs the healing of the wound. Cryophilic bacteria and fungi may also be a problem.

Minor shell defects often require no treatment. Dental repair material may be used in attempts to correct any minor shell defect.

Visceral organs

Reptilian viscera are very friable and must be handled with extreme care. Peters and Coote (1977) reported the problems encountered when suturing the oviduct of a snake due to the extreme friability of this tissue.

Enterotomy/enterectomy may be performed as in mammals. However, in small species a Czerney suture pattern is recommended rather than an inverting suture. In bladders, stomachs and oviducts it may be possible to use a two-layered continuous inverting suture pattern. Fine dissolvable sutures on an atraumatic needle should be used, eg. Vicryl 4/0 – 6/0.

Partial gastrectomy may be performed where indicated, but especially in cases of suspected neoplasia. There is apparently little effect on the animal postoperatively (Wallach and Porter, 1975). A pyloroduodenal resection has also been described for treatment of damage and eventual necrosis of the pyloric region of the stomach (Jacobson and Ingling, 1976). The surgical technique in both these cases is as has been described in mammals, including humans.

SPECIFIC SURGICAL PROCEDURES

Ear abscesses

These are common in chelonians. The tympanic scale or membrane swells to a varying extent and this may be uni- or bilateral. In very severe cases there may be rupture of the tympanic membrane, or inspissated pus may be seen at the back of the mouth, this having travelled down the Eustachian tube.

Under general anaesthesia, a vertical incision is made over the swelling, usually the length of the tympanic scale, but for very large abscesses the incision may have to extend beyond these margins. Once the skin and underlying membrane are opened, the inspissated pus is clearly seen. Using a small Volkmann's currette or a Hobson's ear loop this core of pus should be removed, if possible in one piece. Successful removal yields a comma-shaped mass of pus, the tail of the comma representing the material from the Eustachian tube. The tympanic cavity is then cleaned and flushed with Hartmann's solution, with or without an antibiotic. The Eustachian tube is easily cannulated and should be flushed until fluid comes from the mouth. Failure to remove all the infection at the time of surgery will increase the chances of recurrence postoperatively.

The cavity is then filled with an antibiotic ointment, eg. framomycin (Framycetin Anti-Scour Paste, C-Vet), and the tympanic membrane is closed with fine simple interrupted sutures, eg. Vicryl 6/0.

Where the infection is very severe and the tympanic membrane has ruptured, it may not be possible to repair the damage to the latter. In these cases the wound should be left to granulate as an open wound; however, this may take many months (up to two years) and requires a great deal of attention from the client to prevent recontamination and infection. It is advisable that the owners clean the area two or three times daily with a diluted povidone-iodine solution, even using an intravenous cannula and flushing out the area. Antibiotics may also be used topically for the first month to control infection, after which antiseptics should be relied upon to prevent recontamination.

"Lumpectomy" - tumour removal

The most usual cause of swelling in reptiles is an abscess, although tumours do occur (Elkan, 1981). Other inflammatory processes, eg. steatitis, may also need to be investigated surgically. Where the cause of the swelling is not obvious material should be taken for cytology and/or histopathological evaluation.

In lizards, chelonians and crocodilians, where the mass appears on a limb or a digit, radiography should be carried out first to rule out bone involvement or osteodystrophia fibrosa and a secondary pathological fracture (see "Radiological and Related Investigations"). If, however, osteolysis is occurring the use of antibiotics for several weeks and re-radiographing is recommended, so long as the animal is not septicaemic and the swelling does not increase in size during this time. If the infection or swelling is not resolved, the mass should be removed. However, this often requires amputation of a limb (see later).

Surgery to remove a mass in a reptile is approached slightly differently from mammals. In mammals it is usually advisable to incise around the mass to ensure that a healthy margin of tissue is also removed at the time of surgery. This procedure in reptiles can, however, often result in difficulty in closing the wound due to the lack of elasticity of the skin. Therefore, the majority of incisions are made over the mass or to one side and the mass is removed by blunt or sharp dissection. In the case of an abscess the mass is removed *in toto* by dissection around the outside of the fibrous capsule. The cavity is then flushed where necessary, with or without antibiotics, and sutured as already described.

Baxter and Meek (1988) described the use of cryosurgery in the treatment of a number of cases, including papillomas in a lizard (*Lacerta viridis*), apparently successfully. They stated that cryosurgery is relatively free of pain. This is debatable, however, and patients should always be anaesthetised before this technique is undertaken.

Masses in the abdomen should be investigated by exploratory coeliotomy. Millichamp (1985) described the surgical removal of an ovarian neoplasm in a female sand viper (*Vipera ammodytes montandoni*). Cooper *et al* (1983) reported the use of cryosurgery on five separate occasions for the treatment of a neurilemmal sarcoma in the shell of a tortoise (*Testudo hermanni*); although this appeared to be successful, the tortoise was euthanased ten years postoperatively on account of recurrence of the tumour (Lawton and Cooper, unpublished data).

Limb amputation

The approach to limb amputation is very similar to that used in mammals. Although reptiles follow the basic pentadactyl structure, there are variations in the position of nerves and blood vessels. The prospective surgeon is advised, therefore, to check the anatomy prior to embarking on surgery.

In lizards the amputation, or partial amputation, of digits appears to cause very little handicap to the animal. Partial amputation of the limb, however, other than in small light lizards, eg. geckos, may lead to complications. In iguanas, monitors or other larger lizards total amputation should be considered rather than partial amputation. In these animals partial amputation can result in infection, trauma or disruption of the stump.

In chelonians partial amputation may also result in complications, particularly trauma to the stump; therefore, a high amputation is advised. The authors use the technique of disarticulation of fore and hindlimbs, although this requires working within the shell, which, in some species, eg. *Testudo* spp., is more difficult. However, the results justify the inconvenience at the time of surgery.

Tail injuries in lizards

As a number of species of lizards are capable of autotomy (shedding of the tail), tail injuries or amputation by the veterinary surgeon must be approached with a degree of caution. Often, where the tail is lost, treating the stump conservatively by bathing with antiseptic and/or topical antibiotics without suturing will result in regeneration of the tail to its original length. Suturing the stump will prevent regeneration.

Where there is infection, severe trauma or neoplasia it may be necessary to perform surgical amputation of the tail, in which case this should be approached as for dogs and cats. However, the owner must be forewarned that the tail is unlikely to regenerate.

OPHTHALMIC SURGERY

Some procedures, such as treatment of sub-spectacular abscesses, are described in "Ophthalmology".

Third eyelid flap

This is a useful procedure in which the membrana nictitans is used as a bandage to support and allow healing of any damage to the cornea or prolapse of the globe. In reptiles that have a third eyelid, such as chelonians and lizards, the third eyelid may be sutured under general anaesthesia on to the bulba conjunctivae as described in mammals (Slatter, 1981). The use of Vicryl sutures, 4/0 or 6/0, is recommended, and the flap may have to be left in position for up to 4 – 5 weeks.

Conjunctivoralostomy

This technique has been described (Millichamp *et al*, 1986) for the treatment of excess fluid within the sub-spectacular space following an occluded lacrimal duct. Under general anaesthesia a 30° incision is made on the edge of the spectacle at the medial canthus. An 18G needle is then introduced and pushed from the inferior fornix of the sub-spectacular space down through the roof of the mouth emerging between the palatine and maxillary teeth. Sialistic tubing (0.635mm) is threaded through the needle and the needle is removed. The tubing is sutured to the roof of the mouth and to the face to maintain the patency of the conjunctivoralostomy. The tube is removed one month postoperatively.

Enucleation

Enucleation may be performed, as a last resort, where the globe is severely traumatised or proptosed. In chelonians, crocodilians and lizards it is possible to perform a tarsorropathy afterwards in order to close the orbit. However, following enucleation in snakes, the orbit must be left to granulate by third intention healing.

The technique of enucleation in the snake requires an incision to be made around the spectacle, which is lifted away. The whole eyeball is then grasped with small forceps and the underlying tissue loosened. The optical sheath should be grasped with curved forceps and, after removal of the eye, ligated in order to control haemorrhage (Zwart *et al*, 1973). As the orbit remains opened, it is important to use topical antibiotic drops or powder for up to four weeks postoperatively to prevent infection. Zwart *et al* (1973) stated that scarring will occur within four weeks without complications.

An alternative to carrying out enucleation is evisceration, where the contents of the globe are removed but leaving the sclera and ossicles. It is possible then to close the orbit either by using a third eyelid flap or tarsorropathy in genera with these structures. The advantage of evisceration is the lack of a sunken orbit compared with enucleation.

Enucleation does not appear to affect the reptile's normal behaviour nor its feeding ability (Zwart *et al*, 1973, Ensley *et al*, 1978).

ORTHOPAEDIC SURGERY

The basic principles of orthopaedic surgery in the Reptilia are similar to those in mammals. Internal and external fixation can be used for stabilisation of limb and spinal fractures. Care must be taken, particularly in lizards, before embarking on internal fixation, to make sure that the bones are not osteodystrophic. Placement of intramedullary pins may result in fracture or tearing of the softened bones or failure of the implant to stay *in situ*. If there is any doubt as to the quality of the calcification of the bones an external technique is strongly advised.

External fixation

Hexcelite (Vet Lite) casts are strongly advised rather than plaster of Paris. The majority of reptiles like to enter water, so there is often difficulty in avoiding contact between the cast and water. The lighter weight of Hexcelite compared with plaster of Paris is also an advantage. Lizards with fractured limbs should be treated with a cast, providing it can be applied above and below the fracture site, in order to allow stabilisation. Difficulties arise in osteodystrophic animals where there are multiple fractures. Only instable fracture sites require casting. There are other alternatives to the use of Hexcelite and plaster of Paris. Redisch (1977) described the technique of using a modified Thomas splint in the treatment of a midshaft fracture of the humerus in an *Iguana* spp.

Spinal trauma may be stabilised in snakes and lizards by use of a tube-like cast. Again, Hexcelite is a good material to use. The prognosis following spinal fracture is better than that in mammals (see "Neurological Diseases").

Chelonians with fractured limbs may also be treated with casts; this is particularly useful in giant species. In smaller species, eg. *Testudo* spp., the limb may be placed within the shell and immobilised using epoxyresin and gauze. This is a similar technique to that used in birds with fractured wings where folding the wing and taping it in its natural folded position often allows accurate anatomical alignment and better healing.

Although some authors have reported healing within 6–9 weeks (Redisch, 1977), the authors of this chapter advise that the cast should be left in place for a minimum of 10 weeks. Radiographs should be taken to confirm healing and stable callus formation before the cast is removed. Species that hibernate should not be allowed to do so while healing is taking place.

Internal fixation

The use of intramedullary pins (especially stack pinning), plates, screws, circlage wires and bone grafts, may provide superior support, quicker return of function of the limb and less chance of distortion than would be the case if a cast were used. However, as already mentioned, such techniques are contraindicated if any degree of osteodystrophy pre-exists. This technique is, therefore, virtually ruled out in most pet lizards. Intramedullary pins have been used in the treatment of midshaft humeral fractures in green iguanas (*Iguana* spp.) (Hartman, 1976). The surgical approach was on the lateral surface of the limb over the humerus, and bleeding was reported to be minimal. Following an uneventful recovery and healing the pin was removed 14 weeks postoperatively.

Transverse mandibular fractures may be repaired by inserting pins into the ramus, although the animal has to be large enough to make this technique practicable, eg. the alligator snapping turtle (*Macroclemys temmincki*) as described by Kuehn (1973).

Crane *et al* (1980) described a neutralisation bone plating technique for the repair of a fractured humerus in an Aldabra tortoise (*Megalochelys gigantea*). Because of the size of the tortoise and the fact that the fractured bone had overridden, rigid internal fixation was attempted. Subsequent radiography revealed complete union of the fracture six months postoperatively.

COELIOTOMY

Indications for coeliotomy are eggbinding, peritonitis, gastrointestinal tract obstruction, colopexy in cases of chronic colon prolapse, cystotomy to remove urinary calculi and exploratory surgery.

The procedure varies between species due to the anatomical differences.

Snakes. In snakes the incision should be made between scales to one side of the midline, ideally between the lateral scales and gastropeges. The left side is usually favoured for coeliotomy in order to avoid entering or damaging the air sac, which is commonly found as an extension of the right lung (Brown and Martin, 1990). Ventral incisions are more likely to dehisce because of contact with the ground. There is a large ventral midline abdominal vein which should be avoided.

If coeliotomy is being performed to remove retained eggs a number of incisions through the body wall may be required due to the thinness and friability of the oviduct (Vanderventer and Schmidt, 1977; Millichamp *et al*, 1983; Bennett, 1989b). However, in some cases eggs may be milked down and delivered through a single incision (see Visceral Organs) (Mulder *et al*, 1979; Patterson and Smith, 1979).

Lizards. In lizards a midline and paralumbar approach have both been recommended (Bennett, 1989a). The large ventral midline abdominal vein must be avoided, however. In some lizards and crocodilians the use of "H" flaps improves exposure. Two parallel incisions are made perpendicular to the first incision in order to increase exposure (Frye, 1980).

Chelonians. Chelonian patients present their own problems. In some species where the plastron is reduced, eg. *Trionyx* spp. and *Chelydra* spp., it may be possible to gain entry to the coelom via an incision in the inguinal region; indeed, this is reported in *Xerobates* spp. and was described by Isenburgel and Barandun (1981) in a bastard turtle (*Lepidochelys olivacea*). Where a sub-plastron coeliotomy is carried out, the incision is generally made between the caudal border of the plastron and the cranial border of the femur (Bennett, 1989b). The incision is then extended through the muscles, allowing access into the abdominal cavity where the intestines or oviduct are identified and exteriorised in order to carry out an enterotomy or Caesarean section. However, in the majority of species the plastron has to be opened. This is accomplished under general anaesthesia by using a circular diamond saw to remove a square "trap-door" in the plastron (Frye and Schuchman, 1974; Jordan and Kyzar, 1978; Holt, 1979). This is hinged to one side or removed and kept moist for later replacement. Once the plastron has been traversed the peritoneum can be incised using a scalpel blade. Care must be taken to avoid the paired abdominal ventral veins on either side of the midline.

The plastron is repaired postoperatively using steel wire sutures placed through pre-drilled holes; usually one on each side is sufficient. The join is then made watertight (see Repair of the Shell of Chelonians).

URATE CALCULI

Urate calculi are commonly encountered in all species of reptile. They often cause a partial or complete blockage. Under sedation or anaesthesia it may be possible to break them down *per cloacam* and flush them out, particularly if they are in the cloaca. However, if the calculus is in the bladder (in reptiles which have such an organ) exploratory coeliotomy and removal of the calculus may be required (Frye, 1972).

CLOACAL ORGAN PROLAPSE

The cloaca has openings from the colon, ureters and bladder, although snakes and some species of lizards do not have a bladder. The most likely organ to prolapse is the cloaca and cloacal prolapse is a common presenting sign, especially in chelonians and ophidians. The cause is often constipation following an improper diet or lack of exercise, or "cloacitis" due to a bacterial or parasitic infection.

If the prolapse is fresh and the tissues are viable, cleaning with a dilute povidone-iodine solution and replacement under general anaesthesia may be possible. A purse-string suture is placed around the cloaca leaving a hole wide enough for the reptile to urinate and defecate normally. This is left in place for 7 – 10 days. If the prolapse recurs after suture removal or if the tissues are too badly traumatised to be replaced, the necrotic portion should be amputated leaving healthy tissue. Care must be taken not to close an oviduct, ureter or colon and where possible these structures or their openings should first be identified.

In all cases attempts should be made to determine and rectify the underlying cause of the prolapse.

AMPUTATION OF THE PENIS

This has been briefly described elsewhere (see "Urogenital System"). Most reptiles have two hemipenes, which are muscular structures and do not contain a urethra. Therefore, if recurrent prolapsing or severe infection or trauma occurs, the hemipenes (or one hemipenis) may be amputated with minimal risk of complications. In the first instance it is always advisable to try and replace the hemipenis by use of sedation/anaesthesia, cold compresses and gentle pressure. If this is not possible, amputation is a realistic option.

Prior to amputation a transfixing suture is placed near the base of the hemipenis adjacent to the cloaca. The distal portion of the hemipenis can then be surgically removed. The mucosa is sutured, using a continuous suture pattern, in order to cover the stump.

VASECTOMY

Zwart *et al* (1979) recommended the surgical procedure of vasectomy as the method of choice for controlling fertility in mature garter snakes (*Thamnophis* spp.). Ovariectomy of females is considered more dangerous on account of the numerous blood vessels supplying the ovaries and the uterus; removal of part of the ducti deferentes involves less traumatic surgery.

The technique in garter snakes involves a 5cm incision made from the 45th abdominal scale (from the cloaca) to the 40th abdominal scale. The ducti deferentes are found on either side of the intestines. A 3mm segment is excised from each duct, haemorrhage being controlled by ligation and cauterisation.

DEVENOMATION (LIGATION OF VENOM DUCTS)

This procedure was described by Frye (1980). However, in the UK the Royal College of Veterinary Surgeons considers it to be unacceptable except on medical grounds.

It is impossible to distinguish a devenomated snake from a snake with functional venom glands; therefore, all venomous snakes should be handled with extreme caution.

PLACEMENT OF A PHARYNGOSTOMY TUBE

Supportive therapy of some chelonian species, especially Brown's tortoises (*Manouria emys*), box-turtles (*Terrapene* spp.), *Chinemys* spp. and the more aggressive species such as *Trionyx* spp.

and snappers (*Chelydra* spp.), can be difficult. Instead of extracting the animals' head from within the shell on a daily basis, it is advisable to place a pharyngostomy tube.

The reptile should be anaesthetised. A finger is placed into the pharynx and identified on the left hand side of the neck. An incision is made through the skin and, by gentle blunt dissection, forceps are placed through the lateral skin incision until they emerge into the pharynx. A stomach tube of the correct length is placed into the stomach, the forceps being used to grasp the end of the tube and draw it out of the skin incision. Using a butterfly suture pattern, the tube is secured to the skin of the neck. Sufficient tube is left protruding so that this can be attached to the shell. A suitable stopper is fitted to the opening of the tube.

To use the pharyngostomy tube, a syringe, with or without an adaptor, is connected to the tube opening and the tube is flushed through with a small amount of Hartmann's solution. After the medicant, liquid food or rehydration mixture has been administered, the stopper is replaced.

When supportive therapy is no longer required, the butterfly suture is cut, the tube removed and the skin wound repaired or left to granulate.

REFERENCES

BAXTER, J.S. and MEEK, R. (1988). Cryosurgery in the treatment of skin disorders in reptiles. *Herpetological Journal* **1**, 227.

BENNETT, R.A. (1989a). Reptilian surgery. Part I: basic principles. *Compendium on Continuing Education for the Practicing Veterinarian (North American Edn.)* **11** (**11**), 10.

BENNETT, R.A. (1989b). Reptilian surgery. Part 2: management of surgical diseases. *Compendium on Continuing Education for the Practicing Veterinarian (North American Edn.)* **12** (**2**), 122.

BRAGDON, D.E. (1953). A contribution to the surgical anatomy of the water snake, *Natrix sipedon sipedon*; the location of the visceral endocrine organs with reference to ventral scutellation. *Anatomical Record* **117**, 145.

BROWN, C.W. and MARTIN, R.A. (1990). Dystocia in snakes. *Compendium on Continuing Education for the Practicing Veterinarian (North American Edn.)* **12** (**3**), 361.

CHIODINI, R.J., SUNDBERG, J.P. and CZIKOWSKY, J.A. (1982). Gross anatomy of snakes. *Veterinary Medicine/Small Animal Clinician* **77** (**3**), 413.

COOPER, J.E. (1971). Surgery on a captive gaboon viper. *British Journal of Herpetology* **4**, 234.

COOPER, J.E., JACKSON, O.F. and HARSHBARGER, J.C. (1983). A neurilemmal sarcoma in a tortoise (*Testudo hermanni*). *Journal of Comparative Pathology* **93**, 541.

CRANE, S.W., KURTIS, M., JACOBSON, E.R. and WEBB, A. (1980). Neutralisation bone-plating repair of a fractured humerus in an Aldabra tortoise. *Journal of the American Veterinary Medical Association* **177** (**9**), 945.

CRAWLEY, G.R., BISHOP, C.R. and WOODY, B.J. (1988). Gastrotomy in a Burmese python. *Californian Veterinarian* **42** (**3**), 7.

ELKAN, E. (1981). Pathology and histopathological techniques. In: *Diseases of the Reptilia, Vol. 1.* (Eds. J.E. Cooper and O.F. Jackson). Academic Press, London.

ENSLEY, P.K., ANDERSON, M.P. and BACON, J.P. (1978). Ophthalmic disorders in three snakes. *Journal of Zoo Animal Medicine* **9**, 57.

FRYE, F.L. (1972). Surgical removal of a cystic calculus from a desert tortoise. *Journal of the American Veterinary Medical Association* **16** (**6**), 600.

FRYE, F.L. (1973). Clinical evaluation of a rapid polymerizing epoxyresin for the repair of shell defects in tortoises. *Veterinary Medicine/Small Animal Clinician* **68**, 51.

FRYE, F.L. (1980). Surgery in captive reptiles. In: *Current Veterinary Therapy VII. Small Animal Practice.* (Ed. R.W. Kirk). W. B. Saunders, Philadelphia.

FRYE, F.L. and SCHUCHMAN, S.N. (1974). Salpingotomy and Caesarian delivery of impacted ova in a tortoise. *Veterinary Medicine/Small Animal Clinician* **69**, 454.

HARTMAN, R.A. (1976). Use of an intramedullary pin in repair of a midshaft humeral fracture in a green iguana. *Veterinary Medicine/Small Animal Clinician* **71**, 1634.

HOLT, P.E. (1979). Obstetrical problems in two tortoises. *Journal of Small Animal Practice* **20**, 353.

HOLT, P.E. (1981). Healing of a surgically induced shell wound in a tortoise. *Veterinary Record* **108**, 102.

ISENBURGEL, E. and BARANDUN, G. (1981). Surgical removal of a foreign body in a bastard turtle. *Veterinary Medicine/Small Animal Clinician* **76** (**12**), 1766.

JACKSON, O.F. (1978). Tortoise shell repair over two years. *Veterinary Record* **102**, 184.

JACOBSON, E.R. and INGLING, A.L. (1976). Pyloroduodenal resection in a Burmese python. *Journal of the American Veterinary Medical Association* **169**, 985.

JORDAN, R.D. and KYZAR, C.T. (1978). Intra-abdominal removal of eggs from a gopher tortoise. *Veterinary Medicine/Small Animal Clinician* **73** (**8**), 1051.

KUEHN, G. (1973). Bilateral transverse mandibular fractures in a turtle. *American Association of Zoo Veterinarians Annual Proceedings*, 243.

MILLICHAMP, M.J. (1985). Surgical management of ovarian neoplasia. *Veterinary Medicine/Small Animal Clinician* **80** (**11**), 54.

MILLICHAMP, M.J., JACOBSON, E.R. and DZIEZYC, J. (1986). Conjunctivoralostomy for treatment of an occluded lacrimal duct in a blood python. *Journal of the American Veterinary Medical Association* **189** (**9**), 1136.

MILLICHAMP, M.J., LAWRENCE, K., JACOBSON, E.R., JACKSON, O.F. and BELL, D.A. (1983). Egg retention in snakes. *Journal of the American Veterinary Medical Association* **11**, 1213.

MULDER, J.B., HAUSER, J.J. and PERRY, J.J. (1979). Surgical removal of retained eggs from a king snake (*Lampropeltis getulus*). *Journal of Zoo Animal Medicine* **10**, 21.

NORTHWAY, R.B. (1970). Repair of a fractured shell and lacerated cornea in a tortoise. *Veterinary Medicine/Small Animal Clinician* **65**, 944.

PATTERSON, R.W. and SMITH, A. (1979). Surgical intervention to relieve dystocia in a python. *Veterinary Record* **104**, 551.

PETERS, A.R. and COOTE, J. (1977). Dystocia in a snake. *Veterinary Record* **100**, 423.

REDISCH, R.I. (1977). Management of leg fractures in the iguana. *Veterinary Medicine/Small Animal Clinician* **72**, 1487.

SEDGWICK, C.J. (1980). Anesthesia of reptiles. In: *Current Veterinary Therapy VII. Small Animal Practice.* (Ed. R.W. Kirk). W.B. Saunders, Philadelphia.

SLATTER, D.H. (1981). *Fundamentals of Veterinary Ophthalmology.* W.B. Saunders, London.

SMEAK, D.D. and WENDELBURG, K.L. (1989). Choosing suture materials for use in contaminated or infected wounds. *Compendium of Continuing Education for the Practicing Veterinarian (North American Edn.)* **11** (**4**), 467.

SMITH, D.A., BARKER, I.K. and ALLEN, O.B. (1988). The effect of ambient temperature and type of wound on healing of cutaneous wounds in the common garter snake (*Thamnophis sirtalis*). *Canadian Journal of Veterinary Research* **52**, 120.

VANDERVENTER, T.L. and SCHMIDT, M. (1977). Caesarian section of a western gaboon viper. *International Zoo Yearbook* **17**, 140.

WALLACH, J.D. and PORTER, S. (1975). Partial gastrectomy in a Timor python (*Python timoriens*). *Journal of Zoo Animal Medicine* **6** (**2**), 13.

ZEMAN, W.D., FALCO, F.G. and FALCO, J.J. (1967). Repair of the carapace of a box turtle using a polyester resin. *Laboratory Animal Care* **17** (**4**), 424.

ZWART, P., DORRESTEIN, G.M., STADES, F.C. and BROWER, B.H. (1979). Vasectomy in the garter snake (*Thamnophis sirtalis radix*). *Journal of Zoo Animal Medicine* **10**, 7.

ZWART, P., VERWER, M.A.J., De VRIES, G.A., HERMANDIES-NIJHOF, E.J. and De VRIES, H.W. (1973). Fungal infection of the eyes of the snake *Epicrates chenchria maurus*: enucleation under halothane narcosis. *Journal of Small Animal Practice* **14**, 773.

CHAPTER EIGHTEEN

THERAPEUTICS

Mark A Pokras DVM
Charles J Sedgwick DVM DipACLAM DipACZM
Gretchen E Kaufman DVM

THE IMPORTANCE OF CORE BODY TEMPERATURE AND BODY SIZE WHEN TREATING REPTILES

It is essential to understand two principles when treating reptiles:–

1. To achieve consistent results in medicating such animals, one must consider the dependence of drug dosage, uptake, distribution and excretion on the body temperature of these ectotherms (Frye, 1991).

2. Smaller animals tend to have faster uptake, distribution and excretion of drugs than larger animals.

    This means that, in order to maintain effective serum levels in smaller animals, the dose rate (mg/kg) will be greater and will need to be administered more frequently than in larger animals, assuming that all other factors influencing the pharmacokinetics of a given drug are the same (Kirkwood, 1983a,b; Sedgwick and Pokras, 1988; Mader, 1991). The mathematical technique of allometric scaling can be used to extrapolate accurately appropriate treatment regimens (see later).

The metabolism of reptiles functions most efficiently when the core body temperature is within the preferred body temperature (PBT), which for some species is around 37°C (Hainsworth, 1981). However, all reptiles survive varying periods of time with core body temperatures that are reduced (sometimes drastically), with the animal in a state of torpor or hibernation (Hock, 1964). Trying to bring reptile patients to their PBT is, therefore, essential when medicating such animals. For example, an effective treatment regimen in an aquatic turtle with its core at the animal's PBT would have little or no predictable effect in the same animal if it were chilled to the point of torpor. Mader et al (1985) demonstrated that when treating small gopher snakes with the antibiotic amikacin, raising the ambient temperature from 25°C to 37°C greatly improved the volume distribution and clearance of the drug.

Core body temperature and the heart rate of reptiles

Whether or not a reptile is at its metabolic optimum may be determined by taking the animal's cloacal temperature and noting its heart rate (HR). A reptile with a cloacal temperature around 34°–36°C would probably be considered to have an optimal core body temperature. When at its PBT, a reptile should achieve a definable resting HR (dependent on body size, see later). When cloacal temperatures indicate core body temperature below PBT, heart rates will be lower than those calculated from the allometric formula. However, if a reptile is warmed to its PBT, the heart rate will usually be found to be approximately that calculated for a reptile of its body size. All that is needed to assess heart rate is a knowledge of the animal's mass (kg), an instrument for determining HR in reptiles and an inexpensive electronic calculator capable of scientific functions.

Heart rate may be determined with a Doppler ultrasonic blood flow detector; this can be used in many species and sizes of reptiles. An allometric formula for calculating HR follows (assuming PBT):–

Heart rate (pulse) for reptiles = $34(W_{kg})^{-0.25}$

Examples:

500 gram python
$34(0.5)^{-0.25}$ = 40 beats per min

32 kilogram python
$34(32)^{-0.25}$ = 14 beats per min.

Warming cold reptiles

Very large reptiles, eg, giant tortoises, that have been allowed to become cold, may require days or weeks to regain their PBT when returned to a suitably heated environment. Small snakes can be heated to their PBT in warm water within a few minutes. An ill reptilian patient, warmed to its PBT and determined to have its appropriate heart rate, can be considered optimally prepared for medication, whether the medication is injectable anaesthetic, antibiotic or vermifuge. However, the reptilian patient which is gravely ill (even though it can be warmed temporarily) and found incapable of achieving its appropriate resting state HR, carries a poor prognosis.

Determining drug dosages by allometric scaling

If small reptiles are to maintain serum levels of drugs comparable to those in large reptiles, they require higher doses per unit bodyweight (mg/kg) administered more frequently (Mader, 1991). Dosages (mg/kg) and treatment regimes may be extrapolated from one animal to another by using the ratios of their estimated minimum energy costs (MEC) and specific minimum energy costs (SMEC). To use this information to extrapolate drug doses the clinician needs to calculate the MEC and SMEC of the reptile (see example and Appendix 1). Antimicrobial medications are taken up by the vascular system and delivered in the plasma to discrete sites, eg. those of infected tissues, via myriads of blood capilliaries from which diffusion and filtration occur. The medications or their by-products are later removed from tissues via lymph capilliaries. Medicating interstitial tissues via capilliary membranes is a dynamic model, not a static one. When it is seen that small animals require the highest drug dose rates, administered more frequently in the treatment period, it helps if one understands that smaller animals have greater capillary density (more diffusion/filtration surface area) than larger ones. Smaller animals also have a greater surface area per unit of body mass than larger ones. In fact, in some very small animals it is very difficult to administer a single treatment dose of medication intravenously before most of the dose disappears from the serum (excreted from the animal's body). Small animals of any given taxonomic group have more rapid circulation times and heart rates than do larger ones (Schmidt-Nielson, 1984). Small animals have high intracellular densities of cytochrome-C and mitochondria, and surface area and oxygen requirements per unit of bodyweight are higher than in large animals (Calder, 1984). For a given taxonomic group, small animals have higher hepatocyte densities and glomerular filtration rates than do large ones (Calder, 1984).

It should also be remembered that in cancer chemotherapy, doses of certain particularly toxic therapeutic agents are scaled to the body surface area and not to the mass of the patient. When doses are based on body surface area and are arithmetically converted to dose rates, it can be seen why the highest dose rates relate to small patients and the lowest ones to large patients.

Treatment doses and dose rates scale allometrically to body size (lean body mass in kilograms). MEC is represented by the allometric formula $K(W_{kg})^{0.75}$ and SMEC by $K(W_{kg})^{-0.25}$. For reptiles, the constant K is equal to 10 (Hainsworth, 1981). Appendix 2 for this chapter contains a formulary of antimicrobial drug doses based on metabolic ratios.

As an example, the formulary can be used to calculate treatment regimens using the antimicrobial agent trimethoprim-sulpha for two common iguanas (*Iguana iguana*) of different body size. The small one weighs 300 grams and the large one 2 kilograms.

Small iguana	Large iguana
MEC $= 10(0.3)^{0.75}$	MEC $= 10(2)^{0.75}$
$= 4$ kcal	$= 16.8$ kcal
SMEC $= 10(0.3)^{-0.25}$	SMEC $= 10(2)^{-0.25}$
$= 13.5$ kcal/kg	$= 8.4$ kcal/kg

The allometric dose for trimethoprim is 0.5mg/kcal and the SMEC-Frequency is 0.051 (see the formulary and work-sheet in Appendix 2 and follow the directions provided).

Small iguana	Large iguana
4kcal x 0.5mg/kcal = 2mg	16.8kcal x 0.5mg/kcal = 8.4mg
13.5mg/kg/kcal x 0.5mg/kcal = 6.75mg/kg	8.4mg/kg/kcal x 0.5mg/kcal = 4.2mg/kg
13.5 x 0.051 = 0.7 treatments/24hrs.	8.4 x 0.051 = 0.4 treatments/24hrs.
24/0.7 = 34 (36) hours	24/0.4 = 60 hours

Rx: Trimethoprim-sulpha for 300g iguana
     2mg total dose (or 6.75mg/kg) **every 36 hours**

Rx: Trimethoprim-sulpha for 2kg iguana
     8.4mg total dose (or 4.2mg/kg) **every 60 hours**

Therefore, the small iguana will receive 6.75mg/kg every day and a half while the large iguana will receive 4.2mg/kg once every two and a half days. At the same PBT the smaller animal requires a higher dose rate administered more frequently than in the larger animal.

SUPPORTIVE THERAPY

Warming

Warming a reptile to its optimal temperature **must** be the first therapeutic step. Other than topical treatments and emergency fluids, no food or drugs should be administered to a chilled reptile.

The metabolic processes of reptiles and other ectotherms slow dramatically when their body temperatures are reduced. Cold reptiles cannot properly digest food, nor can they reliably absorb, metabolise or excrete drugs. Force-feeding of cold reptiles may result in food decomposing in the digestive tract, often with fatal results. Medications given repeatedly to a cold reptile will often remain unabsorbed until the reptile is warmed, when all of the doses may suddenly be mobilised with potentially damaging consequences.

Fluids

The indications for fluid administration in reptiles are virtually the same as for mammalian patients. Jarchow (1988) pointed out that dehydration in reptiles may be isotonic, hypotonic or hypertonic. Hypotonic dehydration most often results from prolonged anorexia, while failure to drink is associated with hypertonic dehydration. Mildly dehydrated reptiles may recover well if given large pans of water in which to soak and drink. The water must be warmed to the appropriate temperature and the animal must have adequate access to and from the water. A debilitated animal may well be able to get into the water, but have trouble getting out.

Thorson's (1968) classic study demonstrated that, per unit bodyweight, reptiles have a higher percentage of total body water (63.0 – 74.4%) and a higher percentage of fluid in the intracellular compartment (45.8 – 58%) than do mammals. These values are highest in freshwater reptiles, lower in terrestrial species and lowest in marine forms (Marcus, 1981).

Marcus (1981) stated that the concentration of isotonic saline for most non-marine reptiles is 0.8%. Jarchow (1988) recommended administering slightly hypotonic fluids in cases of significant dehydration to facilitate intracellular diffusion of water. He recommended a solution of two parts 2.5% dextrose in 0.45% sodium chloride and one part Ringer's or equivalent electrolyte solution. He further recommended adding additional potassium chloride in cases of anorexia, emesis or diarrhoea.

Parenteral fluids should always be warmed to the reptile's PBT before administration. The volume of parenteral fluid that should be administered to an ailing reptile is debatable; recommendations vary from 2 – 5% of the animal's bodyweight (Cooper and Jackson, 1981; Jarchow, 1988; Page and Mautino, 1990). The present authors feel that the more cautious approach is appropriate, and that fluids equivalent to 2 – 3% of the animal's bodyweight should be given daily until normal hydration is restored, or until the animal begins drinking on its own. As a general rule one should not try to correct any physiological pathology too rapidly, as overhydration can cause additional harm.

Sites and rates for fluid administration

Lizards often have enough loose skin so that volumes of fluids approaching 1% of bodyweight may be given subcutaneously. In snakes and chelonians this is rarely possible.

Page and Mautino (1990) recommended that when administering intravenous fluids, the rate of administration should be carefully monitored and a rate of 1ml/min be used only for the sickest patients.

Frye (1981), Marcus (1981), Jarchow (1988) and others advocated the use of the intracoelomic route for fluid therapy; this is analogous to the intraperitoneal route in animals with diaphragms, eg. mammals. Only warmed fluids should be used. Reptiles should be restrained in lateral recumbency (on their right side in snakes). A site is chosen on the animal's left side on the ventrolateral aspect of the trunk. A radiograph taken before the procedure may help in assessing the extent of the air sacs as this varies greatly in different snake Families. In general, injections should be administered only at sites in the caudal quarter of the animal's body (to avoid the air sacs in snakes). The needle is directed towards the parietal pleuroperitoneum in order to avoid the abdominal organs. Jarchow (1988) pointed out that fluids given in this manner may result in decreased total lung capacity and tidal volume, especially in chelonians. He recommended extracoelomic administration of fluids in this group. In a 2kg chelonian, a 2.5 – 4cm, 20G – 21G needle is directed caudally and inserted just ventral to the shoulder joint and dorsal to the plastron. Fluids are injected slowly.

The administration of glucose in parenteral fluids may be an important factor in supporting cachectic reptiles. Isotonic (5%) and hypertonic glucose solutions may be useful if fluids are being administered via a large vein. However, one must never give glucose solutions subcutaneously or intramuscularly if they are more concentrated than 2.5%.

Because many reptiles have the ability to absorb significant amounts of fluid from the gastrointestinal tract, the authors regularly use fluids, such as 0.45% saline, via the intracolonic route (enemas) as an adjunct route for rehydrating debilitated reptiles. This location has particular advantages when dealing with animals which are inclined to bite. In many chelonians the authors have found that it is often much easier to gain access to the cloaca than to the mouth.

Vitamins and minerals

Many authors advocate the use of vitamin and mineral supplementation in chronically hypophagic or anorexic reptiles. Depending on the preparation and the health of the animal, these may be given as intramuscular injections or added to parenteral fluids. Injection sites should be alternated to avoid undue pain. Injections in the hindlimbs are not recommended (see Parenteral Administration). Specific recommendations are discussed later.

TECHNIQUES OF DRUG ADMINISTRATION

Oral medication

The use of oral medication is often limited. Sick reptiles are often profoundly anorexic, so placing drugs in the food is generally unreliable. In addition, the gastrointestinal transit time in reptiles can be extremely variable, complicating the task of maintaining therapeutic levels of drugs. However, some animals are difficult or dangerous to handle and some drugs are only available in oral forms. For animals that are eating well, oral medication may also be far less stressful to both patient and clinician. In reptiles that take whole animal prey it is usually quite easy to place or inject medicines into the bodies of food animals.

Administration of medications via drinking water is **not** recommended. Water consumption in reptiles is too sporadic and unpredictable to ensure any reliable therapeutic effects.

Stomach tubing (gavage)

Stomach tubing is often used to medicate or force-feed reptiles, but it must be approached judiciously. Repeatedly having to force open a reptile's mouth is stressful and may easily cause fractures, bruising or mucosal trauma that can predispose to infectious stomatitis. When using a speculum or mouth gag it is important to avoid traumatising oral structures. The tube used for gavage should be blunt-tipped, flexible and well lubricated. It should take little or no force to pass a tube down the oesophagus of most reptiles. The oesophagus of sea turtles is lined with caudally pointing, long, conical, keratinised papillae that can make passing a stomach tube difficult (Jacobson, 1988). Suedmeyer (1991) described a simple technique using a folded piece of radiographic film for a speculum and a soft rubber catheter: this may be useful for administering food or medications to smaller reptiles.

Chemical restraint for stomach tubing may be a useful adjunct in species that are venomous, strong or aggressive: it will permit easy handling.

Snakes are perhaps the easiest group of reptiles to stomach tube, but it must be remembered that snakes are adept at regurgitation, so a tube of sufficient length to reach the stomach is needed (roughly 30% of the animal's length). It is important to avoid handling or otherwise stressing snakes for at least 24 hours after tubing to avoid regurgitation. They should be left warm, quiet and away from disturbance. Snakes can often be forcefed dead, whole, lubricated prey species (with or without medication), although it is relatively easier for these to be regurgitated.

The glottis in snakes is situated in a rostral position and is capable of being closed quite firmly; it is often possible to insert and advance a stomach tube without even opening the animal's mouth, especially if the tube is larger in diameter than the snake's glottis. Padded forceps or a tongue depressor can be used as a speculum if necessary. The teeth of most snakes are quite fragile, so care must be taken to avoid oral trauma. Excessive force in passing the tube may result in oesophageal rupture.

Some chelonians are difficult to stomach tube. It often takes two handlers to extend their necks and open their mouths. In larger species opening the mouth may often not be possible without sedation or anaesthesia. A speculum must be used to keep the mouth open while the stomach tube is passed. The stomach is on the left, about midway between the fore and hindlimbs. Terrapins are difficult to stomach tube as they are slippery, can bite and will not extend their necks.

Because of the potential danger to handlers, crocodilians are usually given medications in their feed. However, if these species can be adequately restrained, stomach tubing is not difficult. A suitably sturdy gag or speculum should be used. Once the mouth has been opened, the animal will readily bite on a gag, and the jaws can be taped closed around the gag. In the caudal oral cavity the dorsal soft palate and the ventral preglottal flap cover the glottis. Once the preglottal flap is depressed, the glottis is readily visible and can be avoided with the stomach tube.

Most lizards are not difficult to restrain manually. Once this is accomplished, they may open their mouths spontaneously or may be encouraged by waving a finger in front of them, or gently tapping on the nasal area or under the jaw. When they attempt to bite, a small gag or speculum may be inserted. If necessary, the lower jaw may be gently pulled open by slow, steady downward tension on the skin below the lower jaw, but care must be taken not to damage the dewlap.

Under anaesthesia, it is possible surgically to insert an oesophagostomy tube in many reptile species. This technique may be most useful in chelonians where opening of the mouth can be a problem. Page and Mautino (1990) described a technique in tortoises that differs little from a standard mammalian pharyngostomy approach (see "Surgery"). Difficulty in maintaining asepsis of the surgical site makes this technique less effective in aquatic species. This may be a useful technique in snakes, although oesophagostomy in snakes or crocodilians does not appear to have been reported.

Parenteral administration

Since reptiles (and other non-mammalian vertebrates) may possess a renal portal venous circulation, it is sometimes recommended that drugs, especially the nephrotoxic aminoglycosides, should not be injected into the caudal portion of the body (Jacobson, 1988). The rationale for this is based on a knowledge of anatomy and physiology, but pharmacokinetic studies have not yet been done. It is believed that drugs injected into caudal areas of the body may go directly to the kidney via the renal portal veins, thus reaching very high concentrations in this tissue and potentially being excreted before travelling to the cranial portion of the animal.

Intramuscular/subcutaneous injections

For many patients intramuscular or subcutaneous injection may be the route of choice. Injections can often be given with minimal manipulation of the animal and they give relatively consistent and predictable uptake, providing that the animal is properly warmed and hydrated. In addition, in less fractious or dangerous animals, owners can often be trained to administer drugs by these routes.

Small volumes of injectable drugs or fluids may be given by the intramuscular route. In very tiny animals, eg. anole or lacotid lizards (*Anolis* spp.), there may not be enough muscle mass to permit administration of more than 0.05ml. In patients weighing less than 100g no more than about 0.2ml should be given at one injection site, whereas several millilitres may be given in tortoises or crocodilians. Clearly, if a drug is known to be very irritating to tissues, another route of administration is advisable.

With adequate restraint, intramuscular injections are not difficult in most reptiles. Snakes have an extensive set of longissimus muscles which run parallel to the spine and small injections can be given in this location. Some chelonians, eg. box-turtles (*Terrapene* spp.), may be able to retract far enough into their shells to make intramuscular injections difficult. Many crocodilians have plates of dermal bone in the cervical region which can prevent or hamper injections.

Intravenous injections

The sites that one may use for obtaining blood or giving intravenous injections to reptiles are described elsewhere in this manual (see "Anaesthesia" and "Laboratory Investigations").
The authors have found that in chelonians the dorsal caudal venous sinus described by Samour

et al (1984) provided excellent access for both obtaining blood samples and giving intravenous injections. This site has been used to place 22G or 24G indwelling catheters in large individuals (>1kg) of several species of chelonians for periods up to a few hours.

Nebulisation

Although not often used in reptile medicine, nebulisation of warm, moist gases and therapeutic drugs may be important in the treatment of bacterial or fungal respiratory diseases in reptiles. Lawrence (1988) described a technique used in birds that may be useful in reptiles.

Topical

Topical administration of some agents may be effective in treating superficial skin or ocular infections in reptiles (see "Integument" and "Ophthalmology"). Many of these are discussed later (see Medications).

MEDICATIONS

Before one can expect any pharmaceutical agents to act predictably in reptiles, the animals' environmental conditions must be adjusted to meet their optimal physiological requirements. This point cannot be overemphasised. The chemical armamentarium can only be expected to aid the animal's natural ability to fight disease. This principle must not be neglected; the animal must first be helped to help itself. Appendix 3 summarises some of the literature on drug dosages recommended for reptiles. Further information may be found in Holt (1981).

ANTIBACTERIAL AGENTS

The use of antibacterial agents in reptiles is based on several factors. First, the identity of the pathogen should be determined by culture. Second, sensitivity testing of the organism isolated will provide a range of possible effective antibiotics. The techniques for culturing reptile pathogens are different from the routine methods used on mammals and birds. One should make sure that the laboratory performing the culture is aware of these differences. Cultures should be incubated at room temperature (22°–25°C) and at 37°C, as well as at a reduced oxygen tension (Frye, personal communication; Needham, 1985). Third, the available route of administration may influence the choice of drug; for example, the majority of medications used in mammals and/or birds in the oral form have not been tested for bioavailability in reptiles and, therefore, may be ineffective. Other medications may be available only in intravenous form and it may be impracticable to administer them repeatedly to most reptile patients.

TOPICAL MEDICATIONS

There are many effective topical antibacterial agents which can be useful in treating superficial skin infections of reptiles, such as infectious stomatitis, rat bites, superficial fungal or bacterial infections and minor wounds.

Triple antibiotic ointments containing a combination of **polymyxin B**, **neomycin** and **bacitracin** (without steroids) are often effective against common reptile pathogens, eg. *Pseudomonas* spp. and *Aeromonas* spp.

Silver sulphadiazine cream (Flamazine, Smith and Nephew) is a topical agent developed to combat infections secondary to severe burns in human medicine. It covers a very broad spectrum including *Pseudomonas* spp. and many fungi (Swaim and Lee, 1987). It can be useful in the treatment of superficial cutaneous infections, including infectious stomatitis, and thermal burns in reptiles. Systemic levels of the sulphadiazine have been detected in humans when a large surface area is being treated and this suggests that toxicity could become a problem. The silver component of the medication is not absorbed and systemic toxicity by this element should not develop (Gilman *et al*, 1980). No toxicity has been recorded in reptiles.

Povidone-iodine solutions (0.05%) and ointments can be used to clean and disinfect wounds.

Animals with generalised cutaneous infections can be soaked in very dilute (0.005%) aqueous povidone-iodine solutions for up to one hour, once or twice daily, to help reduce skin contamination.

Chlorhexidine solutions (0.05%) and ointments can also be used for this purpose (Swaim and Lee, 1987).

Oral antibiotics

Most oral antibiotics have not been studied as repeated stomach tubing can be stressful and may be difficult in some reptiles.

However, the **fluoroquinalones** are being investigated and hold promise as effective antibiotics with good activity against major reptile pathogens, including *Pseudomonas aeruginosa* and many mycobacterial species. They are only minimally active against anaerobes. Currently, few parenteral fluoroquinalones are available and oral formulations are being used only on an experimental basis. The human products include **norfloxacin**, **ciprofloxacin**, **ofloxacin**, **enoxacin** and **perfloxacin**. The veterinary products include an oral form of **enrofloxacin** and a intramuscular injectable form. The injectable form of enrofloxacin is very irritating and the manufacturer cautions against repeated injections (Baytril, package insert). The human product ciprofloxacin is the most potent of the group and clinical experience has shown it to have a better Gram-negative spectrum, especially against *Pseudomonas* spp., than enrofloxacin. Fluoroquinalones are not recommended in children due to cartilage erosions seen in young growing patients (Hooper and Wolfson, 1991). This adverse effect has not yet been reported in reptiles. As this is a new class of antibiotic, newer generation compounds will be forthcoming. The spectrum and low toxicity of this antibiotic suggest that it may prove useful for the treatment of reptile diseases.

Parenteral antibiotics

Parenteral administration is usually the method of choice for systemic treatment of reptiles. Severe infections such as pneumonia, thermal burns or septicaemia require accurate calculation of the appropriate drug dosage and adequate duration of treatment.

Aminoglycosides have historically been used to treat reptiles due to the fact that *Pseudomonas* spp. and *Aeromonas* spp. seem to be the predominant pathogens and are often resistant to other available antibiotics. However, these agents are potentially nephrotoxic to birds, mammals and reptiles (Jacobson, 1976; Montali *et al*, 1979). It has been stated that reptiles are more sensitive to the potential harmful effects of aminoglycosides than other animals. It is perhaps more likely that the dose has been incorrectly determined and that the reptiles have been overdosed.

Amikacin and **gentamicin** are the most frequently used aminoglycosides. They possess a good Gram-negative spectrum but a limited Gram-positive range, with no activity against anaerobes. Of all the aminoglycosides, amikacin possesses the broadest spectrum and is the least likely to encounter resistance (Gilman *et al*, 1980). There is a very narrow safety margin with side effects recorded of nephrotoxicity, ototoxicity, cardiotoxicity and neuromuscular blockade (Jacobson, 1988; Jacobson *et al*, 1988). Toxicity is both dose and frequency related. The toxicity of these agents also depends on the state of hydration of the animal and the resultant functional state of the kidneys. Accurate evaluation of renal function and dehydration are perhaps the most important factors in using these drugs successfully, after the correct dosage has been determined. The state of hydration should always be corrected **before** initiating systemic therapy with aminoglycosides. It is advised that fluid therapy be maintained throughout the course of anitbiotic therapy to help support renal function and to reduce the likelihood of toxicity. Finally, reptiles may be fasted to reduce the nitrogen load (uric acid production) on the system (Jacobson 1988; Jacobson *et al*, 1988). If renal function is compromised, the use of aminoglycosides should be questioned and another antibiotic considered. Clinical evaluation of hydration status and serum uric acid levels will help to assess and monitor renal function.

Another antibiotic which has shown effectiveness against the common reptile Gram-negative pathogens is **carbenicillin**. This drug was developed for its anti-pseudomonal activity. However, resistance may develop quickly and the injection can be irritating (Lawrence *et al*, 1984). The recommended dose is high (Lawrence *et al*, 1984; Lawrence *et al*, 1986) and often requires large volumes to be injected. This may be impracticable in smaller patients. **Piperacillin** is another beta-lactam antibiotic demonstrating excellent activity against *Pseudomonas* spp. and a wide range of other bacteria (Hilf *et al*, 1991a).

Ceftazadime (Tazidime, Lilly) is a third generation cephalosporin which, unlike the other cephalosporins, has been shown to be effective against *Pseudomonas* spp. It possesses minimal nephrotoxicity (Lawrence *et al*, 1984). The drug is available in powder form and must be reconstituted before use. Once reconstituted, the shelf-life (refrigerated) is 10 days. The shelf-life may be extended by freezing the reconstituted solution. Individual doses for a two or three week course of therapy for a particular patient may be drawn up into syringes at the time that the vial is reconstituted. These individual syringes can then be frozen until just prior to use (Kollias, personal communication).

Other cephalosporins, such as **cephalothin** and **cephaloridine**, have also been used in reptiles. Culture and sensitivity information will help select the appropriate drug. Empirical use is not recommended because these agents are generally not effective against *Pseudomonas* spp. and *Aeromonas* spp. However, they do possess good activity against *Klebsiella* spp., *Enterobacter* spp. and *Proteus* spp. (Gilman *et al*, 1980).

Chloramphenicol is a broad-spectrum, bacteriostatic antibiotic which has been used for many years in reptiles. It is effective against many Gram-positive and Gram-negative aerobic and anaerobic bacteria. It is often the most effective agent against *Salmonella* spp. (Clark *et al*, 1985; Booth and MacDonald, 1989). The half-life of this drug is variable among different species of reptiles (Clark *et al*, 1985). Chloramphenicol is recommended specifically for the treatment of septicaemic cutaneous ulcerative disease (SCUD) in turtles, caused by *Citrobacter freundii* (see "Integument"), although other antibiotics have also been found to be effective for this disease (Frye, 1991). Unfortunately, *Pseudomonas* spp. are sometimes resistant to chloramphenicol. No side effects have been reported in reptiles (Jacobson, 1988).

Oxytetracycline has been suggested for use against *Aeromonas* spp. in alligators (Jacobson, 1988).

The use of antibiotics in combination may be required to increase the spectrum, decrease the development of resistance and allow for synergistic or additive effects which would allow reduced dosages and thus reduce any potential toxicity. Combining an **aminoglycoside** plus a **penicillin** or **cephalosporin** can broaden the effective antibiotic spectrum and can be useful in treating mixed infections. Empirical therapy with aminoglycosides alone will not be effective against infections involving Gram-positive or anaerobic organisms (Stewart, 1990). It is suggested that empirical treatments for conditions such as pneumonia should cover the possibility of anaerobic involvement and include a combination such as amikacin plus ampicillin. The use of gentamicin plus carbenicillin is controversial. The advantages cited are questionable and may in fact be detrimental (Jacobson, 1988).

Trimethoprim-sulpha is a relatively non-toxic, broad-spectrum antibiotic combination available in both oral and parenteral form. The oral form has not been evaluated in reptiles for bioavailability; consequently, the parenteral formulation is recommended. The drug is not always effective against *Pseudomonas* spp., in particular *Pseudomonas aeruginosa*, and sensitivity testing should be performed prior to use (Gilman *et al*, 1980).

ANTIFUNGAL AGENTS

Fungal diseases can be difficult to treat in any species. The exact circumstances surrounding the development of this type of infection should be investigated, since fungal diseases are often secondary to some type of immunosuppression or sub-optimal environmental conditions. Improper environmental conditions relating to temperature, humidity and hygiene are frequently implicated (Jacobson, 1980).

Topical medications

Iodine preparations are effective against most superficial fungal skin infections. Dilute **povidone-iodine** baths (a dilution of 1:10 or 0.005%) can be very beneficial in treating superficial conditions as mentioned earlier. Fungal or algal infections in chelonians can be effectively treated by scrubbing the shell with a stiff surgical scrub brush or toothbrush and aqueous povidone-iodine solution. Aquarium preparations of **malachite green**, **gentian violet** and **potassium permanganate**, or **salt**, **coppertox** or **formaldehyde** solutions in water have also been recommended (Jacobson et al, 1983; Murphy and Collins, 1983).

Topical antifungal agents that may be useful include **1% tolnaftate cream** (Jacobson, 1980), **miconazole** preparations and some of the newer **imidazoles** being developed.

Systemic agents

Ketoconazole is still the most commonly used systemic antifungal agent. Unfortunately it is available only in an oral form and must be administered by stomach tube once daily (Page at al, 1988). Hepatoxicity may develop with prolonged use.

Amphotericin B can be used intravenously or may be nebulised in respiratory infections (Jacobson, 1988). It is potentially nephrotoxic, so care should be taken to monitor renal function. It is not absorbed from extravascular sites so nebulisation should avoid potential renal toxicity.

Nystatin can be used for gastrointestinal candidiasis (Jacobson, 1988). It is not absorbed from the gastrointestinal tract and consequently cannot be used for systemic mycoses.

Griseofulvin has not yet been shown to be effective in reptiles (Jacobson, 1988).

ANTIPARASITIC AGENTS

Protozoal infections can be a major cause of anorexia in reptiles. Normal protozoa may proliferate and develop into a pathogenic population, or primary pathogenic protozoa, eg. *Entamoeba invadens*, can cause severe, life-threatening disease. **Metronidazole** is easily administered orally. Toxicity has been seen in indigo snakes (*Drymarchon* spp.) and tricolour ring snakes (*Lampropeltis triangulum*) and care should be taken to calculate the dose accurately (Jacobson, 1988). Other similar compounds, such as **dimetridazole**, have been used, but may no longer be available. Compounds such as **emetine** and **paramomycin**, with or without **diiodohydroxyquin**, have been advocated (Murphy and Collins, 1983).

Coccidiosis in reptiles is not unusual and should be treated as in mammals, ie. when clinical signs are apparent. As in mammals, problems with coccidia are most often related to overcrowding and/ or poor hygiene practices in the captive environment. Treatment may involve the use of oral **sulphadimethoxine, sulphaquinoxaline** or **sulphamethazine**. Injectable **sulphamethoxydiazone** has also been used successfully (Jacobson, 1988). Chelonians have also been treated with **amprolium** (Murphy and Collins, 1983).

Cryptosporidial and haemoprotozoal infections are generally thought to be untreatable except with supportive care.

Trematode infections are not unusual in aquatic species, particularly turtles. Nearly every organ system can be involved depending on the organism. Unfortunately, there are very few safe options for treating such infections. Treatment with **praziquantel** has met with variable success. Usually the encysted forms of the trematode are not reached by this drug. Some of the newer mammalian anti-trematodal drugs, such as **clorsulon** and **albendazole**, may be more effective and need to be investigated (Booth and MacDonald, 1989). Suggested scaled doses are given in Appendix 3.

Cestode infections occur in a wide variety of reptile species, particularly free-living or recently captured animals. **Praziquantel** is the most effective treatment currently available. Older medications such as **bunamidine hydrochloride** and **niclosamide** have been used successfully.

Nematode infections are common and can be very serious. Suggested therapies include compounds from the benzimidazole group, eg. **oxfendazole, thiabendazole** and **fenbendazole, piperazine** and the avermectins, particularly **ivermectin**. Toxicity has been seen with the use of ivermectin in chelonians. Experimentally induced toxicity was seen specifically in red-footed tortoises (*Chelonoidis carbonaria*), leopard tortoises (*Geochelone pardalis*), box-turtles (*Terrapene* spp.) and red-eared terrapins (*Trachemys scripta elegans*) (Teare and Bush, 1983). Pending further research, use of this drug should be avoided in all chelonians (Jacobson, 1988). In other reptiles, treatment with ivermectin should be repeated at approximately two week intervals.

Lungworm infestations with *Rhabdias* spp. may be treated with **levamisole** (Jacobson, 1988). Treatment of lungworm infestations with **fenbendazole** or **ivermectin** is effective in mammals and may also be so in reptiles.

Pentastomes are rarely diagnosed in animals other than reptiles. However, they have been reported to infect mammals, including humans, as an incidental intermediate host. Because of this, proper precautions should be taken when handling infected animals. They can potentially cause significant disease by their migrations during their life-cycle (Hendrix and Blagburn, 1988). There is no effective treatment for all stages, although Jacobson (1988) suggested **ivermectin** as the drug most likely to eliminate gastrointestinal stages.

Acanthocephalan infections may also require treatment. Treatment with **dithiazanine iodide** and **levamisole** has been suggested by Frye (1981).

Arthropod infestations can be difficult to eradicate. Mites and ticks are often seen in collections or on individual animals that have recently been captured or come from a large collection. Historically, treatments have included the use of **dichlorvos** (Vapona, Shell) impregnated strips. These products are potentially toxic. Clinical experience has shown that **ivermectin** appears to be very effective against mites, and can also be effective against ticks when combined with manual removal (Lawton, 1991). Ivermectin should **not** be used in chelonians. Secondary infections, viruses and septicaemias may be induced by arthropod infestations and should not be overlooked. Any treatment regimen must involve elimination of the arthropod from the environment as well as from the individual animal. Cages should be thoroughly cleaned and disinfected and may also be treated with an appropriate insecticide.

NUTRITIONAL THERAPIES

Nutritional deficiencies are commonly encountered in captive reptiles. This is partly due to the poorly understood dietary requirements of this widely diverse group of animals. There is also a distinct lack of information available to the reptile owning public. Nutritional diseases and therapy are discussed elsewhere in this manual (see "Nutritional Diseases").

REFERENCES

BOOTH, N.H. and MacDONALD, L.E. (1989). *Veterinary Pharmacology and Therapeutics.* Iowa State University Press, Ames.

BUSH, M. (1980). Antibiotic therapy in reptiles. In: *Current Veterinary Therapy VII. Small Animal Practice.* (Ed. R.W. Kirk). W.B. Saunders, Philadelphia.

BUSH, T.M., SMELLER, J.M. and CHARACHE, P. (1976). Preliminary studies of antibiotics in snakes. *Proceedings American Association of Zoo Veterinarians.*

BUSH, T.M., SMELLER, J.M., CHARACHE, P. and ARTHUR, R. (1978). Biological half-life of gentamicin in gopher snakes. *American Journal of Veterinary Research* **39 (1)**, 171.

CALDER, W. (1984). *Size, Function and Life History.* Havard University Press, Cambridge.

CALIGIURI, R., KOLLIAS, G.V., JACOBSON, E.R., McNAB, B., CLARK, C.H. and WILSON, R.C. (1990). The effects of ambient temperature on amikacin pharmacokinetics in gopher tortoises. *Journal of Veterinary Pharmacology and Therapeutics* **13**, 287.

CLARK, C.G., ROGERS, E.D. and MILTON, J.L. (1985). Plasma concentrations of chloramphenicol in snakes. *American Journal of Veterinary Research* **46 (12)**, 2654.

COOPER, J.E. and JACKSON, O.F. (1981). Eds. *Diseases of the Reptilia.* Academic Press, London.

FRYE, F.L. (1981). *Biomedical and Surgical Aspects of Captive Reptile Husbandry.* 1st Edn. Veterinary Medical Publishing Company, Edwardsville.

FRYE, F.L. (1991). *Biomedical and Surgical Aspects of Captive Reptile Husbandry.* 2nd Edn. Krieger, Malabar.

GILMAN, A.G., GOODMAN, L.S. and GILMAN, A. (1980). Eds. *The Pharmacological Basis of Therapeutics.* 6th Edn. Pergamon Press, New York.

GINSBERG, M. and TAGER, I. (1980). *Practical Guide to Antimicrobial Agents.* Williams and Wilkins, Baltimore.

HAINSWORTH, F.R. (1981). *Animal Physiology: Adaptations in Function.* Addison-Wesley Publishing Company, Reading.

HENDRIX, C.M. and BLAGBURN, B.L. (1988). Reptilian penastomiasis: a possible emerging zoonosis. *Compendium on Continuing Education for the Practicing Veterinarian (North American Edn.)* **10 (1)**, 46.

HILF, M., SWANSON, D., WAGNER, R. and YU, V.L. (1991a). Pharmacokinetics of piperacillin in blood pythons (*Python curtis*) and *in vitro* evaluation of efficacy against aerobic Gram-negative bacteria. *Journal of Zoo and Wildlife Medicine* **22 (2)**, 199.

HILF, M., SWANSON, D. and WAGNER, R. (1991b). A new dosing schedule for gentamicin in blood pythons (*Python curtis*): a pharmacokinetic study. *Research in Veterinary Science* **50**, 127.

HOCK, R.J. (1964). Terrestrial animals in cold: reptiles. In: *Handbook of Physiology, Section 4: Adaptations to the Environment.* (Eds. D.B. Hill, E.F. Adolph and C.G. Wilber). American Physiological Society, Washington DC.

HOLT, P.E. (1981). Drugs and doses. In: *Diseases of the Reptilia, Vol. 2.* (Eds. J.E. Cooper and O.F. Jackson). Academic Press, London.

HOOPER, D.C. and WOLFSON, J.S. (1991). Fluoroquinolone antimicrobial agents. *The New England Journal of Medicine* 384.

JACOBSON, E.R. (1976). Gentamicin-related visceral gout in two boid snakes. *Veterinary Medicine/Small Animal Clinician* **71**, 361.

JACOBSON, E.R. (1980). Necrotizing mycotic dermatitis in snakes: clinical and pathological features. *Journal of the American Veterinary Medical Association* **177 (9)**, 838.

JACOBSON, E.R. (1988). Use of chemotherapeutics in reptile medicine. In: *Exotic Animals.* (Eds. E.R. Jacobson and G.V. Kollias). Churchill Livingstone, New York.

JACOBSON, E.R., KOLLIAS, G.V. and PETERS, L.J. (1983). Dosages for antibiotics and parasiticides used in exotic animals. *Compendium on Continuing Education for the Practicing Veterinarian (North American Edn.)* **5 (4)**, 315.

JACOBSON, E.R., BROWN, M.P., CHANG, M., VLIET, K. and SWIFT, R. (1988). Serum concentrations and disposition kinetics of gentamicin and amikacin in juvenile American alligators. *Journal of Zoo Animal Medicine* **19 (4),** 188.

JARCHOW, J.L. (1988). Hospital care of the reptile patient. In: *Exotic Animals.* (Eds. E.R. Jacobson and G.V. Kollias). Churchill Livingstone, New York.

KIRKWOOD, J.K. (1983a). Influence of body size on animals in health and disease. *Veterinary Record* **113**, 287.

KIRKWOOD, J.K. (1983b). Bodyweight and drug dosage rates. *British Veterinary Zoological Society Newsletter* **16**, 5.

LAWRENCE, K. (1984). A preliminary study on the use of carbenicillin in snakes. *Journal of Veterinary Pharmacology and Therapeutics* **7**, 119.

LAWRENCE, K. (1988). Therapeutics. In: *Manual of Parrots, Budgerigars and other Psittacine Birds.* (Ed. C.J. Price). BSAVA, Cheltenham.

LAWRENCE, K., MUGGLETON, P.W. and NEEDHAM, J.R. (1984). Preliminary study on the use of ceftazidime, a broad-spectrum cephalosporin antibiotic, in snakes. *Research in Veterinary Science* **36**, 16.

LAWRENCE, K., PALMER, G.H. and NEEDHAM, J.R. (1986). Use of carbenicillin in two species of tortoise *(Testudo graeca and T. hermanni). Research in Veterinary Science* **40**, 413.

LAWTON, M.P.C. (1991). Lizards and snakes. In: *Manual of Exotic Pets.* New Edn. (Eds. P.H. Beynon and J.E. Cooper). BSAVA, Cheltenham.

LEWIS, B.P. and WILKIN, L.O. (1982). *Veterinary Drug Index.* W.B. Saunders, Philadelphia.

MADER, D.R. (1991). Antibiotic therapy. In: *Biomedical and Surgical Aspects of Captive Reptile Husbandry.* 2nd Edn. (Ed. F.L. Frye). Krieger, Malabar.

MADER, D.R., CONZELMAN, G.M. and BAGGOT, J.D. (1985). Effects of ambient temperature on the half-life and dosage regimen of amikacin in the gopher snake. *Journal of the American Veterinary Medical Association* **187 (11)**, 1134.

MARCUS, L.C. (1981). *Veterinary Biology and Medicine of Captive Amphibians and Reptiles.* Lea and Febiger, Philadelphia.

MONTALI, R.J., BUSH, M. and SMELLER, J.M. (1979). The pathology of nephrotoxicity of gentamicin in snakes. *Veterinary Pathology* **16**, 108.

MORGAN, R. (1988). *Angell Memorial Animal Hospital Formulary.* Boston.

MURPHY, J.B. and COLLINS, J.T. (1983). *A Review of the Diseases and Treatment of Captive Turtles.* AMS Publishing, Lawrence.

NEEDHAM, J.R. (1985). Laboratory aspects of reptilian infections. In: *Reptiles: Breeding, Behaviour and Veterinary Aspects.* (Eds. S. Townson and K. Lawrence). British Herpetological Society, London.

PAGE, C.D., MAUTINO, M., MEYER, J.R. and MECHLINSKI, W. (1988). Preliminary pharmacokinetics of ketoconazole in gopher tortoises *(Gopherus polyphemus). Journal of Veterinary Pharmacology and Therapeutics* **11**, 397.

PAGE, C.D. and MAUTINO, M. (1990). Clinical management of tortoises. *Compendium on Continuing Education for the Practicing Veterinarian (North American Edn.)* **12 (2)**, 221.

PAGE, C.D., MAUTINO, M., DERENDORF, H. and MECHLINSKI, W. (1991). Multiple-dose pharmacokinetics of ketoconazole administered orally to gopher tortoises *(Gopherus polyphemus). Journal of Zoo and Wildlife Medicine* **22 (2)**, 191.

RAPHAEL, B., CLARK, C.H. and HUDSON, R. Jr. (1985). Plasma concentration of gentamicin in turtles. *Journal of Zoo Animal Medicine* **16**, 136.

SAMOUR, H.J., RISLEY, D., MARCH, T., SAVAGE, B., NIEVA, O. and JONES, D.M. (1984). Blood sampling techniques in reptiles. *Veterinary Record* **114,** 472.

SCHMIDT-NIELSON, K. (1984). *Scaling: Why is Animal Size so Important.* Cambridge University Press, London.

SEDGWICK, C.J. and POKRAS, M.A. (1988). Extrapolating rational drug doses and treatment periods by allometric scaling. *Proceedings of the American Animal Hospital Association 55th Annual Meeting.*

STEWART, J.S. (1990). Anaerobic bacterial infections in reptiles. *Journal of Zoo and Wildlife Medicine* **21 (2)**, 180.

SUEDMEYER, W.K. (1991). A simple method for administering oral medications to reptiles. *Journal of Small Exotic Animal Medicine* **1 (1)**, 43.

SWAIM, S.F. and LEE, A.H. (1987). Topical wound medications: a review. *Journal of the American Veterinary Medical Association* **190 (12)**, 1588.

TEARE, J.A. and BUSH, M. (1983). Toxicity and efficacy of ivermection in chelonians. *Journal of the American Veterinary Medical Association* **183 (11)**, 1195.

THORSON, T.B. (1968). Body fluid partitioning in the Reptilia. *Copeia* **3**, 592.

APPENDIX 1

Extrapolating drug doses MEC/SMEC-dose, SMEC-frequency work sheet.

Drug:–
    Reference:–
    Serum Inhibitory Concentration (μg/ml):–
        Peak _____ Trough _____
Date _____ Clinician _____

---

1. Control species _____ Weight (W_{kg}) _____
2. Dose (control) = _____ mg (dose rate _____ mg/kg)q _____ h (interval as hours).
3. Minimum Energy Cost (MEC) = $(K[W_{kg}]^{0.75})$ _____
4. Specific Minimum Energy Cost (SMEC) = $(K[W_{kg}]^{-0.25})$ _____
5. Dose (MEC/SMEC) is the control's treatment dose in mg divided by its MEC or its dose rate

   (mg/kg) divided by its SMEC; either, gives the same number {

MEC/SMEC-DOSE = _____

---

6. Frequency (number of treatment intervals per 24 hours) _____
7. SMEC frequency (frequency divided by SMEC)= _____

SMEC-FREQUENCY = _____

---

8. Patient species _____ Bodyweight (kg)
9. MEC $(K[W_{kg}]^{0.75})$ _____
10. SMEC $(K[W_{kg}]^{-0.25})$ _____
11. MEC x **MEC/SMEC-DOSE** = mg _____
12. SMEC x **MEC/SMEC-DOSE** = mg/kg _____
13. SMEC x **SMEC-FREQUENCY** = treatments (frequency) per day _____
14. Divide 24 treatments per day to obtain treatment interval as hours _____

    Patients dose = _____ mg [dose rate _____ mg/kg] every _____ hours

    Control's dose = _____ mg (dose rate _____ mg/kg) every _____ hours

    Energy constants (K):

Passerine bird	129
Non-passerine	78
Placental mammal	70
Marsupial	49
Reptile	10

Example:–

Drug:– *Cephalexin*
    Reference:– *Angell formulary*
    Serum Inhibitory Concentration (µg/ml):–
        Peak _____ Trough _____
Date   *01/9/91*   Clinician *Sedgewick / POKRAS*

---

1. Control species *dog*   Weight (W_{kg}) *20kg*
2. Dose (control) = *220 – 440* mg (dose rate *11 – 22* mg/kg)q *8* h (interval as hours).
3. Minimum Energy Cost (MEC) = $(K[W_{kg}]^{0.75})$ *662kcal*
4. Specific Minimum Energy Cost (SMEC) = $(K[W_{kg}]^{-0.25})$ *33kcal/kg*
5. Dose (MEC/SMEC) is the control's treatment dose in mg divided by its MEC <u>or</u> its dose rate

    (mg/kg) divided by its SMEC; either, gives the same number   { *0.33 – 0.66mg/kcal*

MEC/SMEC-DOSE = *0.33 – 0.66mg/kcal*

---

6. Frequency (number of treatment intervals per 24 hours) *3*
7. SMEC frequency (frequency divided by SMEC)= *0.08*

SMEC-FREQUENCY = *0.08*

---

8. Patient species *Bell python*   *3* Bodyweight (kg)
9. MEC $(K[W_{kg}]^{0.75})$ *22.8kcal*
10. SMEC $(K[W_{kg}]^{-0.25})$ *7.6kcal/kg*
11. MEC x **MEC/SMEC-DOSE** = mg *7.5 – 15mg*
12. SMEC x **MEC/SMEC-DOSE** = mg/kg *2.5 – 5mg/kg*
13. SMEC x **SMEC-FREQUENCY** = treatments (frequency) per day *0.6*
14. Divide 24 treatments per day to obtain treatment interval as hours *40 hours*

    Patients dose = *7.5 – 15* mg [dose rate *2.5 – 5* mg/kg] every *40* hours

    Control's dose = *220 – 440* mg (dose rate *11 – 22* mg/kg) every *8* hours

    Energy constants (K):

	Passerine bird	129
	Nonpasserine	78
	Placental mammal	70
	Marsupial	49
	Reptile	10

APPENDIX 2

Drugs/dose (MEC/SMEC), frequency (SMEC).

Directions:—
Multiply a patient's MEC by dose *(MEC/SMEC)* to get a treatment in mg; or, (optionally) multiply a patient's SMEC by dose (MEC/SMEC) to get a treatment in mg/kg. Multiply a patient's SMEC by frequency *(SMEC)* to get treatments per day (24 hours). Divide 24 by treatments per day to get the number of hours (interval) between treatments.

Antimicrobial	Reference control	Dose (MEC/SMEC)	Frequency (SMEC)
Albendazole (po)	6	0.617	------
Amikacin (i/m, i/v)	1, 2	0.127	0.076
Ampicillin (i/m, i/v)	1, 2	0.3-2.5	0.076
Ampicillin (po)	1, 2	2.5	0.1
Amoxicillin (po)	1	0.254	0.076
Amoxicillin/Clavulanic Acid (po)	1	0.349	0.051
Amprolium		3 – 6	sid 5 days
Carbenicillin (i/m, i/v)	1	0.381	0.1
Carbenicillin (i/v - serious infection)	2	3.34	0.2
Ceftizoxime (i/m, /iv)	1, 2	0.5	0.1
Cephalexin (i/m, i/v, po)	1	0.254	0.076
Cephalothin (i/m, i/v, po)	1	0.5	0.076
Chloramphenicol (i/v)	1	0.6	0.076
Chlortetracycline (po)	1	0.51	0.076
Clindamycin (i/m, i/v)	1	0.254	0.076
Doxycycline (po)	3	0.254	0.025
Enrofloxacin (i/m, po)	4	0.12	0.05
Erythromycin (po)	1	0.254	0.076
Fenbendazole		0.34 – 0.68	sid 3 days
Gentamicin (i/m, i/v)	1	0.07	0.076
Ivermectin (s/c)	5	0.01	once
Kanamycin (i/m, i/v)	1	0.19	0.051
Metronidazole		1.8	sid
Neomycin (po)	1	0.05	0.076
Nitrofurantoin (po)	1	0.1	0.076
Praziquantel		0.3	once
Ticarcillin (i/v)	2	2.5	0.134
Tobramycin (i/m)	2	0.05	0.1
Trimethoprim sulfa (i/m, i/v, po)	1	0.5	0.051

References
1. Morgan (1988).
2. Ginsberg and Tager (1980).
3. Lewis and Wilkin (1982).
4. Package insert.
5. Package insert.
6. Package insert.

APPENDIX 3

Appendix formulary of empirical and pharmacokinetically derived dosages.

Note that these doses are **not** scaled, but are recorded from the literature. Most of the studies involving pharmacokinetics with reptiles did not include the temperature at which the animals were kept for the duration of the study, and/or did not record the bodyweights of the experimental subjects involved.

Scaled doses for many of the same drugs listed below are given in Appendix 2.

Dosages for chelonians are based on the entire weight of the animal, including the shell. It has been recommended to use the snake dosage for tortoises, rather than the water turtle dose, unless a tortoise dose is specifically listed (Bush, 1980).

* = Pharmacokinetic information available
NS = Species not specified

Medication	Species	Dosage	Comments
Amikacin	Gopher snake (*Pituophis* spp.)	5mg/kg i/m (loading dose), then 2.5mg/kg every 72 hours.*	Mader *et al*, 1985.
	Alligator	2.25mg/kg i/m every 72 – 96 hours.*	Jacobson *et al*, 1988.
	Gopher tortoise (*Gopherus* spp.)	5mg/kg i/m every 48 hours (30°C).*	Caligiuri *et al*, 1990.
Amphotericin B	NS	5mg in 150ml NaCl, nebulised 1h bid for 7 days.	Jacobson, 1988. (Treatment for fungal pneumonia.)
Ampicillin	Tortoise	20mg/kg i/m every 24 hours for 7 – 14 days.	Page and Mautino, 1990.
Carbenicillin	Snakes	400mg/kg i/m every 24 hours for 2 weeks.*	Lawrence, 1984.
	Tortoises	400mg/kg i/m every 48 hours for 2 weeks.*	Lawrence *et al*, 1986.
Cefotaxime	Tortoises	20 – 40mg/kg i/m every 24 hours for 7 – 14 days.	Page and Mautino, 1990. (May be used with aminoglycosides.)
Cephaloridine	NS	10mg/kg i/m every 12 hours for 10 days.	Frye, 1991.

Medication	Species	Dosage	Comments
Cephalothin	NS	20 – 40mg i/m every 12 hours for 10 days.	Frye, 1991.
Chloramphenicol	Gopher snakes (*Pituophis* spp.)	40mg/kg s/c every 24 hours for 2 weeks.*	Bush *et al*, 1976.
	Indigo, rat and king snakes	50mg/kg s/c every 12 hours for 2 weeks.*	Clark *et al*, 1985.
	Boid and moccasin snakes	50mg/kg s/c every 24 hours for 2 weeks.*	Clark *et al*, 1985.
	Rattlesnake (*Crotalus* spp.)	50mg/kg s/c every 48 hours for 2 weeks.*	Clark *et al*, 1985.
	Nerodia snake (*Nerodia* spp.)	50mg/kg s/c every 72 hours for 2 weeks.*	Clark *et al*, 1985.
	Tortoises	20mg/kg i/m or orally every 12 hours for 7-14 days.	Page and Mautino, 1990.
Dimetridazole	Snakes	40mg/kg orally every 24 hours for 5 days.	Jacobson, 1988.
Dithiazanine	NS	20mg/kg orally every 24 hours for 10 days.	Frye, 1981.
Doxycycline	NS	10mg/kg.	
Emetine HCL	NS	0.5mg/kg i/m or s/c every 12 – 24 hours for 10 days.	Frye, 1981.
Fenbendazole	NS	50 – 100mg/kg orally repeat in 2 weeks.	Jacobson, 1988.
Gentamicin	Gopher and bull snakes (*Pituophis* spp.)	2.5mg/kg i/m every 72 hours for 2–3 weeks.*	Bush *et al*, 1978 and Bush, 1980. (Use snake dose for desert tortoises.)
	Blood pythons (*Python curtis*)	2.5mg/kg i/m, then 1.5mg/kg every 96 hours.*	Hilf *et al*, 1991b.
	Red-eared slider turtles (*Trachemys* spp.)	6mg/kg i/m every 72 hours for 3 weeks.*	Raphael *et al*, 1985.
	American alligator	1.75mg/kg i/m every 96 hours for 3 weeks.*	Jacobson *et al*, 1988.

Medication	Species	Dosage	Comments
Ivermectin	All except chelonians	200mcg/kg orally or s/c, repeat in 2 weeks.	(DO NOT USE in chelonians.) Teare and Bush, 1983.
Kanamycin	NS	10 – 15mg/kg divided i/v or i/m every 12 – 24 hours.	Frye, 1991.
Ketoconazole	Gopher tortoise (*Gopherus polyphemus*)	15mg/kg orally every 24 hours for 2 – 4 weeks.*	Page *et al*, 1991. (Tablets crushed and suspended in water for gavage.)
Levamisole	Snake	10mg/kg i/p, repeat in 2 weeks.	Jacobson, 1988. (Treatment for *Rhabdias* spp. infection.)
Mebendazole	NS	20 – 25mg/kg orally, repeat in 2 weeks.	Jacobson, 1988.
Metronidazole	Snakes	100 – 275mg/kg orally, repeat in 2 weeks.	Jacobson, 1988.
	Indigo snakes (*Drymarchon* spp.)	40mg/kg orally, repeat in 2 weeks.	(CAUTION in indigo snakes and tricolour ring snakes.)
	King snakes (*Lampropeltis* spp.)	100mg/kg orally, repeat in 2 weeks.	
	Tortoises	250mg/kg orally, repeat in 2 weeks.	Page and Mautino, 1990.
Nystatin	Python	100,000 units/kg orally every 24 hours for 10 days.	Jacobson, 1988.
Oxfendazole	NS	3ml/kg orally.	Lawton, 1991.
Oxytetracycline	NS	6 – 10mg/kg i/v or i/m every 24 hours.	Frye, 1991.
	Alligators	10mg/kg orally every 24 hours.	Jacobson, 1988.
Paramomycin	NS	33 – 110mg/kg orally every 24 hours for up to 4 weeks.	Jacobson, 1988.
Penicillin (Potassium G)	NS	10,000 – 20,000 units/kg i/m or s/c every 6 – 8 hours.	Frye, 1991.
Piperacillin	Blood pythons (*Python curtis*)	100mg/kg i/m every 48 hours.*	Hilf *et al*, 1991a.

Medication	Species	Dosage	Comments
Piperazine	Alligators and crocodiles	50mg/kg orally, repeat in 2 weeks.	Jacobson et al, 1983.
Praziquantel	Green turtles (*Chelonia* spp.)	7.5mg/kg orally, repeat in 2 weeks.	Jacobson, 1988.
Sulphadiazine	Lizards and snakes	25mg/kg orally every 24 hours for 1 week.	Jacobson et al, 1983.
Sulphadimethoxine	NS	90mg/kg orally day 1, then 45mg/kg orally for 5 days.	Jacobson et al, 1988.
Sulphamerazine	Lizards and snakes	25mg/kg orally every 24 hours for 1 week.	Jacobson et al, 1983.
Sulphamethazine	NS	75mg/kg orally day 1, then 40mg/kg orally for 5 – 7 days.	Jacobson, 1988.
Sulpha-methoxydiazone (20% injectable solution)	NS	80mg/kg s/c or i/m day 1, then 40mg/kg every 24 hours for 4 days.	Jacobson, 1988.
Sulphaquinoxaline	NS	75mg/kg orally day 1, then 40mg/kg orally every 24 hours for 5 – 7 days.	Jacobson, 1988.
Tetracycline	NS	10mg/kg orally every 24 hours for 10 – 14 days.	Jacobson et al, 1983.
Thiabendazole	NS	50mg/kg orally, repeat in 2 weeks.	Frye, 1981.
Tobramycin	NS	2mg/kg i/m every 24 hours (26°C).	Lawton, 1991.
Trimethoprim-sulpha	NS	15mg/kg s/c every 24 hours for 10 days.	Jacobson, 1988.
Tylosin	Lizards and snakes	5mg/kg i/m every 24 hours for 10 days.	Jacobson et al, 1983.
Vitamin A	Tortoises	11,000 IU/kg i/m.	Page and Mautino, 1990.
Vitamin B complex	NS	0.5ml/kg i/v, i/m or s/c.	Frye, 1981.
Vitamin C	NS	10 – 20mg/kg i/m every 24 hrs.	Frye, 1991.
Vitamin D_3	Tortoises	1,650 IU/kg i/m.	Page and Mautino, 1990.
Vitamin K	NS	0.25 – 0.75mg/kg i/m.	Frye, 1991.

APPENDIX ONE
Useful Addresses

The following addresses may prove helpful to those who are dealing with reptiles. While every effort has been made to check that the list is correct at the date of publication, no responsibility can be accepted for mistakes or omissions. The inclusion of an organisation on the list should not necessarily be taken to imply recommendation or endorsement by the British Small Animal Veterinary Association or by the editors and authors of this manual.

SOCIETIES, CLUB AND ORGANISATIONS

Animal Welfare Foundation
c/o British Veterinary Association, 7 Mansfield Street, London W1M 0AT.

Association of Amphibian and Reptilian Veterinarians
c/o Dr M Frahm DVM, Gladys Porter Zoo,
500 Ringhold Sreet, Brownsville, Texas 78520, USA.

Association of British Wild Animal Keepers (ABWAK)
2A Northcote Road, Clifton, Bristol BS8 3HB.

Association for the Study of Reptiles and Amphibians (ASRA)
Cotswold Wildlife Park, Burford, Oxfordshire OX8 4JW.

British Association of Tortoise Keepers (BATK)
c/o Egbaston Hotel, 323 Hagley Road,
Birmingham B17 8ND.

British Chelonia Group (BCG)
PO Box 2163, London NW10 5HW.

British Herpetological Society (BHS)
c/o Zoological Society of London, Regent's Park, London NW1 4RY.

British Small Animal Veterinary Association
Kingsley House, Shurdington, Cheltenham, Gloucestershire GL51 5TQ.

British Veterinary Association (BVA)
7 Mansfield Street, London W1M 0AT.

British Veterinary Zoological Society (BVZS)
c/o 7 Mansfield Street, London W1M 0AT.

British Wildlife Rehabilitation Council
c/o RSPCA, The Causeway, Horsham
West Sussex RH12 1HG
and 1 Pemberton Close, Aylesbury,
Buckinghamshire HP21 7NY.

East Sussex Herpetological Society
Mr P R Martin, 20 Silverlands Road, St Leonards-on-Sea, East Sussex TN37 7DE.

Fauna and Flora Preservation Society (The)
78-83 North Street, Brighton, East Sussex BN1 1ZA.

Glasgow and West of Scotland Society for the Protection of Animals
15 Royal Terrace, Glasgow G3 7NY.

Institute of Biology
20 Queensberry Place, London SW7 2DZ.
International Herpetological Society (IHS)
65 Broadstone Avenue, Walsall, West Midlands WS3 1JA.

Leicester Herpetological Society
Mr J Hayward, 11 Ledbury Green, Leicester, Leicestershire LE4 2LY.

People's Dispensary for Sick Animals (PDSA)
Whitechapel Way, Priorslee, Telford,
Shropshire TF2 9PQ.

Pet Health Council (PHC)
4 Bedford Square, London WC1B 3RA.

Portsmouth Reptile and Amphibian Society
c/o Southsea Community Centre, King Street, Southsea, Hampshire PO5 4EE.

Royal College of Veterinary Surgeons (RCVS)
32 Belgrave Square, London SW1X 8QP.

Royal Society for the Prevention of Cruelty to Animals (RSPCA)
The Causeway, Horsham, West Sussex RH12 1HG.

Scottish Society for the Protection of Animals (SSPCA)
19 Melville Street, Edinburgh EH3 7PL.

Scottish Wildlife Trust
25 Johnston Terrace, Edinburgh EH1 2NH.

Society of Reptiles, Amphibians and Snakes (SRAS)
c/o Iris Close, Springfield, Chelmsford, Essex CM1 5XS.

South Eastern Herpetological Society
Sutton Valance, Westerham, Kent.

South Western Herpetological Society
59 St Marychurch Road, Torquay, Devon TQ1 3HG.

Thames and Chilterns Herpetological Society
Mr M Matthewson, 18 Haynes Road,
Berkhampstead, Hertfordshire.

Ulster Society for the Prevention of Cruelty to Animals (USPCA)
Knockdeen, 11 Drumview Road, Lisburn,
Co Antrim BT27 6YF.

Universities Federation for Animal Welfare (UFAW)
8 Hamilton Close, South Mimms, Hertfordshire EN6 3BD.

Wildlife Department
RSPCA Headquarters, The Causeway, Horsham,
West Sussex RH12 1HG.
Wildlife Disease Association
PO Box 886, Ames, Iowa 50010, USA.

Wildlife Hospitals Trust
1 Pemberton Close, Aylesbury, Buckinghamshire HP21 7NY.

World Association of Wildlife Veterinarians
c/o Zoological Society of London, Regent's Park,
London NW1 4RY.

World Society for the Protection of Animals (WSPA)
106 Jermyn Street, London SW1Y 6EE.

Zoological Society of London
Regent's Park, London NW1 4RY.

EQUIPMENT AND SERVICES

Arnolds Veterinary Products Limited
Cartmel Drive, Harlescott, Shrewsbury SY1 3TB.
(*veterinary equipment*)

BCG Tortoise Registration
c/o British Chelonia Group, PO Box 2163,
London NW10 5HW.
Brindus Industrial Services Limited
Dartford Industrial Trading Estate, Dartford, Kent.
(*gloves*)

Chas F. Thackray Limited
PO Box 171, Park Street, Leeds, West Yorkshire
(*endoscopes*)

Distinject
Peter Ott AGG, Postfach, CH-4007, Basel, Switzerland.
(*capture equipment*)

International Market Supplies (IMS)
Dane Mill, Broadhurst Lane, Congleton, Cheshire CW12 1LA.
(*snake sexing probes*)

3M Health Care, Electro Medical Products
Morley Street, Loughborough, Leicestershire LE11 1EP.
(*kitecko pads*)

MD Components
Hamelin House, 211-213 Hightown, Luton,
Bedfordshire LU2 0BZ.
(*Jackson-Lawton tortoise measurer, anaesthetic chambers*)

Medical Diagnostic Services Inc
PO Box 1441, Brandon, Florida 34299, USA.
(*diagnostic instruments including endoscopes*)

Norfine Nets
15 The Drive, Fakenham, Norfolk NR21 8EE.
(*nets*)

Olympus Optical Company Limited
Keymed House, Stock Road, Southend on Sea,
Essex SS2 5QH.
(*endoscopes*)

Richard Wolf UK Limited
PO Box 47, Micham, Surrey CR4 4TT.
(*endoscopes*)

Temple Cox Limited
Cray Avenue, Orpington, Kent BR5 3TT.
(*capture equipment including nets and pole syringes*)

Vetark Animal Health
PO Box 60, Winchester, Hampshire SO23 9XN.
(*specialist vitamin/mineral mixes and disinfectants*)

JOURNALS AND MAGAZINES

Biologist (The)
c/o Institute of Biology, 20 Queensbury Place, London SW7 2DZ.

British Journal of Herpetology
Available from British Herpetological Society
Journal of Wildlife Diseases
Available from Wildlife Disease Association
Journal of Zoo and Wild Animal Medicine
Available from Executive Director of American Association of Zoo Veterinarians, 34th Street and Girard Avenue, Philadelphia, Pennsylvania 19104, USA.

Journal of Small Exotic Animal Medicine (The)
11552 Hartsook Street, Hollywood, California 91603, USA.

Testudo (BCG Journal)
PO Box 2163, London NW10 5HW

INFORMATION AND ADVICE

Association for the Study of Reptiles and Amphibians (ASRA)
Cotswold Wildlife Park, Burford, Oxfordshire OX8 4JW.

British Association of Tortoise Keepers (BATK)
c/o Egbaston Hotel, 323 Hagley Road, Birmingham B17 8ND.

British Chelonia Group (BCG)
PO Box 2163, London NW10 5HW.

British Herpetological Society (BHS)
c/o Zoological Society of London, Regent's Park, London NW1 4RY.

Natural History Museum
Cromwell Road, London SW7 5BD.
(*identification of specimens including parasites*)

British Veterinary Zoological Society (BVZS)
c/o 7 Mansfield Street, London W1M 0AT.

Commonwealth Institute of Parasitology
395A Hatfield Road, St Albans, Hertfordshire AL4 0XU.
(*identification of helminth parasites*)

Focus Publications
PO Box 235, Harston, Cambridge CB2 5TL.
(*literature on exotic pets*)

Joint Nature Conservation Committee
Monkstone House, City Road, Peterborough, Cambridgeshire PE1 1JY.

Pedigree Petfoods Education Centre
National Office, Waltham on the Wolds, Leicestershire LE14 4RS.
(*general*)

Society of Reptiles, Amphibians and Snakes (SRAS)
c/o Iris Close, Springfield, Chelmsford, Essex CM1 5XS.

Zoological Society of London
Regent's Park London NW1 4RY.
(*advice on care of reptiles*)

GOVERNMENT DEPARTMENTS

Agricultural Department (Welsh Office)
Crown Buildings, Cathays Park, Cardiff CF1 3NQ.

Department of Agriculture and Fisheries for Scotland (DAFS)
Pentland House, 47 Robb's Loan, Edinburgh EH14 1TW.

Department of the Environment (DOE) (Wildlife Division)
Tollgate House, Houlton Street, Bristol BS2 9DJ.

Ministry of Agriculture Fisheries and Food (MAFF)
(Animal Exports and Imports)
Animal Health Division, Hook Rise South, Surbition, Surrey KT6 7NF.

(*for Scotland see DAFS; for Wales see Agricultural Department (Welsh Office)*)

APPENDIX TWO
Clinical Examination and Post-Mortem Sheets

CLINICAL EXAMINATION OF REPTILES

Clinical ref. no:– .... Laboratory ref. no:– ....
Species:– .... Sub-species:– .... Sex:– .... Age:– ....
Owner/Origin:– .... Time in captivity:– ....
Management:– ....
Veterinary surgeon:– ....
Indications for examination:– ....
Date when first examined:– .... Weight when first examined:– ....
Measurements:– ....

HISTORY AND RELEVANT DATA

RESULTS OF CLINICAL EXAMINATION

FURTHER TESTS, RADIOGRAPHY

LABORATORY RESULTS

SUBSEQUENT INVESTIGATIONS

COMMENTS

(attach additional sheets as necessary)   .... MRCVS

Sequential weights/measurements

REPTILE POST-MORTEM REPORT

Clinical ref. no:– .... Pathology ref. no:– ....
Owner:– .... Veterinary surgeon:– ....
Date of examination:– .... Performed by:– ....
Died/killed (date):– .... Storage prior to examination:– ....
Species:– .... Sub-species:– ....
Sex:– .... Age:– ....

HISTORY
Period in captivity:– .... Origin:– ....
Management, including records of feeding, sloughing etc:–
Clinical history:–
(attach separate sheets as necessary)

POST-MORTEM FINDINGS
Radiography:–   **Jackson's ratio** (where appropriate)

Weight:–   **Measurements** rostrum – cloaca:–   **Condition** 1/2/3/4
                                     cloaca – tail tip:–
                                     carapace length:–

External
Skin:–   Cloaca:–   Other:–

Internal
Respiratory system:–
Liver:–
Spleen:–
Kidney and urinary tract:–
Reproductive tract:–
Gastrointestinal tract:–
Cardiovascular system:–
Central nervous system:–
Other organs:–

LABORATORY INVESTIGATIONS
Bacteriology:–
Parasitology:–
Cytology:–
Histopathology:–
Other:–

SAMPLES STORED:–
SAMPLES SUBMITTED ELSEWHERE:–
DIAGNOSIS/COMMENTS:–   .... MRCVS

APPENDIX THREE
Haematological and Biochemical Data

Haematological and Biochemical Data:– Chelonia and Crocodylia

Parameter	Units	Chelonia	Crocodylia
Haematological data			
RBC	$10^6/mm^3$	0.15 – 0.98	0.6 – 1.48
WBC	$10^3/mm^3$	6 – 48	6.4 – 10.2
Haemaglobin	g/dl	3.5 – 11.3	5.9 – 12
Haematocrit (PCV)	l/l	0.20 – 0.35	0.20 – 0.35
Biochemical data			
Total protein	g/l	29 – 66	51 – 65
Uric acid	µmol/l	88 – 707	244 – 265.5
Urea	mmol/l	0.2 – 16	0
Creatinine	µmol/l	....	....
Glucose	mmol/l	1.8 – 8.2	2.8 – 5.5
Calcium	mmol/l	1.3 – 17	2.6 – 7.9
Phosphorus	mmol/l	0.8 – 3.5	....
Lactate dehydrogenase (LDH)	u/l	10 – 136	....
Gamma glutamyltransferase (GGT)	u/l	0 – 15	....
Alkaline phosphatase (ALP)	u/l	....	....
Aspartate aminotransferase (SGOT)	u/l	10 – 136	....
Alanine aminotransferase (ALT)	u/l	....	....
Triglycerides	µmol/l	....	....
Cholesterol	mmol/l	....	....

Haematological and Biochemical Data:– Serpentes and Sauria

Parameter	Units	Serpentes	Sauria
Haematological data			
RBC	$10^6/mm^3$	0.4 – 2.5	0.4 – 2.1
WBC	$10^3/mm^3$	0.4 – 2.5	12 – 22.5
Haemaglobin	g/dl	5.2 – 12	4.6 – 11.9
Haematocrit (PCV)	l/l	0.16 – 0.45	0.16 – 0.35
Biochemical data			
Total protein	g/l	29 – 80	30 – 81
Uric acid	µmol/l	60 – 600	160 – 475
Urea	mmol/l	0.17 – 1.87	0.17 – 1.99
Creatinine	µmol/l	0 – 45	5 – 13
Glucose	mmol/l	0.5 – 6	3 – 11
Calcium	mmol/l	2.5 – 5.5	1.9 – 2.5
Phosphorus	mmol/l	0.9 – 1.85	0.6 – 1.66
Lactate dehydrogenase (LDH)	u/l	30 – 600	250 – 1000
Gamma glutamyltransferase (GGT)	u/l	0 – 15	0 – 10
Alkaline phosphatase (ALP)	u/l	80 – 145	60 – 99
Aspartate aminotransferase (SGOT)	u/l	5 – 35	5 – 105
Alanine aminotransferase (ALT)	u/l	260	....
Triglycerides	µmol/l	0.6 – 2	0.6 – 1.2
Cholesterol	mmol/l	1.3 – 3.6	1.2 – 3.62

APPENDIX FOUR
Legal Aspects

Both captive and free-living (wild) reptiles are subject to various laws in many countries. Veterinary surgeons who are dealing with reptiles should have some familiarity with this legislation and know how to obtain further information, since they may be asked for guidance by owners or potential purchasers.

The Table outlines some of the main legislation that is relevant to reptiles in the UK. Those seeking information about other countries should consult the references and also make enquiries of the appropriate Embassy, High Commission or Government Department.

While every effort has been made to state the legislation correctly, this Appendix should not be relied upon as the ultimate authority on any aspect of the law. The current state of the law must always be ascertained and any person who has a problem involving a point of law should seek help from a solicitor or legal advice centre.

TABLE

Some UK legislation which is relevant to reptiles.

Title and provisions	Species affected and relevant references	Comments
Wildlife and Countryside Act 1981 (as amended 1988, 1991). Illegal to kill. Illegal to injure. Illegal to take. Illegal to possess. Illegal to sell.	Applies only to certain UK species. For details see legislation, Anon (1988) and Cooper and Cooper (1991).	There may be exceptions - for example, under licence, taking and possession of sick and and injured specimens, killing of severely sick or injured specimens.
Dangerous Wild Animals Act 1976 and (Modification) Order 1984. Licence to keep.	Covers certain species of venomous snakes (native and exotic), both species of poisonous lizards and all crocodilians. See legislation, Cooper (1978) and RCVS (1991).	A licence is not required if the reptile is undergoing veterinary treatment or is in, for example, a licensed pet shop, zoo or research laboratory.

Title and provisions	Species affected and relevant references	Comments
Zoo Licensing Act 1981. Licence to display.	All species of reptiles are covered. See legislation and Cooper (1983).	A licence is only needed if display is open to the public and for more than seven days in a year.
Protection of Animals Acts 1911-1988. Illegal to cause unnecessary suffering.	All species of reptiles are covered if in captivity. See legislation and Cooper (1987).	Can include failure to feed and to provide water or veterinary treatment.
Animals (Scientific Procedures) Act 1986. Licence to perform research.	All species of reptiles are covered. See legislation and Cooper (1987).	Applies to research in the laboratory and in the field.
Pet Animals Act 1951. Licence to sell as pets.	All species of reptiles are covered. See legislation and RCVS (1991).	Applies to any such business, not only pet shops.
EEC Regulation 3626/82 (as amended).	Certain species of reptiles are covered. See Regulation and Department of the Environment (1987).	Implements Convention on International Trade in Endangered Species of wild Fauna and Flora (CITES). Permits required for import/export and sale/exchange of specified species.
Endangered Species (Import and Export) Act 1976 (as amended by Wildlife and Countryside Act Schedule 10 and elsewhere).	Certain species of reptiles are covered. See legislation and Department of the Environment (1987).	Controls import/export and sale of certain species not covered by CITES.
Transit of Animals (General) Order 1973 (as amended 1988). Animals must be transported without unnecessary suffering.	All species of reptiles are covered. See legislation and Cooper (1987).	Imposes a duty on the carrier as well as the consignor.
Veterinary Surgeons Act 1966. Diagnosis and treatment must be by a veterinary surgeon.	All species of reptiles are covered. See legislation and RCVS (1991).	Exemptions exist for minor medical treatment by owner and emergency first aid.

REFERENCES

ANON. (1988). *Protecting Britain's Wildlife*. The Department of the Environment, Bristol.

COOPER, M.E. (1978). The Dangerous Wild Animals Act 1976. *Veterinary Record* **102**, 457.

COOPER, M.E. (1983). The Zoo Licensing Act 1981. *Veterinary Record* **112**, 564.

COOPER, M.E. (1987). *An Introduction to Animal Law*. Academic Press, London.

COOPER, M.E. and COOPER, J.E. (1991). Snakes and snake-bite. *Veterinary Record* **129**, 203.

DEPARTMENT OF THE ENVIRONMENT. (1987). *Controls on the Import and Export of Endangered and Vulnerable Species (with supplements)*. Department of the Environment, Bristol.

ROYAL COLLEGE OF VETERINARY SURGEONS (1991). *Legislation Affecting the Veterinary Profession in the United Kingdom*. 6th Edn. (with supplement). Royal College of Veterinary Surgeons, London.

Acknowledgement

The editors are indebted to **Mrs. M.E. Cooper LLB FLS** for preparing this Appendix.

APPENDIX FIVE
Conversion Tables

Metric and Temperature Conversions

Length		**Volume**	
1 millimetre	= 0.03937 inch	1 litre	= 1.76 pints
1 centimetre	= 0.394 inch	4.546 litres	= 1 gallon
1 metre	= 1.094 yards		
Area		**Weight**	
1 square centimetre	= 0.155 square inch	1 gram	= 0.03527 ounce
1 square metre	= 1.1960 square yards	1 kilogram	= 2.205 pounds

Temperature Comparisons

Centigrade	Fahrenheit	Centigrade	Fahrenheit
10	50	38.1	101
12.5	55	38.6	101.5
15.5	60	38.9	102
18.2	65	39.2	102.5
21	69.8	39.4	103
26.6	80	39.7	103.5
28	82.4	40	104
32.2	90	40.3	104.5
35	95	40.6	105
35.5	96	40.8	105.5
36.1	97	41.1	106
36.6	98	41.4	106.5
37.2	99	41.7	107
37.7	100	41.9	107.5
38	100.4	42.1	108

INDEX

Abscesses
    in "Integument" 75
Aestivation 9, 19
Ageing
    in "Cardiovascular System" 85
    in "Miscellaneous" 153
Allometric scaling 194, 195, 207, 208
Alphaxalone/alphadolone 177
Amyloidosis 153
Analgesia 170
Anaesthetic chamber 178
Anaesthetic circuits 178
Anaesthetic requirements 170
Anaphylactic shock 155
Animals (Scientific Procedures) Act 222
Anophthalmia 167
Anorexia/cachexia in snakes 144
Antibacterial agents 200, 201
Antibiotic toxicity 132
Antifungal agents 202, 203
Antiparasitic agents 203
Aortic valvular stenosis 82
Arcus lipoides corneae 165
Arteriosclerosis 153
Arthritis 118
Atresia of follicular cells 125
Atrophic gastritis 107
Ausculation 81, 96
Avulsed spectacle 161

Bacterial diseases
    in "Cardiovascular System" 84
    in "Gastrointestinal System" 111
    in "Respiratory System" 90
Beak and claw deformities
    in "Gastrointestinal System" 105
    in "Integument" 75
Biochemical data 219, 220
Biochemical evaluations 53
Biopsies (digestive system) 55
Biopsies (integument) 54, 55
Biotin deficiency
    in lizards 146
Blepharoedema 162
Blindness 130, 134
Blister disease 17, 75
Blood sampling 50, 51
Blood supply
    in "Ophthalmology" 159
Boid inclusion disease 132
Breeding 25, 151
Bronchoscopy 57
Burns 75

Caesarian section 190
Calcium/phosphorous ratios 148
Canker (see Stomatitis)
Cannabalism 153
Cardiomegaly 82
Cardiomyopathy 83, 85, 131
Caseous plaque 164
Cataracts 166
Cerebrocorticonecrosis 131
Chlamydial diseases
    in "Cardiovascular System" 84
Classification 7, 11, 12, 13
Clinical examination
    of chelonians 33
    in general 32
    of lizards 38
    in "Ophthalmology" 157
    of snakes 35
Clinical examination sheet 217
Cloacal calculi 126
Cloacal prolapse
    in "Surgery" 191
    in "Urogenital System" 126
Cloacal tumours 126
Cloacitis 126, 191
Coccidiosis
    in "Therapeutics" 203
Coeliotomy 170, 187, 190
Colic 108, 111
Colonic washes 60
Colony problems 48
Complete blood counts 52
Computerised tomography 67, 82
Congestive heart failure 83
Congenital abnormalities 85
Conjunctivitis 164
Conjunctivoralostomy 164, 188
Constipation 111, 147, 191
Conus papillaris 157, 159
Conversion tables 224
Convulsions
    in "Anaesthesia" 176
    in "Neurological Diseases" 128
    in "Respiratory System" 92
Core body temperature 8, 155, 194
Cyrosurgery 187
Cyanosis
    in "Respiratory System" 90
    in "Therapeutics" 194
Corneal ulceration 165
Cyclopia 167
Cytodiagnostics 56

Damaged shell 75
Dangerous Wild Animals Act 8, 20, 221
Decapitation 154
Degenerative changes
    in "Cardiovascular System" 85
Depression
    in "Respiratory System" 90
Dermatitis (scale rot) 75, 77
Dermatophilosis 76
Devonomation 191
Drinking
    in "Nutritional Diseases" 139
Drowning 96, 147
Dysecdysis 17, 76, 155

Ear abscess 33, 186
Ecdysis 19
Ectoparasites 76
Ectothermic 8, 97, 128, 138, 170
EEC Regulations 222
Eggbinding 121, 134
Electroanaesthesia 180
Electrocardiograms 81, 85, 181
Electron microscopy 56,
Endangered Species Act 222
Endocrinological disorders 154
Endoscopy 34, 36, 118, 126
Enemas 198
Enophthalmus 163
Enteritis
    in lizards 111
Enucleation 188
Enterotomy/enterectomy 186, 190
Ether 179
Etorphine 175
Evisceration (of the globe) 189
Eurythermic 140
Euthanasia 154
Exophthalmia 167
Eye enucleation 188
Eyelids 158, 160

Faecal examinations 60, 120
Fatty liver
    in chelonians 112, 134
Feeding
    in general 16, 17
    of chelonians 18, 142
    of crocodilians 18
    of lizards 18
    in "Nutritional Diseases" 139
    of snakes 19
Feeding frenzy 147, 153, 160
Fluids
    in "Anaesthesia" 182
    in "Therapeutics" 197
Flukes
    in "Gastrointestinal System" 106
    in "Respiratory System" 93
Follicular rupture 126
Foreign bodies
    in Gastrointestinal System" 107, 108, 109, 111
    in "Ophthalmology" 162
    in "Respiratory System" 94
Foreign-body dermatitis 77
Formulary 210
Fracture fixation 189
Fractures
    in chelonians 135
    in iguanas 151
    in lizards 133
    in snakes 130
    in "Surgery" 189
    in "Therapeutics" 198
Freeze damage
    in chelonians 134, 166
Fundus 167
Fungal diseases
    in "Cardiovascular System" 84
    in "Gastrointestinal System" 113
    in "Integument" 77
    in "Ophthalmology" 165
    in "Respiratory System" 94
    in "Therapeutics" 202

Gasping 89
Gastrectomy 186
Gastric ulceration
    in snakes 106, 110
Gastritis
    in snakes 106
Gavage (see Stomach tubing)
Gout 117, 132, 145, 153
Great vessels 80

Haematological data 219, 220
Haematological evaluations 52
Halothane 179
Handling
    in "Anaesthesia" 170, 172
    of chelonians 21, 32
    of crocodilians 22
    in general 20, 153
    of lizards 22, 38
    of snakes 23, 34
Harderian gland 159, 163
Health monitoring 48
Heart 80
Heart rate 181, 194
Heat sources 15
Heliotherm 138
Hepatitis 112
Hexamitiasis
    in chelonians 119
Hibernation 9, 19
Hides 14, 144
Housing 14, 16
Humidity 16, 17, 19, 88, 97, 98, 202
Hyperplastic pancreatic regeneration 155
Hypervitaminosis D_3 145
Hyphaema 134, 166
Hypocalcaemia 151
Hypocalcaemic tetany 133
Hypoglycaemia
    in crocodilians 147
Hypopyon
    in "Ophthalmology" 166
    in "Respiratory System" 90
Hypothermia
    in "Anaesthesia" 180
    in "Miscellaneous" 154
Hypovitaminosis A 95, 117, 139, 146, 163
Hypovitaminosis B_1 130, 139, 146
Hypovitaminosis B_{12} 146
Hypovitaminosis C 139, 146
Hypovitaminosis E 139, 147
Hypovitaminosis K 147

Immunological disorders 155
Incubation 27
Incubation temperatures 26

Injection sites
    in "Therapeutics" 197
Intracoelomic injection 197
Intramuscular injection 199
Intravenous injection 172, 174, 175, 197, 199
Intubation 179
Intussusception
    in lizards 112
Iodine deficiency 145, 147, 155
Iris 160
Isoflurane 180
Isolation 98

Jackson's ratio 33, 112, 135, 143, 150

Keratitis 165
Keratitis sicca 163
Ketamine hydrochloride 176

Lacrimal glands 163
Legislation 8, 9, 17, 41, 48, 176, 221
Lens luxation 167
Lighting 15
Light intensity 124, 138
Light sources 17
Limb amputation 135, 188
Live feeding 140, 147

Magnetic resonance imaging 67, 82
Maladaptation syndrome 156
Membrana vasculosa retinae 159
Meningoencephalitis 131
Metabolic diseases
    in "Cardiovascular System" 85
Methoxyflurane 180
Metomodate 175
Microbiology 58
Microphthalmia 167
Minimum energy costs 195
Misting system 17
Mouth rot (see Stomatitis)
Muscle relaxation 171
Mydriatics 157
Myocarditis 84
Myiasis 77

Nasal discharge
    in "Respiratory System" 89
Nasolacrimal system 159, 163
Nebulisation 200
Negative-staining electron microscopy 57
Neoplasms
    in "Cardiovascular System" 86
    of eyelids 162
    in "Integument" 77
Nephritis 117
Neurological examination
    of chelonians 134
    of snakes 128
Nitrates 145
Nitrous oxide 179
Nomenclature 9
Nutritional therapies 204

Obesity 18, 98, 134, 145
Oesophagostomy tube 199
Ophthalmic surgery 188
Ophthalmoscope 157

Oral medication 198
Organophosphorus poisoning
    in snakes 131
Orthopaedic surgery 189
Ossicles 158
Osteodystrophy 187
    in crocodilians 136
    in iguanas 151
    in "Integument" 77
    in lizards 133
    in "Neurological System" 133
    in "Nutritional Diseases" 148
Osteolysis
    in chelonians 135, 187
Ovarian disturbances 124, 125
Ovarian inflammation 125
Oxalate toxicity 135

Palpatation 81
Panophthalmitis 162, 165
Papillomatosis 77, 78
Paramyxo-like virus 92
Paraphymosis 121
Parasitic diseases
    in "Cardiovascular System" 84
    in "Gastrointestinal System" 105, 107, 109, 113
    in "Neurological Diseases" 131
    in "Ophthalmology" 162
    in "Respiratory System" 93
    in "Therapeutics" 203
Parenteral administration 199, 200, 202
Paresis
    in alligators 145
    in chelonians 134
    in lizards 134
    in snakes 130
Parietal eye (see Third eye)
Pathological evaluations 56
Penis amputation 121, 191
Pentastomes
    in "Gastrointestinal System" 105
    in "Respiratory Sytems" 94
    in "Therapeutics" 204
Pentobarbitone 154, 173
Pericarditis 84, 85
Pet Animals Act 222
Pharyngostomy tube 191
Plugging (of hemipenes) 121
Pneumonia 89, 91, 166
Poikilothermic 8, 170
Post-hibernation anorexia 143
Post-hibernation jaundice 112
Post-mortem examination
    in "Cardiovascular System" 83
    in general 40
Post-mortem sheet 218
Postoperative care
    in "Anaesthesia" 182
Posture
    in "Respiratory System" 89
Preferred body temperature 8, 15, 16, 97, 128, 138, 155, 170, 171, 194, 195, 197
Preferred optimum temperature zone 9, 138
Premedication 171
Propofol 177
Protozoa
    in "Respiratory System" 93
    in "Therapeutics" 203
    in "Urogenital System" 119
Protection of Animals Act 222
Pseudogout 118
Pseudopregnancy 125
Pyloroduodenal resection 186

Radiography
    of chelonians 65, 97
    in general 63
    of lizards 64, 97
    of snakes 64, 84, 97
Regurgitation
    in chelonians 107
    in lizards 107
    in snakes 106
    in "Therapeutics" 198
Renal portal system 80, 199
Renal trematodiasis 119
Reproductive cycle 124
Reproductive scanning 70
Retained spectacle 161
Retention of slough 78
Retina 160
Rhinitis 89, 90
Rostral abrasions 78

Salmonellosis
    in chelonians 108
Salpingitis 85, 123
Salpingotomy 123
Schirmer tear test 163
Scale rot (see dermatitis)
Sendai virus 92
Septicaemic cutaneous ulcerative disease 78, 202
Serology 54
Sexing
    of chelonians 24, 34, 160
    of crocodilians 25
    in general 24, 28, 74
    of lizards 25, 39, 120
    of snakes 25, 36, 120, 126
Shell growth 148
Shell repair 186
Shock 155
Skin lacerations 77
Space-occupying lesions
    in "Respiratory System" 95
Specific minimum energy costs 195
Spinal abscesses 132
Spinal fluid collection 61
Starvation 156
Steatitis 147, 187
Stenothermic 140
Stomach tubing 198
Stomatitis
    in chelonians 102
    in general 102
    in lizards 104
    in snakes 104
    in "Therapeutics" 198
Stress 155
Subcutaneous injections 197, 199
Sub-spectacular abscess
    in "Integument" 75
    in "Laboratory Investigations" 56
    in "Ophthalmology" 163, 164
    in "Respiratory System" 90
Succinylcholine 176
Supplements 142, 198
Suturing 185
Systemic medication 200, 202

Tail injuries
    in lizards 188
Tarsorropathy 188, 189
Tear film 159
Tear staining syndrome 163
Temperature
    in "Anaesthesia" 181
    in "Management in Captivity" 15
    in "Respiratory System" 97
Therapeutics
    in "Cardiovascular System" 83
    in general 194
    in "Respiratory System" 97
Thermoregulation 8, 15, 160
Thiamine deficiency
    in chelonians 135, 167
    in "Managment in Captivity" 18
    in snakes 129, 130, 139
Thigmotherm 138
Thiopentone 175
Third eye (in snakes) 159
Third eyelid 158
Third eyelid flap 165, 188
Tiletamine hydrochloride 176
Topical administration 200, 202
Torticollis 130
Toxicity
    in chelonians 120, 135
    in crocodilians 136
    in snakes 131
Tracheal washes 57, 91, 93, 94, 97
Transit of Animals (General) Order 222
Transmission electron microscopy 57
Trauma 189
    in chelonians 134, 153
    of eyelids 160
    in "Respiratory System" 95
    in "Therapeutics" 198
Tricaine methanesulphonate 176
Tubulonephrosis 118
Tumour removal 187

Ulcerative shell disease 78
Ultra-violet light 15, 17, 64, 151
Ultrasonography 69, 70, 82, 84
Urate calculi 190
Urine collection 60
Uveitis 166

Vasectomy 191
Ventricular hypertrophy 82
Veterinary Surgeons Act 222
Viral diseases
    in "Cardiovascular System" 83
    in "Gastrointestinal System" 111
    in "Integument" 78
    in "Miscellaneous" 156
    in "Neurological System" 132
    in "Respiratory System" 92
Visceral organs
    in "Surgery" 186
Vitamin A deficiency 146, 162, 163, 164, 167
Vitamin B deficiencies
    in lizards 134

Warming
    in "Therapeutics" 195, 196
Water
    in "Management in Captivity" 19
    in "Nutritional Diseases" 139
Wildlife and Countryside Act 221
Wound healing 185

Xerophthalmia 163
Xylazine 177

Zoo Licensing Act 222
Zoonoses 20